Manufacturing Technology and Industrial Engineering

Manufacturing Technology and Industrial Engineering

Edited by **Michelle Vine**

WILLFORD PRESS

New York

Published by Willford Press,
118-35 Queens Blvd., Suite 400,
Forest Hills, NY 11375, USA
www.willfordpress.com

Manufacturing Technology and Industrial Engineering
Edited by Michelle Vine

International Standard Book Number: 978-1-68285-053-4 (Hardback)

Contents

Preface

This book elucidates the concepts and innovative models around prospective developments with respect to industrial engineering and manufacturing technology. The objective of the topics covered in this book is to give a general view of the different areas of this field such as operations research, production planning, supply chain management and logistics, ergonomics, etc. Scientists and students actively engaged in this field will find this book full of crucial and unexplored concepts.

This book is the end result of constructive efforts and intensive research done by experts in this field. The aim of this book is to enlighten the readers with recent information in this area of research. The information provided in this profound book would serve as a valuable reference to students and researchers in this field.

At the end, I would like to thank all the authors for devoting their precious time and providing their valuable contributions to this book. I would also like to express my gratitude to my fellow colleagues who encouraged me throughout the process.

Editor

Performance enhancement for crystallization unit of a sugar plant using genetic algorithm technique

P. C. Tewari[1], Rajiv Khanduja[2*] and Mahesh Gupta[3]

Abstract

This paper deals with the performance enhancement for crystallization unit of a sugar plant using genetic algorithm. The crystallization unit of a sugar industry has three main subsystems arranged in series. Considering exponential distribution for the probable failures and repairs, the mathematical formulation of the problem is done using probabilistic approach, and differential equations are developed on the basis of Markov birth-death process. These equations are then solved using normalizing conditions so as to determine the steady-state availability of the crystallization unit. The performance of each subsystem of crystallization unit in a sugar plant has also been optimized using genetic algorithm. Thus, the findings of the present paper will be highly useful to the plant management for the timely execution of proper maintenance decisions and, hence, to enhance the system performance.

Keywords: Performance enhancement, Crystallization unit, Genetic algorithm

Background

The sugar industry comprises of large complex engineering systems arranged in series, parallel, or a combination of both. Some of these systems are feeding, crushing, refining, steam generation, evaporation, crystallization, etc. The crystallization unit is one of the most important functionary units of a sugar plant where the sugar crystals are formed. The concentrated juice available in the form of thick syrup from refining unit is heated slowly for long time at low temperature condition resulting into the formation of crystals called crystallization process. The semi-solid juice from the cooking pans of refining unit is first fed to the crystallizers arranged in parallel. Now, the juice mixture consisting of yellowish sugar crystals is suspended in a semi solid mass (molasses or magma). This mixture is processed in centrifuges to separate the sugar crystals from magma. These yellowish sugar crystals are treated chemically to yield white crystals, whereas crystal-free magma is recycled through sulphitors for more recovery. The sugar crystals are then sent to the grading unit, which comprises of a hopper, elevator, cooler, and grader, arranged in series. It grades the sugar crystals according to their shape and size.

Literature review

The available literature reflects that several approaches have been used to analyze the system performance in terms of reliability and availability. These include reliability block diagram, Monte Carlo simulation, Markov modeling, failure mode and effect analysis, fault tree analysis, and Petri nets (Misra and Weber 1989; Singer 1990; Bradley and Dawson 1998; Modarres et al. 1999; Gandhi et al. 2003; Adamyan and Dravid 2004; Panja and Ray 2007; Bhamare et al. 2008). Dhillon and Singh (1981) have frequently used the Markovian approach for the availability analysis, using exponential distribution for failure and repair times. Kumar et al. (1988, 1989, 1993) used the Markov modeling in the analysis and evaluation of the performances of sugar and urea fertilizer plants. Srinath (1994) has explained a Markov model to determine the availability expression for a simple system consisting of only one component. Gupta et al. (2005) have evaluated the reliability parameters of butter manufacturing system in a dairy plant considering

* Correspondence: rajiv_khanduja@rediffmail.com
[2]Department of Mechanical Engineering, Seth Jai Parkash Mukand Lal Institute of Engineering and Technology (JMIT), Radaur, Yamuna Nagar, Haryana135133, India
Full list of author information is available at the end of the article

Table 1 Availability matrices of the subsystems for crystallization unit

Availability matrices of the three subsystems

Availability matrices of crystallizer subsystem for crystallization unit

β_{22}	0.01	0.02	0.03	0.04	0.05	Parameter constraints
α_{22}						
0.01	0.6491	0.8312	0.8953	0.86283	0.9403	
0.02	0.4310	0.6619	0.7770	0.83522	0.8761	
0.03	0.3171	0.5384	0.6736	0.80444	0.8119	$\alpha_{23} = 0.02; \beta_{23} = 0.10;$
0.04	0.2496	0.4501	0.5808	0.77276	0.7493	$\alpha_{24} = 0.02; \beta_{24} = 0.10$
0.05	0.2053	0.3853	0.5224	0.6224	0.6944	

Availability matrices of centrifuge subsystem for crystallization unit

β_{23}	0.01	0.02	0.03	0.04	0.05	Parameter constraints
α_{23}						
0.04	0.6491	0.6603	0.6609	0.6611	0.6615	
0.06	0.6216	0.6528	0.6578	0.6591	0.6595	
0.08	0.5887	0.6426	0.6533	0.6566	0.6580	$\alpha_{22} = 0.02; \beta_{22} = 0.10;$
0.10	0.5532	0.6301	0.6475	0.6534	0.6500	$\alpha_{24} = 0.02; \beta_{24} = 0.10$
0.12	0.5177	0.6157	0.96471	0.6495	0.6536	

Availability matrices of sugar grader subsystem for crystallization unit

β_{24}	0.01	0.02	0.03	0.04	0.05	Parameter constraints
α_{24}						
0.02	0.4574	0.5387	0.5741	0.5936	0.6059	
0.04	0.3465	0.4547	0.5075	0.5387	0.5594	
0.06	0.2801	0.3933	0.4547	0.4932	0.5195	$\alpha_{22} = 0.01; \beta_{22} = 0.02;$
0.08	0.2348	0.3465	0.4118	0.4547	0.4849	$\alpha_{23} = 0.04; \beta_{23} = 0.10$
0.10	0.2022	0.3097	0.3764	0.4218	0.4547	

exponentially distributed failure rates of various components. The reliability of the system is determined by forming the differential equations with the help of transition diagram using Markovian approach and then solving these differential equations with the help of fourth-order Runge–Kutta method. They applied the recursive method for calculating long-run availability and mean time between failure (MTBF) using numerical technique. Kumar et al. (2007) dealt with the simulated availability of CO_2 cooling system in a fertilizer plant.

Gupta et al. (2008) developed the performance models and decision support system for a feed water unit of thermal power plant with the help of mathematical formulation based on Markov birth-death process using probabilistic approach. In this way, the decision matrices are developed which provide the various performance levels for different combinations of failure and repair rates for all subsystems. The model developed helps in to decide about correct and orderly execution of proper maintenance in order to enhance the performance of the feed water unit of the thermal power plant. Khanduja et al. (2008a, b) have discussed availability analysis of bleaching unit of a paper plant. They also developed the performance evaluation system of screening unit in a paper plant. For long-run failure-free operation of the bleaching and screening units, the expression of steady-state availability has been developed, and behavior of each sub-system has also been analyzed.

Deb (1995) has explained the optimization techniques and how they can be used in the engineering problems. Tewari et al. (2000, 2005) dealt with the development of decision support system of refining system of sugar plant. They determined the availability for the refining system with elements exhibiting independent failures and repairs or the operation with standby elements for sugar industry. They also dealt with mathematical modeling and behavioral analysis for a refining system of a sugar industry using genetic algorithm. Ying-Shen et al. (2008) proposed a genetic algorithm-based optimization model to optimize the availability for a series–parallel system. The objective is to determine the most economical policy of component's MTBF and mean time to repair.

In this paper, the mathematical (availability) model has been developed to evaluate the performance of crystallization unit of a sugar plant on the basis of certain assumptions. After that, the performance optimization using genetic algorithm technique (GAT) is done, which gives the optimum unit availability levels for different combinations of failure and repair rates of the subsystems of crystallization unit for improving the performance of the sugar plant. Thus, the findings of the present paper will be highly useful to the plant management in futuristic maintenance planning and control to enhance the unit performance.

The crystallization unit

Crystallization unit consists of three subsystems in series configuration with the following description:

- Subsystem A_i ($i = 1$ to 6): It consists of six crystallizer units connected in parallel. The failure of any one reduces the capacity of the system and, hence, loss in production. Complete failure occurs when more than one unit fail at a time.

- Subsystem A_j ($j = 1$ to 19): It consists of nineteen centrifuge units connected in parallel. Complete failure occurs when more than two units fail at a time.

- Subsystem A_k ($k = 1$ to 4): It consists of four sugar grader units connected in series. The failure of any one causes the complete failure of the system.

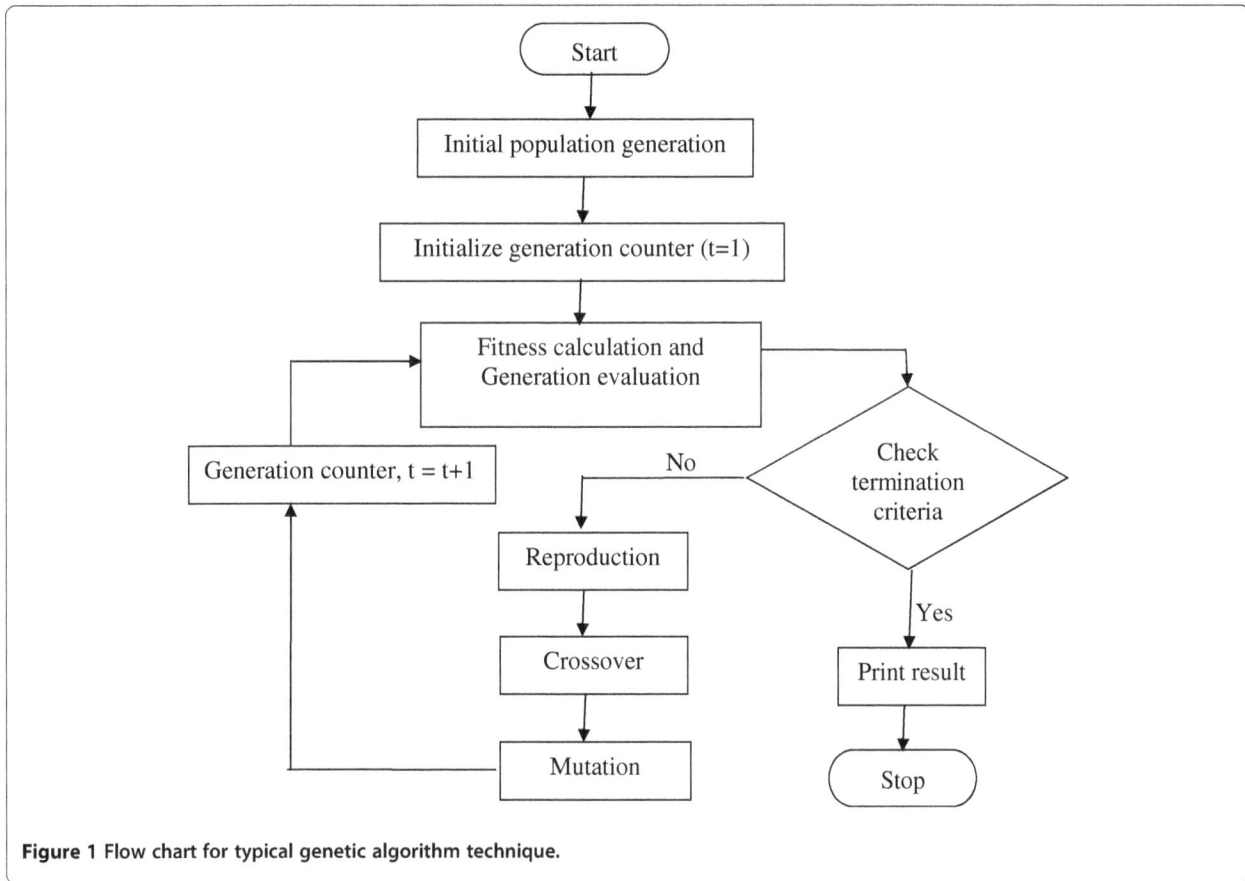

Figure 1 Flow chart for typical genetic algorithm technique.

Assumptions

The assumptions used in the probabilistic model are the following:

1. Failure/repair rates are constant over time and statistically independent.
2. A repaired unit is as good as new and performance wise for a specified duration.
3. Sufficient repair facilities are provided, i.e., no waiting time to start the repairs.
4. Standby units (if any) are of the same nature and capacity as the active units.
5. System failure/repair follows exponential distribution.
6. Service includes repair and/or replacement.
7. System may work at a reduced capacity/efficiency.
8. There is no simultaneous failure among the system. However, simultaneous failure may occur among various subsystems in a system/unit.

Notations

The following notations are associated with the crystallization unit:

- α_i, β_i
  $i = 22, 23, 24$
  Respective failure and repair rates of various subsystems
- $P_i(t)$
  Probability function that the unit is in a particular state at time 't'
- $P_i'(t)$
  Derivative of probability function $P_i(t)$

Performance modeling

The mathematical modeling is carried out and done using simple probabilistic considerations and differential equations which are developed on the basis of Markov birth-

Table 2 Effect of population size on availability of the crystallization unit using genetic algorithm

Population size	Availability	α_{22}	β_{22}	α_{23}	β_{23}	α_{24}	β_{24}
20	0.9458	0.0387	0.3737	0.0209	0.4537	0.0202	0.4717
40	0.9455	0.0132	0.2172	0.0290	0.4974	0.0209	0.4721
60	0.9474	0.0249	0.3668	0.0208	0.4384	0.0205	0.4940
80	0.9480	0.0178	0.2555	0.0229	0.4117	0.0206	0.4995
100	**0.9491**	**0.0205**	**0.2823**	**0.0207**	**0.4406**	**0.0203**	**0.4905**
120	0.9491	0.0205	0.2823	0.0207	0.4406	0.0203	0.4905

Mutation probability = 0.015; number of generation = 150; crossover probability = 0.875.

death process. These equations are further solved for determining the steady-state availability of crystallization unit. Various probability considerations give the following differential equations associated with the crystallization unit:

- State 0 - full capacity working with no standby
- State 1 to 5 - reduced capacity working
- State 6 to 16- represents the system in failed state

$$P_0'(t) + \sum \alpha_r P_0(t) = \sum \beta_j P_k(t) \tag{1}$$

$$P_1'(t) + \sum \alpha_r P_1(t) = \sum \beta_j P_k(t) \tag{2}$$

$$P_2'(t) + \sum (\alpha_r \beta_{23}) P_2(t) = \sum \beta_j P_5(t) + \alpha_{23} P_0(t) \tag{3}$$

$$P_3'(t) + \sum (\alpha_r + \beta_m) P_3(t) = \sum \beta_j P_k(t) + \alpha_{23} P_1(t) + \alpha_{22} P_2(t) \tag{4}$$

$$P_4'(t) + \sum (\alpha_r \beta_{23}) P_4(t) = \sum \beta_j P_k(t) + \alpha_{23} P_2(t) \tag{5}$$

$$P_5'(t) + \sum (\alpha_r \beta_m) P_5(t) = \sum \beta_j P_k(t) + \alpha_{22} P_4(t) + \alpha_{23} P_3(t) \tag{6}$$

$$P_i'(t) + \beta_m P_i(t) = \alpha_m P_1(t) \tag{7}$$

By putting $d/dt = 0$ as $t \to \infty$ in Equations 1 to 7, the steady-state probabilities are given as follows:

$$\sum \alpha_r P_0 = \sum \beta_i P_k$$

$$\sum \alpha_r P_1 = \sum \beta_i P_k$$

$$\sum (\alpha_r + \beta_m) P_3 = \sum \beta_i P_k + \alpha_{23} P_1 + \alpha_{22} P_2$$

$$\sum (\alpha_r + \beta_m) P_5 = \sum \beta_i P_k + \alpha_{22} P_4 + \alpha_{23} P_3$$

$$P_i = (\alpha_m / \beta_m) P_1$$

The probability of full capacity working $viz.$ P_0 is determined by normalizing condition, i.e.,

$$\sum_{i=0}^{16} P_i = 1$$

Substituting the values of P_1 to P_{16} in terms of P_0 into normalizing condition, we get

$$P_0 N = 1$$

Let

$$A = \alpha_{22} / \beta_{22}, B = \alpha_{23} / \beta_{23}, C = \alpha_{24} / \beta_{24}$$

$$X_1 = \alpha_{22} + \alpha_{23} - (\alpha_{22} * \beta_{23} / (\alpha_{22} + \beta_{23}))$$

$$X_2 = \beta_{22} + (\alpha_{23} * \beta_{22} / (\alpha_{22} + \beta_{23}))$$

$$X_3 = X_1 / X_2$$

$$X_4 = (\alpha_{22} + \alpha_{23} - \beta_{22} * X_3) / \beta_{23}$$

$$X_5 = (\alpha_{23} / \beta_{23})^2 * \beta_{23} + \alpha_{23} * X_4.$$

Then,

$$N = 1 + X_3 + X_4 + B * X_3$$
$$+ X_5 + B^2 * X_3 + B * X_5 + B^3 * X_3 + C * B^2 * X_3 + A * B^2 *$$
$$X_3 + A * B * X_3 + C * B * X_3 + A * X_3 + C * X_3 + C + C$$
$$* X_4 + C * X_5$$

Now, the steady-state availability of the crystallization unit may be obtained as the summation of all the working state probabilities, i.e.,

Table 3 Effect of number of generation on availability of the crystallization unit using genetic algorithm

Number of generations	Availability	α_{22}	β_{22}	α_{23}	β_{23}	α_{24}	β_{24}
100	0.895029	0.01070	0.04914	0.04345	0.38526	0.02055	0.48939
150	0.895665	0.01030	0.04658	0.04026	0.44370	0.02075	0.47348
200	0.896551	0.01125	0.04951	0.04052	0.37082	0.02028	0.49569
250	0.902337	0.01013	0.04999	0.04123	0.45499	0.02076	0.47859
300	**0.903933**	**0.01001**	**0.04978**	**0.04033**	**0.46868**	**0.02049**	**0.47530**
350	0.903933	0.01001	0.04978	0.04033	0.46868	0.02049	0.47530

Mutation probability = 0.015; population size = 150; crossover probability = 0.875.

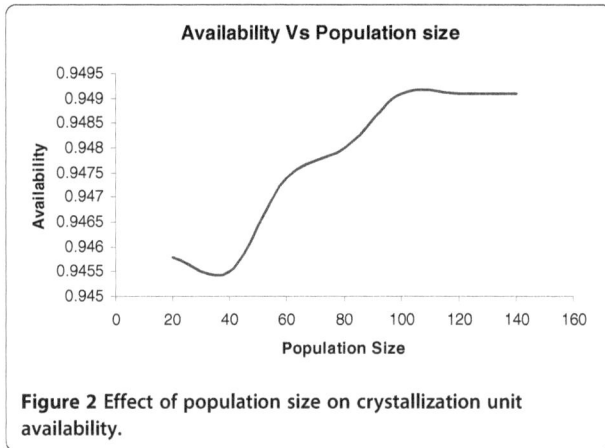

Figure 2 Effect of population size on crystallization unit availability.

$$\text{Availability} = \sum_{i=0}^{5} P_i$$

or

$$\text{Availability(Av.)} = (1/N)[1 + X_3 + X_4 + X_5 + B * X_3 \ ] + B^2 * X_3$$

Performance analysis

From the maintenance history sheet of crystallization unit of sugar plant and the detailed discussions with the plant personnel, appropriate failure and repair rates of all the subsystems are taken, and availability matrices (performance values) are prepared accordingly by putting these failure and repair rate values in expression of availability for P_0. This deals with the quantitative analysis of all the factors *viz.* courses of action and states of nature, which influence the maintenance decisions associated with the crystallization unit. These availability models are developed under the real decision-making environment, i.e., decision making under risk (probabilistic model) and used to implement the proper maintenance decisions for the crystallization unit of sugar plant.

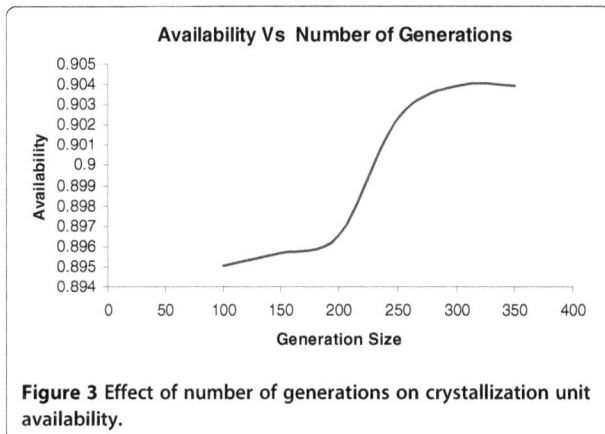

Figure 3 Effect of number of generations on crystallization unit availability.

Table 1 represents the availability matrices for various subsystems of the crystallization unit. These matrices simply reveal the various performance levels for different combinations of failure and repair rates/priorities. It also depicts the effect of failure/repair rate of all the subsystems on crystallization unit performance. On the basis of analysis, one may select the best possible combinations (α_i, β_i) to increase the unit availability. Table 1 shows optimal availability level for all the subsystems (for crystallizer is 0.9403; for centrifuge, 0.6615; for sugar grader, 0.6059) which can be optimized using genetic algorithm technique.

Genetic algorithm technique

Genetic algorithms (GA) are computerized search and optimization algorithms based on the mechanics of natural genetics and natural selection (Figure 1). Genetic algorithms have become important because they are found to be potential search and optimization techniques for complex engineering optimization problems. The action of GAT for parameter optimization in the present problem can be stated as follows:

1. Initialize the parameters of the genetic algorithm.
2. Randomly generate the initial population and prepare the coded strings.
3. Compute the fitness of each individual in the old population.
4. Form the mating pool from the old population.
5. Select two parents from the mating pool randomly.
6. Perform the crossover of the parents to produce two off springs.
7. Mutate if required.
8. Place the child strings to new population.
9. Compute the fitness of each individual in new population.
10. Create best-fit population from the previous and new population.
11. Repeat the steps 4 to 10 until the best individuals in new population represent the optimum value of the performance function (unit availability).

The performance behavior of the crystallization unit is highly influenced by the failure and repair parameters of each subsystem. These parameters ensure high performance of the crystallization unit. GAT is hereby proposed to coordinate the failure and repair parameters of each subsystem for stable system performance, i.e., high availability. Here, the number of parameters is six (three failure parameters and three repair parameters). The design procedure is described as follows: To use GAT for solving the given problem, the chromosomes are to be coded in real structures. Here, concatenated, multi-parameter, mapped, fixed-point coding is used. Unlike, unsigned fixed-point integer coding parameters are mapped to a specified interval $[X_{min}, X_{max}]$, where X_{min} and X_{max} are the maximum and minimum values of system parameters. The maximum value of the availability function corresponds to the optimum values of system parameters. These parameters are optimized according to the performance index, i.e., desired availability level. To test the proposed method, failure and repair rates are determined simultaneously for optimal value of unit availability. Effects of population size and number of generations on the availability of crystallization unit are shown in Tables 2 and 3. To specify the computed simulation more precisely, trial sets are also chosen for GA and system parameters. The performance (availability) of the crystallization unit is determined by the designed values of the unit parameters.

Failure and repair rate parameter constraints

$$(\alpha_{22}, \beta_{22}, \alpha_{23}, \beta_{23}, \alpha_{24}, \beta_{24})$$
$$\alpha_{22}, [A1] \, \epsilon(0.01, 0.05) \quad \alpha_{23}, \epsilon(0.04, 0.12) \quad \alpha_{24}, \epsilon(0.02, 0.10)$$
$$\beta_{22}, \epsilon(0.01, 0.05) \quad \beta_{23}, \epsilon(0.01, 0.05) \quad \beta_{24}, \epsilon(0.01, 0.05)$$

Here, real-coded structures are used. The simulation is done to a maximum number of population size, which is varying from 20 to 120. The effect of population size on availability of the crystallization unit is shown in Figure 2. The optimum value of unit's performance is 94.91%, for which the best possible combination of failure and repair rates is $\alpha_{22} = 0.0205$, $\beta_{22} = 0.2823$, $\alpha_{23} = 0.0207$, $\beta_{23} = 0.4406$, $\alpha_{24} = 0.0203$, and $\beta_{24} = 0.4905$ at population size 100 as given in Table 2.

Now, the simulation is done to a maximum number of generations, which is varying from 100 to 350. The effect of number of generations on availability of the crystallization unit is shown in Figure 3. The optimum value of unit's performance 90.39%, for which the best possible combination of failure and repair rates is $\alpha_{22} = 0.01001$, $\beta_{22} = 0.04978$, $\alpha_{23} = 0.04033$, $\beta_{23} = 0.46868$, $\alpha_{24} = 0.02049$, and $\beta_{24} = 0.47530$ at generation size 300 as given in Table 3.

Conclusions

The performance optimization of crystallization unit of a sugar plant is discussed in this paper. Genetic algorithm technique is hereby proposed to select the various feasible values of the unit failure and repair parameters. Then, GAT is successfully applied to coordinate simultaneously these parameters for an optimum level of unit performance. Besides, the effect of GA parameters such as population size and number of generations on unit performance, i.e., availability, has also been discussed. The findings of this paper are discussed with the concerned sugar plant management. Such results are found highly beneficial for the purpose of performance enhancement of a crystallization unit in the sugar plant concerned.

Author details
[1]Department of Mechanical Engineering, National Institute of Technology, Kurukshetra, Haryana136119, India. [2]Department of Mechanical Engineering, Seth Jai Parkash Mukand Lal Institute of Engineering and Technology (JMIT), Radaur, Yamuna Nagar, Haryana135133, India. [3]Department of Mechanical Engineering, National Institute of Technology, Kurukshetra, Haryana136119, India.

References

Adamyan A, Dravid H (2004) System failure analysis through counters of Petri nets. J Qual Reliability Int 20:317–335

Bhamare SS, Yaday OP, Rathore A (2008) Evolution of reliability engineering discipline over the last six decades: a comprehensive review. Int J Reliab Saf 1 (4):377–410

Bradley ML, Dawson R (1998) The cost of unreliability: a case study. J Qual Maint Eng 4(3):212–218

Deb K (1995) Optimization for engineering design: algorithms and examples. Prentice Hall of India, New Delhi, India

Dhillon BS, Singh C (1981) Engineering reliability: new techniques and applications. John Willey and Sons, New York

Gandhi OP, Sehgal R, Angra S (2003) Failure cause identification of tribo-mechanical system. Reliab Eng Syst Saf 65:259–270

Goldberg DE (2001) Genetic algorithm in search, optimization and machine learning. Pearson Education Asia Ltd., New Delhi, India

Gupta P, Lal A, Sharma R, Singh J (2005) Numerical analysis of reliability and availability of the series processes in butter oil processing plant. Int J Qual Reliability Manage 22(3):303–316

Gupta S, Kumar A, Sharma R, Tewari PC (2008) A performance modeling and decision support system for a feed water unit of a thermal power plant. S A J Ind Eng 19(2):125–134

Khanduja R, Tewari PC, Kumar D (2008a) Development of performance evaluation system for screening unit of a paper plant. Int J Appl Eng Res 3(3):451–460

Khanduja R, Tewari PC, Kumar D (2008b) Availability analysis of bleaching system of paper plant. J Ind Eng, Udyog Pragati, NITIE Mumbai (India) 32(1):24–29

Kumar D, Singh IP, Singh J (1988) Reliability analysis of the feeding system in the paper industry. Microelectron Reliab 28(2):213–215

Kumar D, Singh J, Pandey PC (1989) Availability analysis of the washing system in the paper industry. Microelectron Reliab 29:775–778

Kumar D, Singh J, Pandey PC (1993) Operational behavior and profit function for a bleaching and screening system in the paper industry. Microelectron Reliab 33:1101–1105

Kumar S, Tewari PC, Sharma R (2007) Simulated availability of CO_2 cooling system in a fertilizer plant. Ind Eng J (Indian Inst Ind Eng, Mumbai) 36(10):19–23

Misra KB, Weber GG (1989) A new method for fuzzy fault tree analysis. Microelectron Reliab 29:195–216

Modarres M, Kaminsky M, Kriotsov V (1999) Reliability engineering and risk analysis: a practical guide. Marcel Dekker, New York

Panja SC, Ray PK (2007) Reliability analysis of track circuit of Indian railway signaling system. Int J Reliab Saf 1(4):428–445

Singer D (1990) A fuzzy set approach to fault tree and reliability analysis. J Fuzzy Sets Syst 34:145–155

Srinath LS (1994) Reliability engineering, 3rd edn. East–west Press Pvt. Ltd, New Delhi, India

Tewari PC, Joshi D, Sreenivasa Rao M (2005) Mathematical modeling and behavioral analysis of a refining system using genetic algorithm. In: Proceedings of national conference on Competitive Manufacturing Technology and Management for Global Marketing, Chennai, 2005

Tewari PC, Kumar D, Mehta NP (2000) Decision support system of refining system of sugar plant. J Inst Eng (India) 84:41–44

Ying-Shen J, Shui-Shun L, Hsing-Pei K (2008) A knowledge management system for series–parallel availability optimization and design. J Expert Syst Appl 34:181–193

2

Design of supply chain in fuzzy environment

Kandukuri Narayana Rao[1][*], Kambagowni Venkata Subbaiah[2] and Ganja Veera Pratap Singh[3]

Abstract

Nowadays, customer expectations are increasing and organizations are prone to operate in an uncertain environment. Under this uncertain environment, the ultimate success of the firm depends on its ability to integrate business processes among supply chain partners. Supply chain management emphasizes cross-functional links to improve the competitive strategy of organizations. Now, companies are moving from decoupled decision processes towards more integrated design and control of their components to achieve the strategic fit. In this paper, a new approach is developed to design a multi-echelon, multi-facility, and multi-product supply chain in fuzzy environment. In fuzzy environment, mixed integer programming problem is formulated through fuzzy goal programming in strategic level with supply chain cost and volume flexibility as fuzzy goals. These fuzzy goals are aggregated using minimum operator. In tactical level, continuous review policy for controlling raw material inventories in supplier echelon and controlling finished product inventories in plant as well as distribution center echelon is considered as fuzzy goals. A non-linear programming model is formulated through fuzzy goal programming using minimum operator in the tactical level. The proposed approach is illustrated with a numerical example.

Keywords: Supply chain, Fuzzy goal programming, Performance vector, Continuous review policy, Strategic level, Tactical level

Introduction

Globalization itself is a great change in business environment which causes great trend towards global trade and competition. It became important for enterprises to develop long-term strategic relations between suppliers and customers. Also, due to the introduction of new products with shorter life cycles and the heightened expectations of customers, it became a must for business organizations to focus on their supply chains. The supply chain, which is also referred to as the network of suppliers, manufacturing centers, warehouses, distribution centers, and customers as well as logistic information systems connected by an organization's suppliers and customers of its customers. Basing on the organization's competitive strategy, efficient or responsive supply chains are modeled by integrating business plans at strategic and tactical levels.

Cohen and Lee (1988) developed a model of material requirement policy for every shop in a production system using cost-based stochastic sub-models, namely material control sub-model, production sub-model, and distribution sub-model, to predict the performance of alternative manufacturing strategies. Robinson et al. (1993) designed an integrated distribution system for a two-echelon, uncapacitated distribution location problem as a mixed integer programming problem and illustrated with a case study. Pyke and Cohen (1994) developed a supply chain model by considering multiple products, with independent demand and expedited batches, and optimized the total cost of supply chain subjected to the service levels for all products. Petrovic and Roy (1998) developed a simulation model of supply chain in an uncertain environment, with customer demand and supply of raw materials as vague, which are represented with fuzzy sets. Beamon (1998) presented an overview and evaluation of the performance measures used in supply chain models.

Sabri and Beamon (2000) developed a supply chain model that facilitates simultaneous strategic and operational planning using an interactive method. Vorst (2000) formulated supply chains for agriculture products through linear programming approach. Chen and Tzeng

---
* Correspondence: nr_kandukuri@rediffmail.com
[1]Department of Mechanical Engineering, Govt. Polytechnic, Visakhapatnam 530008, India
Full list of author information is available at the end of the article

(2000) adopted fuzzy multi-objective approach in order to reduce the computational complexity of the integrated supply chain model. Tsiakis et al. (2001) developed a mixed integer linear programming model for the design of multi-product, multi-echelon supply chain networks consisting of manufacturing sites, warehouses, distribution centers, and customer zones. Yu et al. (2003) proposed a strategic production-distribution mixed integer programming model for supply chain design by incorporating logical constraints to represent the BOM. Lee et al. (2002) presented a hybrid methodology that combines the analytic and simulation models for an integrated production-distribution model in supply chain. Li (2002) proposed a method of building a supply chain management system that determines production plan purchasing plan, inventory plan, and distribution plan by minimizing the total cost. Chen et al. (2003) formulated a multi-objective mixed integer non-linear programming problem and adopted a two-phase fuzzy decision-making method to solve the supply chain model involving manufacturing plants, distribution centers, and retailers. Eskigun et al. (2005) modeled a supply chain network design and proposed Lagrangian heuristic method to obtain strategic decision of the model. Amiri (2006) developed a mixed integer programming model and proposed a heuristic procedure for a supply chain network design. The paper addresses network design problem in a supply chain system that involves locating production plants and distribution warehouses and determining the best strategy for distributing the product from plants to warehouses and from the warehouses to the customers. Liang (2008) developed a fuzzy multi-objective linear programming model with piecewise linear membership function to solve integrated multi-product and multi-time-period production-distribution planning decision problems with fuzzy objectives.

Farahani and Elahipanah (2008) developed a mixed integer linear programming model with two objective functions, namely minimizing costs and minimizing the sum of backorders and surpluses of products. Peidro et al. (2009) proposed a fuzzy mixed integer linear programming model where data are ill known and modeled by triangular fuzzy numbers for supply chain planning under supply, process, and demand uncertainties. Troncoso et al. (2011) adopted a mixed integer programming method for integrated strategy and compared this strategy with decoupling strategy. Narayana Rao and Venkatasubbaiah (2011) developed an integrated procurement, production, and distribution supply chain model in fuzzy environment. Bouzembrak et al. (2011) developed a green supply chain network design problem with environmental concerns. In the study, the authors considered warehouse and distribution center locations, building technology selection, and processing-

distribution planning as the strategic decisions in the model.

Supply chain models developed, except for a few ones, ignored vagueness, but in the real world, supply chains operate in a vague or uncertain environment. In this context, uncertainty may be related to the specification of objectives, constraints, or variables. In this paper, a supply chain model is developed through simultaneous strategic and tactical planning in fuzzy environment.

In fuzzy environment, a fuzzy goal programming method is adopted to incorporate the inherent vagueness in supply chain cost and volume flexibility at strategic level. In tactical level also, a fuzzy goal programming method is developed to incorporate vagueness present in three objectives, *viz.* cost of controlling raw material inventory at the supplier echelon and cost of controlling products at plant and distribution center echelons.

A numerical illustration with four echelons, to produce and distribute four types of products using four types of raw materials from five suppliers, is presented. The mixed integer programming problem is formulated at strategic level and non-liner programming problem is formulated at tactical level, and both are solved using LINGO 8.0 optimization solver.

Strategic level

The strategic level considers the design of an integrated procurement, production, and distribution supply chain network. The objective functions (supply chain cost and volume flexibility) and the constraints considered in the model are discussed below.

Supply chain cost

The supply chain cost (SCC) is a mixed integer linear function. This linear objective function contains various components. These components are raw material purchase price, transportation cost of raw materials shipped from suppliers to plants, fixed cost associated plant and distribution center operations, transportation cost of products shipped from plants to distribution centers, and transportation cost of products from distribution centers to customer zones:

$$\text{SCC} = \sum_{njk} \left(a_{njk} + C1_{nj} \right) \times R_{njk} + \sum_k f1_k \times q2_k + \sum_l f2_l \times q3_l + \sum_{ikl} S_{ikl} \times \text{BC}_{ikl} + \sum_{ilm} d_{ilm} \times D_{im} \times Y_{lm} \tag{2.1}$$

Volume flexibility

It is the linear objective function that comprises plant volume flexibility VF and distribution volume flexibility. Plant volume flexibility is measured as the difference between plant capacity and capacity utilization. Distribution volume flexibility is measured as the difference between the available throughput and demand requirements. The following function represents volume flexibility of the supply chain:

$$\text{VF} = \left[\sum_k \left(q2_k \times P_k - \sum_i E2_{ik} \times X_{ik} \right) \right]$$
$$+ \left[\sum_l \left(q3_l \times (T3_l)_H - \sum_{im} E3_{il} \times D_{im} \times Y_{\text{lm}} \right) \right]$$
$$(2.2)$$

The various constraints governing the supply chain model are shown below.

- Raw material availability
  Raw material requirement at the plants should be within the limits of raw material availability at the supplier.

  $$\sum_k R_{njk} \leq R_{nj} \qquad \forall n, j \qquad (2.3)$$

- Plant production capacity
  Total production quantities at each plant should not exceed the plant capacity.

  $$\sum_i E2_{ik} \times X_{ik} \leq P_k \times q2_k \qquad \forall k \qquad (2.4)$$

- Raw material supply
  Shipping of raw materials from the ranked suppliers to the plants should be sufficient to meet the production requirement of the products at the plants.

  $$\sum_i u_{ni} \times X_{ik} \leq \sum R_{njk}$$

  $$\forall n, k \qquad R_{njk} = \text{RA}_{nj} \times \sum_i u_{ni} \times X_{ik} \qquad \forall n, j, k$$
  $$(2.5)$$

- Minimum and maximum bounds on production quantity

  $$(P2_{ik})_L \times q2_k \leq X_{ik} \leq (P2_{ik})_H \times q2_k \qquad \forall i, k \quad (2.6)$$

- Minimum and maximum bounds on throughput capacities of distribution centers

  $$(T3_l)_L \times q3_l \leq \sum_{im} E3_{il} \times D_{im} \times Y_{lm} \leq (T3_l)_H \times q3_l \quad \forall l$$
  $$(2.7)$$

- Assignment of each customer zone to exactly one distribution center

  $$\sum_l Y_{\text{lm}} = 1 \qquad \forall m \qquad (2.8)$$

- Production requirement
  Quantity of the products available at the plant ensures the shipping quantity from the plant.

  $$X_{ik} = \sum_l \text{BC}_{ikl} \qquad \forall i, k \qquad (2.9)$$

- Shipping quantity of products

Total shipments to customer zones should be equal to the demand at the customer zones.

$$\sum_{k,l} \text{BC}_{ikl} = \sum_m D_{im} \qquad \forall i \qquad (2.10)$$

- Demand requirement at each distribution center
  Shipping quantity of the product should satisfy the demand of the product at the distribution center.

  $$\sum_k \text{BC}_{ikl} = \sum_m Y_{\text{lm}} \times D_{im} \qquad \forall i, l \qquad (2.11)$$

- Non-negativity constraints
  Ensure the following variables to be non-negative.

  $$X_{ik}, \text{BC}_{ikl}, R_{njk} \geq 0 \qquad \forall n, i, j, k, l \qquad (2.12)$$

- Binary variable restriction
  Ensure the following variables to be binary.

  $$q2_k, q3_l, Y_{\text{lm}} = 0 \text{ or } 1; \qquad \forall k, l, m \qquad (2.13)$$

Tactical level

In the tactical level, a non-linear programming model is formulated at supplier, plant and distribution center echelons to minimize the cost of controlling raw materials and finished products at plant and distribution center echelons. Continuous review inventory policy is assumed at the echelons. The models at these echelons are discussed below.

Supplier echelon model

In the supplier echelon, the total cost of controlling raw material n required at plant k under continuous review policy is given by the following equation:

$$\text{TCS}_{nk} = q2_k \left[\frac{\text{MD1}_{nk}}{\text{Q1}_{nk}} \times K1_{nk} + H1_{nk} \left(\frac{\text{Q1}_{nk}}{2} + s1_{nk} \right) \right. \quad (3.1)$$
$$\left. + \left(\frac{\text{MD1}_{nk} \times P1_{nk}}{\text{Q1}_{nk}} \right) \times (\sigma1_{nk} \times \text{LI1}_{nk}) \right],$$

where TCS_{nk} total cost of controlling raw material n required at plant k, MDI_{nk} mean demand of raw material n required at plant k, ML1_{nk} expected demand of raw material n during lead time at plant k, $\text{LI1}_{nk} = I_N((u_s)_{nk})$, loss integral representing the expected number of units out of stock during an order cycle of raw material n at plant k, and $(u_s)_{nk}$ standardized variable at the supplier echelon.

The reorder point of raw material n required at plant k can be determined from the following relationship:

$$s1_{nk} = \text{ML1}_{nk} + (u_s)_{nk} \times \sqrt{\text{VL1}_{nk}}. \qquad (3.2)$$

The expected demand over lead time is determined from the following relationship:

$$\text{ML1}_{nk} = \left[\sum_i u_{ni} \times X_{ik}\right] \times T4_{nk}; \qquad \forall n, k \quad (3.3)$$

The average total lead time of raw material n at plant k is calculated as the sum of the raw material lead time and delay time considering all the suppliers, as shown below (Sabri and Beamon 2000):

$$T4_{nk} = \frac{\sum_v \left(T1_{njk} + T2_{nj} \times \left(1 - A1_{nj}\right)\right)}{N} \quad (3.4)$$

where N is the number of vendors.

The following relation gives the lead-time demand variance and standard deviation since production demand (X_{ij}) is fixed:

$$\text{VL1}_{nk} = \left[\sum_i u_{ni} \times X_{ik}\right]^2 \times \text{Var}(T4_{nk}), \quad (3.5)$$

$$\sigma 1_{nk} = \sqrt{\text{VL1}_{nk}} \quad (3.6)$$

The following relation gives variance of total lead time of raw material n at plant k:

$$\text{Var}(T4_{nk}) = \max_j(V1_{njk}) + \max_j\left[(V2_{nj})\left(1 - A1_{nj}\right)\right]$$

$$+ \max_j\left[V2^2_{nj}\left(1 - A1_{nj}\right)\left(A1_{nj}\right)\right]. \quad (3.7)$$

Customer service level of raw material n at plant k is given by the following relation (Ballou 2004):

$$F1_{nk} = 1 - \left(\frac{\text{LI1}_{nk} \times \sigma 1_{nk}}{Q1_{nk}}\right). \quad (3.8)$$

Determine the optimal lot size by minimizing the total cost of controlling raw material n required at plant k subjected to the given customer service level of raw material n at plant k.

Plant echelon model

In the plant echelon, the cost function to be minimized consists of setup costs, processing costs, and work-in-process carrying costs and cost related to the finished product stockpile. The finished product stockpile cost comprises stockpile holding cost, transportation holding cost, and backorder cost. The optimum order quantity, reorder point, and service level of product i at plant k are obtained by optimizing the cost function. The following equations represent the total cost of production

and cost of controlling finished product stockpile, respectively:

$$\text{TCP}_{ik} = q2_k \times \left[K2_{ik} \times \frac{X_{ik}}{Q2_{ik}} + \text{PC}_{ik} \times X_{ik} + \text{HP}_{ik} \times X_{ik} \times T5_{ik}\right] \quad (3.9)$$

$$\text{TCF}_{ik} = q2_k \times \left\{ \begin{array}{l} H2_{ik}\left(\dfrac{Q2_{ik}}{2} + s2_{ik}\right) + \sum_l \text{CH}_{ikl} \\ \times \text{BC}_{ikl}(\text{BN}_{ikl} \times F2_{ik} + \text{BE}_{ikl} \times (1 - F2_{ik})) \\ + \left\{\dfrac{\text{MD2}_{ik} \times F2_{ik}}{Q2_{ik}}\right\} \times \sigma 2_{ik} \times \text{LI2}_{ik} \end{array} \right\} \quad (3.10)$$

$$\text{TC}_{ik} = \text{TCP}_{ik} + \text{TCF}_{ik}, \quad (3.11)$$

where TC_{ik} total cost of controlling finished product i at plant k, TCP_{ik} cost of production of product i at plant k, TCF_{ik} total cost of controlling finished product stockpile, MD2_{ik} mean demand of product i at plant k, ML2_{ik} expected demand of product i during leadtime at plant k, $\text{LI2}_{ik} = I_N(u_p)_{ik}$, loss integral representing the expected number of units out of stock during an order cycle of product i at plant k, and $(u_p)_{ik}$ standardized variable of product i at plant echelon.

Reorder point is determined from the following relationship:

$$s2_{ik} = \text{ML2}_{ik} + (u_p)_{ik} \times \sqrt{\text{VL2}_{ik}}. \quad (3.12)$$

Expected demand of product i over production lead time at plant k is determined from the following relationship:

$$\text{ML2}_{ik} = X_{ik} \times T5_{ik}; \qquad \forall i, k. \quad (3.13)$$

Determination of various parameters is given by the following relations (Sabri and Beamon 2000):

The total production lead time of product i at plant k is given as the sum of setup time and waiting time at the work stations, processing times, and material delay times:

$$T5_{ik} = S_{ik} + P_{ik} + w_{ik} + T6_{ik} \qquad \forall i, k \quad (3.14)$$

The material delay time can be determined from the following relationship:

$$T6_{ik} = \max(T4_{nk} \times (1 - F1_{nk})) \qquad \forall i, k. \quad (3.15)$$

The following relation gives variance of the production lead time:

$$\text{Var}(T5_{ik}) = \text{Var}(w_{ik}) + \text{Var}(T6_{ik}). \quad (3.16)$$

Variance of the material delay time is determined from the following relation:

$$\mathrm{Var}(T6_{ik}) = \max[\mathrm{Var}(T4_{nk}) \times (1-F1_{nk}) \quad (3.17)$$
$$+ (T6_{ik})^2 \times (1-F1_{nk}) \times (F1_{nk})]$$

Variance and standard deviation of demand during lead time are determined from the following equations:

$$\mathrm{VL2}_{ik} = (X_{ik})^2 \times \mathrm{Var}(T5_{ik}) \quad (3.18)$$

$$\sigma 2_{ik} = \sqrt{\mathrm{VL2}_{ik}}. \quad (3.19)$$

Customer service level of product i at plant k is given by the following relation (Ballou 2004):

$$F2_{ik} = 1 - \left(\frac{\mathrm{LI2}_{ik} \times \sigma 2_{ik}}{Q2_{ik}} \right) \quad (3.20)$$

The expected replenishment lead time for product i from plant k to distribution center l is given by the following relation:

$$T_{ikl} = \mathrm{BN}_{ikl} \times F2_{ik} + (T5_{ik} + \mathrm{BE}_{ikl}) \\ \times (1-F2_{ik}). \quad (3.21)$$

Variance of lead time for product i from plant k to distribution center l is calculated from the following relation:

$$\mathrm{Var}\, T_{ikl} = F2_{ik} \times (1-F2_{ik}) \times \{T_{ikl} - (T5_{ik} + \mathrm{BE}_{ikl})\}^2 \quad (3.22)$$

Determine the optimal lot size ($Q2_{ik}$) by minimizing total cost of production and cost of controlling finished product stockpile subjected to customer service level of product.

Distribution center echelon model

In the distribution echelon, the total cost of distribution of product i at the distribution center (DC)l under dynamic continuous review policy is given by the following equation:

$$\mathrm{TCD}_{il} = q3_l \times \left[\begin{array}{l} \frac{K3_{il} x MD3_{il}}{Q3_{il}} + H3_{il} \times \left\{ \frac{Q3_{il}}{2} - ML3_{il} + s3_{il} \right\} \\ + \left\{ \frac{H3_{il} x ML3_{il}}{2x Q3_{il}} + \frac{MD3_{il} x P3_{il}}{Q3_{il}} \right\} \times \sigma 3_{il} \times \mathrm{LI3}_{il} \end{array} \right] \quad (3.23)$$

where TCD_{il} total cost of controlling finished product i at distribution center l, $MD3_{il}$ mean demand of product i at distribution center l, $ML3_{il}$ expected demand of product i during lead time at distribution center l, $\mathrm{LI3}_{il} = I_N$ $((u_d)_{il})$ loss integral representing the expected number of units out of stock during an order cycle of product i at

distribution center l, and $(u_d)_{il}$ standardized variable at distribution center echelon.

The reorder point of product i at distribution center l is determined from the following relationship:

$$s3_{il} = ML3_{il} + (u_d)_{il} \times \sqrt{\mathrm{VL3}_{il}} \quad (3.24)$$

Find out the expected and variance of transportation lead time for product i to distribution center l from the following relation (Sabri and Beamon 2000):

$$T7_{il} = \frac{\Sigma_k q2_k \times T_{ikl}}{\Sigma_k q2_k} \forall i, l. \quad (3.25)$$

$$\mathrm{Var}(T7_{il}) = \max(q2_k \times \mathrm{Var}(T_{ikl})). \quad (3.26)$$

Determine lead time demand variance and standard deviation of product i at distribution center l from the following relation:

$$\mathrm{VL3}_{il} = \left[\sum_m Y_{lm} \times D_{im} \right]^2 \times \mathrm{Var}(T_{ikl}), \quad (3.27)$$

$$\sigma 3_{il} = \sqrt{\mathrm{VL2}_{il}}. \quad (3.28)$$

Customer service level of product i at distribution center l is calculated from the following relation (Ballou 2004):

$$F3_{il} = 1 - \left(\frac{\mathrm{LI3}_{il} \times \sigma 3_{il}}{Q3_{il}} \right). \quad (3.29)$$

Determine the optimum order quantity ($Q3_{il}$) by minimizing total cost of controlling finished product i at distribution center l subjected to the given customer service level of product i at distribution center l.

Problem formulation in fuzzy environment

In this section, formulation of fuzzy goal programming with minimum operator in strategic and tactical level planning is presented.

Fuzzy goal programming

Fuzzy set theory in goal programming (GP) was first considered by Narasimhan (1982). Ramik (2000), Mohamed (2000), and Abd El-Wahed and Abo-Sinna (2001) have investigated various aspects of decision-making problems using FGP theoretically. In fuzzy goal programming, membership functions are formulated for the objectives. After considering the aspiration levels of the objectives and the nature of the objectives, 'approximately less than or equal to,' and 'approximately greater than or equal to,' the membership functions can be developed for each objective as follows:

- For approximately less than or equal to:

$$\mu_{z_m}(x) = \begin{cases} 1 & ; \quad Z_m(x) \leq l_m \\ \dfrac{u_m - Z_m(x)}{u_m - l_m} & ; \quad l_m < Z_m(x) \leq u_m \\ 0 & ; \quad Z_m(x) > u_m \end{cases} \qquad (4.1)$$

- For approximately greater than or equal to:

$$\mu_{z_k}(x) = \begin{cases} 1 & ; \quad Z_k(x) > l_k \\ \dfrac{Z_k(x) - l_k}{u_k - l_k} & ; \quad l_k < Z_k(x) \leq u_k \\ 0 & ; \quad Z_k(x) \leq l_k \end{cases} \qquad (4.2)$$

$Z_m(x)$ = mth objective function
l_m = lower aspiration level of mth objective
u_m = higher aspiration level of mth objective
$\mu_{z_m}(x)$ = membership function of mth objective

Fuzzy goal programming with minimum operator

Using the approach of Bellman and Zadeh (1970), the feasible fuzzy solution set is obtained by the intersection of all membership functions representing the fuzzy goals. This solution set is then characterized by its membership $\mu_F(x)$ which is,

$$\mu_F(x) = \mu_{Z1}(x) \cap \mu_{Z2}(x) ... \cap \mu_{Zk}(x) \qquad (4.3)$$

$$= \min \left[\mu_{Z1}(x), \mu_{Z2}(x), ..., \mu_{Zk}(x) \right].$$

Then, the optimum decision can be determined to be the maximum degree of membership for the fuzzy decision:

$$\underset{x \in F}{Max}\, \mu_F(x) = \max_{x \in F}\ \min \left[\mu_{Z1}(x), \mu_{Z2}(x), ..., \mu_{Zk}(x) \right]. \qquad (4.4)$$

By introducing the auxiliary variable λ, which is the overall satisfactory level of compromise, the following conventional mathematical programming problem can be formulated:

$$\left. \begin{array}{ll} \text{Maximize} & \lambda \\ \text{Subject to} & \lambda \leq \mu_{zk} \qquad k = 1, ..., K \\ & \lambda \in [0, 1] \end{array} \right\} . \qquad (4.5)$$

Strategic level

These objective functions (supply chain cost and volume flexibility) and constraints are discussed in the 'Methodology' section. The supply chain model that is developed at strategic level with uncertainty in the objectives is taken care by specifying aspiration levels for the objectives. The decision-maker is able to specify the aspiration levels for each objective. Linear membership

functions are defined for each objective depending on their nature. The nature may be approximately less than or equal or greater than or equal to the specified value. Considering the nature of the fuzzy parameters, linear membership functions of either non-increasing or non-decreasing are formulated. Membership function of total supply chain cost is assumed as non-increasing, and membership function of volume flexibility is assumed as non-decreasing.

Formulation of the membership functions of the fuzzy variables are shown below:

1. Membership function for supply chain cost (μ_{SCC})

$$\begin{aligned} \mu_{SCC} &= 1 & & \text{if } SCC \leq C1_{min} \\ &= \frac{(C1_{max}) - SCC}{(C1_{max} - C1_{min})} & & \text{if } C1_{min} < SCC < C1_{max} \\ &= 0 & & \text{if } SCC \geq C1_{max} \end{aligned}$$

$$(4.6)$$

2. Membership function for volume flexibility (μ_{VF})

$$\begin{aligned} \mu_{VF} &= 0 & & \text{if } VF \leq C2_{min} \\ &= \frac{VF - (C2_{min})}{(C2_{max} - C2_{min})} & & \text{if } C2_{min} < VF < C2_{max} \\ &= 1 & & \text{if } VF \geq C2_{min} \end{aligned}$$

$$(4.7)$$

where $C1_{min}$ and $C2_{min}$ are the minimum aspiration levels of SCC and VF and $C1_{max}$ and $C2_{max}$ are the maximum aspiration levels of SCC and VF.

According to Zimmermann (1978), Zadeh's minimum operator is used in aggregating the objectives to determine the optimal solution at the strategic level. The following are the equations in a fuzzy goal programming approach at the strategic level with minimum operator:

$$\begin{array}{ll} \text{Maximize } \lambda \\ \text{Subject to } \lambda \leq \mu_{SCC} \\ \qquad\quad \lambda \leq \mu_{VF} \end{array} \qquad (4.8)$$

and also subject to the crisp constraints shown in (2.3) to (2.13), where

$$\lambda = \min(\mu_{SCC}, \mu_{VF})$$

Tactical level

In tactical level, the total cost of controlling raw materials, cost of controlling finished products at the plant echelon, and cost of controlling finished products at the distribution center echelon are assumed as fuzzy goals. Service levels at the respective echelons are assumed as constraints. Formulation of the membership functions of the fuzzy goals are shown below:

1. Membership function of the fuzzy goal (cost of controlling raw materials)

$$\mu_{TCS} = 1 \qquad\qquad \text{if } TCS \leq C3_{min}$$
$$= \frac{(C3_{max}) - TCS}{(C3_{max} - C3_{min})} \quad \text{if } C3_{min} < TCS < C3_{max}$$
$$= 0 \qquad\qquad \text{if } TCS \geq C3_{max}$$

$$(4.9)$$

2. Membership function of the fuzzy goal (cost of controlling products at plant echelon)

$$\mu_{TCP} = 1 \qquad\qquad \text{if } TCP \leq C4_{min}$$
$$= \frac{TCP - (C4_{min})}{(C4_{max} - C4_{min})} \quad \text{if } C4_{min} < TCP < C4_{max}$$
$$= 0 \qquad\qquad \text{if } TCP \geq C4_{min}$$

$$(4.10)$$

3. Membership function of the fuzzy goal (cost of controlling products at distribution center echelon)

$$\mu_{TCD} = 1 \qquad\qquad \text{if } TCD \leq C5_{min}$$
$$= \frac{TCD - (C5_{min})}{(C5_{max} - C5_{min})} \quad \text{if } C5_{min} < TCD < C5_{max}$$
$$= 0 \qquad\qquad \text{if } TCD \geq C5_{min}$$

$$(4.11)$$

where $C3_{min}$, $C4_{min}$, and $C5_{min}$ are the minimum aspiration levels of TCS, TCP, and TCD and $C3_{max}$, $C4_{max}$, and $C5_{max}$ are the maximum aspiration levels of TCS, TCP, and TCD.

The following equations are obtained in fuzzy goal programming approach at tactical level with minimum operator:

Maximize λ
Subject to $\lambda \leq \mu_{TCS}$
$$\lambda \leq \mu_{TCD} \qquad\qquad (4.12)$$
$$\lambda \leq \mu_{TCP}$$
$$0 \leq \lambda \leq 1$$

and the following service level constraints:

$$0.85 \leq F1_{nk} \leq 0.99 \qquad \forall n, k,$$
$$0.85 \leq F2_{ik} \leq 0.99 \qquad \forall i, k, \qquad (4.13)$$
$$0.85 \leq F3_{il} \leq 0.99 \qquad \forall i, l.$$

Methodology

A supply chain model in fuzzy environment is formulated, and the decision variables at strategic and tactical levels are determined by combining strategic level and tactical level models through the following iterative procedure:

1. Stage 1: Solve the supply chain modeling problem in crisp environment and obtain a payoff table.
   (a) Step 1: Formulate the mixed integer linear programming problem at strategic level with total supply chain cost as objective function and equations given in (2.3), (2.4), (2.5),... (2.13) as constraints and solve to determine supply chain cost and the volume flexibility.
   (b) Step 2: Formulate tactical level models discussed in the 'Tactical level' section and solve to determine total cost of controlling raw material at the supplier echelon, total cost of production and cost of controlling finished product stockpile at the plant echelon, and total cost of controlling products at the distribution center echelon.
   (c) Step 3: Formulate the mixed integer linear programming problem with volume flexibility as objective function and constraints given in Equations 2.3, 2.4, 2.5,... 2.13.
   (d) Step 4: Repeat step 2.
   (e) Step 5: Prepare the payoff table which contains extreme values of the objectives in strategic and tactical levels.
2. Stage 2: Formulate the strategic level model in fuzzy environment and solve.
   (a) Step 1: Formulate membership functions of the fuzzy goals (supply chain cost and volume flexibility) using the aspiration levels of supply chain cost and volume flexibility from the payoff table obtained in the crisp environment.
   (b) Step 2: Formulate tactical level models (strategic, plant, and distribution center echelon models) as discussed in the 'Tactical level' section and solve to determine total cost of controlling raw material at the supplier echelon, total cost of production and cost of controlling finished product stockpile at the plant echelon, and total cost of controlling products at the distribution center echelon.

Numerical illustration

As a numerical illustration, a supply chain with four echelons, namely supplier, plant, distribution center, and customer zone, is considered. In the supply chain, it is assumed that four types of raw materials flow between five suppliers and four plants. In addition, it is assumed that four types of products flow from plants to four customer zones through four distribution centers.

The input variables required for developing and analyzing the supply chain model under study are shown in the Appendix. The procedure discussed in the methodology is used to determine the extreme solutions. These extreme solutions are useful to formulate membership functions of the fuzzy objectives in strategic and tactical levels.

Model formulation in strategic level

A fuzzy goal programming problem is formulated in strategic level as discussed in the 'Strategic level' subsection and is given below.

Maximize λ
Subject to the following constraints :
$$\lambda \le \frac{176,200 - SCC}{176,200 - 175,700}$$
$$\lambda \le \frac{VF - 10432}{10,832 - 10,432} \tag{6.1}$$
$$0 \le \lambda \le 1,$$

and also subjected to the constraints given in Equations 2.3 to 2.13.

Model formulation in tactical level

A fuzzy goal programming problem in tactical level is formulated as discussed in the 'Tactical level' subsection and is given below.

Maximize λ
Subject to the following constraints :
$$\lambda \le \frac{5,300 - TCS}{5,300 - 5,000}$$
$$\lambda \le \frac{4,000 - TCP}{4,000 - 3,600} \tag{6.2}$$
$$\lambda \le \frac{1,050 - TCD}{1,050 - 1,000}$$
$$0 \le \lambda \le 1,$$

and also the service level constraints.

Results and discussions

The supply chain model is solved in crisp environment by implementing the methodology for the numerical example given in the 'Numerical illustration' section. LINGO code is developed to solve the mixed integerprogramming problem formulated at the strategic level and the non-linear programming problems at the three echelons. The extreme solutions are obtained by implementing step 1 to step 11 of stage 1 of the 'Methodology' section.

The extreme solutions shown in Table 1 indicate that the performance measures have two strategies, namely efficient strategy and responsive strategy. The supply chain cost at strategic level in efficient strategy is 0.28% less than the supply chain cost in responsive strategy. The cost of controlling raw materials at the supplier echelon in perspective 2 is 4.22% less than that of the responsive strategy. The cost of controlling products at the plant echelon in perspective 1 is 12.66% less than that of the responsive strategy. The cost of controlling products at the distribution center echelon in both

perspectives is almost the same. In the case of volume flexibility, the responsive strategy is 2.94% more than that obtained in efficient strategy. Further, the total supply chain cost in the efficient strategy is 0.4% less than that of the responsive strategy.

The models developed in strategic and tactical levels in fuzzy environment are solved using LINGO 9.0. Table 2 shows the values of supply chain cost, volume flexibility, total cost of controlling raw materials at the supplier echelon, total cost of controlling products at the plant echelon, and total cost of controlling products at the distribution center echelon in tactical level.

The minimum operator assures minimum satisfaction of all the goals: supply chain cost (175,758.2), volume flexibility (10,627.1), total cost of controlling raw materials (5,184.1) at the supplier echelon, total cost of controlling products at the plant echelon (3,767.8), and total cost of controlling products (1,024.7) at the distribution center echelon obtained with the minimum operator lying between the extreme solutions. The above performance measures obtained in fuzzy environment indicate the compensation strategy. Comparison of the performance measures (normalized values) of the three strategies, namely efficient (EF), responsive (RS), and compensation (CS), is shown in Figure 1.

In the case of heavy industries like integrated steel plants, they can choose perspective 1 as they adopt efficient (cost-effective) supply chain strategy. On the other hand, companies involved in the manufacture of electronic goods like computers, cell phones, etc. can choose responsive supply chain strategy. Today, companies involving manufacturing of volatile and unforeseeable products like apparel and automotive must pioneer in compensation strategy.

Conclusions

In this paper, a multi-objective-oriented approach is adopted for supply chain modeling in fuzzy environment. Aspiration levels of the fuzzy objectives are derived from extreme solutions obtained by modeling in crisp environment. The results revealed that supply chain cost is low with low volume flexibility in the case of efficient supply chain strategy. In responsive strategy,

Table 1 Extreme solutions

Objective function	SCC	VF	TCS	TCP	TCD
Minimize SCC	175,700	10,432	4,800	3,600	1,000
Maximize VF	176,200	10,832	5,300	4,000	1,050

Table 2 Supply chain performance measures

Performance measures	Value
SCC at strategic level	175,758.2
VF	10,627.1
SCC at tactical level	
Total cost of controlling raw materials at the supplier echelon (TCS)	5,184.1
Total cost of controlling products at the plant echelon (TCP)	3,767.8
Total cost of controlling products at the distribution center echelon (TCD)	1,024.7

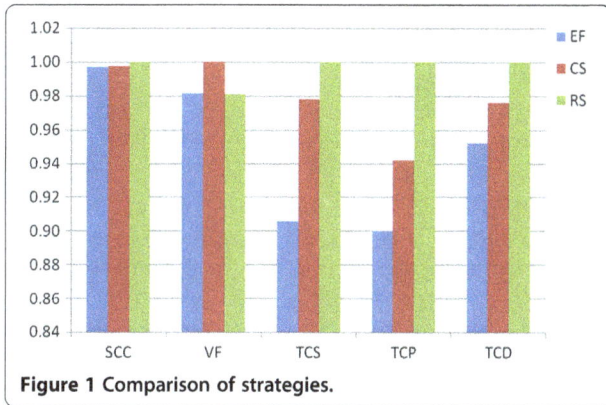

Figure 1 Comparison of strategies.

volume flexibility is high with high total supply chain cost. In fuzzy environment, the total supply chain cost and volume flexibility which lie between the values obtained with the two perspectives indicate the compensation strategy. Implementing the proposed methodology may generate more satisfactory solutions, and the developed model is robust to evaluate different supply chain strategies. Further, fuzzy goal programming techniques provide feasible solutions with flexible model formulation in decision-making problems, which involve human judgments in decision-making.

This paper focuses on how supply chain is designed by implementing a structured methodology which integrates strategic planning and tactical planning. This study can be extended to develop interactive user-friendly application software for supply chain planning in an organization. Also, investigation of integrating, operational level planning is an interesting research area.

Appendix
Input data

i, product index; j, vendor index; k, plant index; l, distribution center index; m, customer zone index; n, raw material index

$f1_k$, fixed charges for plant k (Rs/period) (4,750, 4,450, 4,550, 5,000)

$f2_l$, fixed charges for distribution center l (Rs/period) (120, 150, 120, 150)

P_k, production capacity of plant k (unit/period) (3,750, 2,600, 2,500, 3,000)

$(T3_l)_L$, minimum throughput of DC l (unit/period) (100, 100, 50, 50)

$(T3_l)_H$, maximum throughput of DC l (unit/period) (300,500,300,300)

R_{nj}, raw material n availability at vendor j (2,500, 2,500, 2,500, 2,500, 2,500, 2,500, 2,500, 3,000, 3,500, 3,500, 2,000, 2,500, 2,000, 2,000, 2,500, 2,000, 2,500, 2,500, 2,000, 2,000)

$C1_{nj}$, unit cost of raw material n at vendor j (Rs/unit) (5, 4, 5, 6, 6, 4, 5, 5, 6, 3, 4, 5, 4, 6, 5, 5, 4, 3, 4, 5)

u_{ni}, utilization rate of each raw material n per unit of product i(1.3, 1.2, 1.2, 1.3, 1.2, 1.2, 1.3, 1.2, 1.2, 1.2, 1.1, 1.2, 1.2, 1.2, 1.2, 1.3)

$E2_{ik}$, equivalent units at plant k per unit of product i (2, 2, 5, 5, 2, 2, 5, 4, 5, 2, 2, 2, 5, 2, 5, 2)

$E3_{il}$, equivalent units at DC l per unit of product i (3, 2, 2, 3, 2, 2, 3, 2, 3, 2, 2, 3, 2, 2, 3, 3)

D_{im}, average demand for product i at customer zone m (unit/period) (20, 30, 20, 20, 40, 50, 40, 30, 40, 30, 40, 20, 20, 30, 40, 40)

$(P2_{ik})_L$, minimum production volume of product i at plant k (unit/period) (5, 10, 5, 10, 5, 5, 12, 5, 5, 12, 15, 5, 15, 15, 15, 15)

$(P2_{ik})_H$, maximum production volume of product i at plant k (unit/period) (50, 50, 100, 50, 50, 50, 50, 50, 50, 100, 100, 50, 50, 50, 100, 50)

a_{njk}, unit transportation cost from vendor j to plant k of raw material n (Rs/unit) (1, 1.25, 1.45, 1.4, 1.85, 1.6, 1.5, 1.9, 1, 1, 1.4, 1.4, 1.65, 1.15, 1.5, 1.5, 1, 1.25, 1, 1.4, 1.5, 1.6, 1.7, 1.8, 1, 1.25, 1.4, 1, 1.15, 1.5, 1.5, 1.5, 1.6, 1.6, 1.6, 1.6, 1.5, 1.65, 1.65, 1.65, 1.5, 1, 1, 1, 1.15, 1.5, 1.5, 1.5, 1.4, 1.4, 1.25, 1.25, 1.5, 1.5, 1.5, 1.15, 1.25, 1.4, 1.3, 1.4, 1.5, 1.6, 1.5, 1.5)

$T2_{nj}$ ($\times 10^{-3}$), expected delay time of raw material n at vendor j (period) (13, 17, 14, 14, 11, 12, 14, 14, 16, 15, 16, 12, 14, 13, 14, 15)

$V2_{nj}$ ($\times 10^{-4}$), variance of lead time for raw material n from vendor j to plant k (period^2) (41, 28, 35, 33, 39, 34, 41, 39, 38, 40, 48, 48, 36, 32, 42, 35)

d_{ilm}, unit transportation cost of product i from DC l to customer zone m (Rs/unit) (1.66, 1.3, 1.86, 1.62, 1.94, 1.38, 1.96, 1.74, 1.72, 1.38, 1.54, 1.16, 1.3, 1.24, 1.64, 1.68, 1.02, 1.7, 1.56, 1.1, 1.9, 1.32, 1.18, 1.88, 1.14, 1.14, 1.28, 1.72, 1.02, 1.58, 1.82, 1.08, 1.26, 1.66, 1.72, 1.86, 1.44, 1.98, 1.48, 1.1, 1.16, 1.75, 1.46, 1.56, 1.2, 1.3, 1.38, 1.14, 1.74, 1.46, 1.92, 1.88, 1.62, 1.38, 1.72, 1.72, 1.72, 1.84, 1.04, 1.24, 1.36, 1.62, 1.02, 1.82)

$T1_{njk}$ ($\times 10^{-3}$), expected lead time for raw material n from vendor j to plant k (period) (27, 27.8, 27.2, 27.6, 27.6, 27.6, 27.6, 27.2, 27.1, 27.6, 27.3, 27.5, 27.2, 27.1, 27.6, 27.3, 27.5, 27.2, 28.1, 27.8, 27.4, 27.8, 28.1, 27.4, 27.4, 26.8, 27.7, 27.5, 28.1, 27.4, 28.3, 27.4, 27.6, 27.9, 26.3, 27.4, 27.3, 27.5, 28.4, 26.7, 26.9, 26.8, 27.7, 28.1, 26.8, 28.3, 27.1, 27.4, 27.4, 27.7, 27.1, 27.2, 27.3, 26.9, 28.1, 27.8, 27.8, 26.5, 26.7, 27.7, 27, 26.7, 27.3, 27.3, 27.3, 27.2, 27.1, 27.5, 26.9)

$V1_{njk}$ ($\times 10^{-4}$), variance of lead time for raw material n from vendor j to plant k (period^2) (26, 23, 14, 13, 25, 33, 36, 22, 16, 31, 26, 45, 26, 25, 26, 26, 23, 26, 25, 29, 13, 20, 257, 19, 25, 28, 23, 25, 30, 30, 35, 18, 36, 26, 26, 25, 41, 12, 20, 22, 26, 41, 26, 27, 13, 28, 26, 23, 30, 20, 19, 22, 25, 20, 17, 15, 27, 31, 21, 26, 24, 19, 39, 21)

$K1_{nk}$, order setup cost of replenishing raw material n required at plant k (Rs) (35, 38, 33, 38, 38, 37, 33, 35, 32, 35, 35, 39, 33, 37, 32, 32)

$H1_{nk}$, holding cost of raw material n required at plant k (Rs/period/unit) (1, 2, 5, 2, 5, 2, 1, 2, 5, 3, 4, 1, 4, 5, 2, 2)

$P1_{nk}$, backorder cost for shortage of raw material n required at plant k (Rs/unit) (17, 15, 20, 17, 25, 19, 18, 16, 28, 19, 21, 26, 25, 25, 14, 18)

$K2_{ik}$, production setup cost of product i at plant k (Rs) (54, 50, 53, 55, 50, 55, 56, 57, 55, 51, 52, 53, 53, 55, 54, 56)

$H2_{ik}$, holding cost of product i at plant k (Rs/period/unit) (1.6, 1.2, 1.5, 1.1, 1.3, 1.1, 1.3, 1.7, 1.2, 1.3, 1.5, 1.3, 1.6, 1.4, 1.1, 1.5)

$P2_{ik}$, backorder cost for shortage of product i required at plant k (Rs/unit) (24, 30, 32, 45, 30, 32, 38, 39, 38, 36, 34, 44, 37, 38, 33, 39)

$K3_{il}$, order setup cost of product i at DC l (Rs) (44, 40, 43, 44, 40, 45, 45, 46, 45, 41, 42, 43, 43, 45, 43, 44)

$H3_{il}$, holding cost of product i at DC l (Rs/period/unit) (1.5, 1.6, 1.11, 1.5, 1.3, 1.3, 1.6, 1.2, 1.6, 1.2, 1.2, 1.4, 1.4, 1.3, 1.5, 1.6)

$P3_{il}$, backorder penalty cost for shortage of product i at DC l (Rs/unit) (30, 32, 31, 30, 31, 29, 28, 23, 29, 30, 27, 28, 27, 26, 25, 29)

PC_{ik}, processing cost of product i at plant k (Rs/unit) (1, 1.2, 1, 0.8, 1.2, 0.9, 1.1, 1.2, 0.9, 1, 1.3, 1.3, 0.85, 0.75, 1.2, 1)

HP_{ik}, work-in-process holding cost of product i at plant k (Rs/period/unit) (5, 2, 9, 6, 2, 5, 2, 8, 1, 9, 5, 9, 9, 1, 3, 7)

S_{ik} ($\times 10^{-3}$), production setup time for product i at plant k (period) (10, 20, 30, 40, 20, 10, 20, 10, 30, 30, 10, 20, 10, 20, 10, 10)

P_{ik} ($\times 10^{-3}$), production processing time for product i at plant k (period) (40, 30, 20, 10, 30, 20, 50, 40, 60, 50, 70, 50, 30, 40, 30)

$A1_{nj}$ (%), availability of raw material n at vendor j (90, 85, 95, 90, 90, 95, 95, 90, 90, 85, 95, 90, 90, 85, 95, 90, 90, 85, 95, 90)

BN_{ikl} ($\times 10^{-3}$), normal transportation lead time for product i from plant k to DC l (period) (6, 3, 7, 6, 2, 3, 7, 7, 2, 7, 5, 8, 5, 6, 4, 5, 4, 5, 5, 4, 4, 7, 5, 5, 7, 6, 5, 5, 4, 4, 4, 3, 76, 5, 7, 6, 8, 5, 5, 5, 1, 5, 8, 8, 6, 1, 7, 8, 6, 7, 7, 9, 4, 5, 10, 3, 9, 6, 6, 6, 9, 3, 5)

BE_{ikl} ($\times 10^{-3}$), expedited transportation lead time of product i from plant k to DC l (period) (5, 2, 6, 5, 1, 2, 6, 6, 1, 6, 4, 7, 4, 5, 3, 4, 3, 4, 4, 3, 3, 6, 4, 4, 6, 5, 4, 4, 3, 3, 3, 2, 6, 5, 4, 6, 5, 7, 4, 4, 4, 1, 4, 7, 5, 6, 6, 7, 5, 6, 6, 8, 3, 4, 9, 2, 8, 5, 5, 5, 8, 2, 4)

w_{ik} ($\times 10^{-3}$), waiting time of the product i at plant k (period) (5, 2, 2, 2, 0.3, 4, 2, 3, 2, 2, 4, 2, 3, 5, 4, 2)

Var w_{ik} ($\times 10^{-3}$), 1, 4, 8, 7, 2, 7, 7, 6, 3, 3, 8, 7, 5, 6, 9, 5

s_{ikl}, unit transportation cost from plant k to DC l of product i (Rs/unit), from tactical model (2, 3, 1, 1, 5, 3, 3, 1, 3, 5, 1, 1, 3, 1, 3, 3, 3, 2, 2, 1, 2, 3, 2, 1, 4, 3, 2, 1, 3, 4, 2, 2, 1, 4, 4, 1, 3, 4, 3, 1, 4, 1, 3, 1, 1, 4, 1, 2, 4, 4, 1, 2, 4, 3, 2, 2, 4, 3, 4, 2, 3, 4, 3, 1)

CH_{ikl}, holding for i product from plant k to DC l (Rs/period/unit) (0.5, 0.7, 0.8, 0.8, 0.6, 0.11, 0.2, 0.3, 0.3, 0.13, 0.1, 0.19, 0.6, 0.12, 0.13, 0.17, 0.9, 0.1, 0.11, 0.1, 0.14, 0.18, 0.14, 0.18, 0.5, 0.14, 0.18, 0.5, 0.14, 0.17, 0.16, 0.11, 0.1, 0.12, 0.11, 0.1, 0.15, 0.11, 0.02, 0.2, 0.12, 0.12, 0.11, 0.1, 0.1, 0.11, 0.1, 0.1, 0.11, 0.1, 0.13, 0.16, 0.14, 0.12, 0.17, 0.1, 0.11, 0.19, 0.16, 0.11, 0.13, 0.14, 0.18, 0.12, 0.11, 0.1.17, 0.13, 0.14, 0.10)

Competing interests
The authors declare that they have no competing interests.

Authors' contributions
NR coined the problem and its formulation. VS helped in the manuscript preparation. PS helped in problem solving. All authors read and approved the final manuscript.

Acknowledgement
The authors are very much thankful to the anonymous reviewers for their constructive comments and useful suggestions.

Author details
[1]Department of Mechanical Engineering, Govt. Polytechnic, Visakhapatnam 530008, India. [2]Department of Mechanical Engineering, Andhra University, Visakhapatnam 530003, India. [3]Department of Mechanical Engineering, GITAM University, Visakhapatnam 530045, India.

References
Abd El-Wahed WF, Abo-Sinna MA (2001) A hybrid fuzzy goal programming approach to multiple objective decision making problems. Fuzzy Set Syst 119(1):71–85

Amiri A (2006) Design a distribution network in a supply chain system formulation and efficient solution procedure. Eur J Oper Res 171:567–576

Ballou RH (2004) Business logistics/supply chain management. Pearson Education, New Delhi

Beamon BM (1998) Supply chain design and analysis: models and methods. Int J Prod Econ 55:281–294

Bellman RE, Zadeh LA (1970) Decision-making in a fuzzy environment. Manag Sci 17:141–164

Bouzembrak Y, Allaoui H, Goncalves G, Bouchriha H (2011) A multi-objective green supply chain network design. In: 2011 4th International conference on logistics (LOGISTIQUA). Hammamet, . 31 May 3 June2011, pp 357–361

Chen YW, Tzeng GH (2000) Fuzzy multi-objective approach to the supply chain model. Int J Fuzzy Syst 1(3):220–227

Chen CL, Wang BW, Lee WC (2003) Multi-objective optimization for a multi-enterprise supply chain network. Ind Eng Chem Res 42:1879–1889

Cohen MA, Lee HL (1988) Strategic analysis of integrated production-distribution systems. Model Meth Oper Res 36(2):216–228

Eskigun E, Uzsoy R, Preckel PV, Beaujon G, Krishnan S, Tew JD (2005) Outbound supply chain network design with mode selection, leadtimes and capacitated vehicle distribution centers. Eur J Oper Res 165(1):182–206

Farahani RZ, Elahipanah M (2008) A genetic algorithm to optimize the total cost and service level for just-in-time distribution in a supply chain. Int J Prod Econ 111:229–243

Lee HL, Kim SH, Moon C (2002) Production-distribution planning in supply chain using a hybrid approach. Prod Plann Contr 13(1):46–75

Li H-L (2002) SCM system in Taiwan electronic industry. Productivity 42(4):574–581

Liang TF (2008) Fuzzy multi-objective production/distribution planning decisions with multi-product and multi-time period in supply chain. Comput Ind Eng 55:676–694

Mohamed W (2000) Chance constrained fuzzy goal programming and right-hand side uniform random variable coefficients. Fuzzy Set Syst 109:107–110

Narasimhan R (1982) Geometric averaging procedure for constructing super transitive approximation to binary comparison matrix. Fuzzy Set Syst 8:53

Narayana Rao K, Venkatasubbaiah K (2011) Electronic supply network coordination in intelligent and dynamic environments. IGI Global, Hershey, pp 146–167

Peidro D, Mula J, Poler R, Verdegay JL (2009) Fuzzy optimization for supply chain planning under supply, demand and process uncertainties. Fuzzy Set Syst 160(18):2640–2657

Petrovic D, Roy R (1998) Modeling and simulation of a supply chain in an uncertain environment. Eur J Oper Res 109:299–309

Pyke DF, Cohen MA (1994) Multiproduct integrated production–distribution systems. Eur J Oper Res 74:18–49

Ramik J (2000) Fuzzy alternatives in goal programming problems. Fuzzy Set Syst 111:81–86

Robinson PE, Goa LL, Muggenborg ST (1993) Designing an integrated distribution system at Dowbrands, Inc. Interfaces 23(3):107–117

Sabri EH, Beamon BM (2000) A multi-objective approach to simultaneous strategic and operational planning in supply chain design. Omega 28(5):581–595

Troncoso J, D'Amours S, Flisberg P, Ronnqvist M, Weintraub A (2011) A mixed integer programming model to evaluate integrating strategies in the forest value chain – a case study in the Chilean forest industry. CIRRELT 28:1–29

Tsiakis P, Shah N, Pantelidas CC (2001) Design of multi-echelon supply chain networks under demand uncertainty. Ind Eng Chem Res 40:3585–3604

van der Vorst JGAJ (2000) Effective food supply chains:generating modeling and evaluating supply chain scenarios. Wageningen University, Wageningen

Yu Y, Zu H, Cheng TCE (2003) A strategic model for supply chain design with logical constraints: formulation and solution. Comput Oper Res 30:2135–2155

Zimmermann HJ (1978) Fuzzy programming and linear programming with several objective functions. Fuzzy Set Syst 1:45–55

3

Stochastic extension of cellular manufacturing systems: a queuing-based analysis

Fatemeh Fardis[1], Afagh Zandi[1] and Vahidreza Ghezavati[2*]

Abstract

Clustering parts and machines into part families and machine cells is a major decision in the design of cellular manufacturing systems which is defined as cell formation. This paper presents a non-linear mixed integer programming model to design cellular manufacturing systems which assumes that the arrival rate of parts into cells and machine service rate are stochastic parameters and described by exponential distribution. Uncertain situations may create a queue behind each machine; therefore, we will consider the average waiting time of parts behind each machine in order to have an efficient system. The objective function will minimize summation of idleness cost of machines, sub-contracting cost for exceptional parts, non-utilizing machine cost, and holding cost of parts in the cells. Finally, the linearized model will be solved by the Cplex solver of GAMS, and sensitivity analysis will be performed to illustrate the effectiveness of the parameters.

Keywords: Cellular manufacturing system; Stochastic arrival rate and service rate; Average waiting time; Queuing theory

Introduction

Cellular manufacturing system (CMS) is an application of the group technology concept, which classifies parts with closest features and processes into the part families and assigns machines into the cells, with the goal of increasing production efficiency while decreasing the unit cost. Some advantages of CMS such as simplification of material flow, reduction of transportation and queuing times, reduction of material handling cost and setup times, and the increase of machine utilization and throughput rates are declared in literatures (Muruganandam et al. 2005; Olorunniwo and Udo 2002; Wemmerlov and Hyer 1989). The four major decisions in the implementation of cellular manufacturing systems are the following:

1. Cell formation: grouping parts with the similar processes and features into part families and allocating machines to the cells (Mahdavi et al. 2007; Muruganandam et al. 2005; Yasuda et al. 2005)

2. Group layout: laying out cells and machines within each cell (Mahdavi and Mahadevan 2008; Tavakkoli-Moghaddam et al. 2007; Wu et al. 2007a)

3. Group scheduling: operating and managing the cell operation (Mak and Wang 2002; Tavakkoli-Moghaddam et al. 2010)

4. Resource allocation: assigning resources, such as tools, materials, and human resources, to the cells Cesani and Steudel (2005; Mahdavi et al. 2010)

Wu et al. (2007b) considered cell formation, group layout, and group scheduling decisions simultaneously in their model, which minimize the makespan. They presented a hierarchical genetic algorithm to solve it. Logendran (1993) developed a mathematical programming model to minimize part inter-cell and intra-cell movements and proposed a heuristic algorithm to solve it. Chen (1998) proposed an integer programming model to minimize material handling and machine cost and reconfiguration cost to design a sustainable cellular manufacturing system in a dynamic environment. Most of the developed models in cellular manufacturing systems are cost-based, but there are some models in which machine reliability is considered simultaneously with different cost types. A multi-objective mixed

* Correspondence: v_ghezavati@azad.ac.ir
[2]Faculty of Industrial Engineering, Islamic Azad University, South Tehran Branch, Tehran, Iran
Full list of author information is available at the end of the article

integer programming model was presented by Das et al. (2006) that minimizes variable cost of machine and penalty cost of non-utilizing machine as well as inter-cell material handling cost, and maximizes system reliability with minimizing failure rate. Machine breakdown cost (Jabal-Ameli and Arkat 2007) and inverse of the reliability of the system (Das et al. 2007) are two more objective components in order to maximize machine reliability, which will develop cell performance.

Cellular manufacturing problems can be under static or dynamic conditions. In static conditions, cell formation is done for a single-period planning horizon, where product mix and demand rate are constant. However, in dynamic conditions, the planning horizon is considered as a multi-period planning horizon where the product mix and demand rate differ from one period to another. In order to reach the best efficiency, there will be different cell formations for each period. Some recent

Table 1 Summarized cellular manufacturing system review

Publication	Objective	Problem definition	Solution method	Output
Sarker and Li (1997)	Min INTER + OC	SO, MIP	B&B	CF, PR
Wicks and Reasor (1999)	Min INTER + RCC + CI	SO, DP, MPP, MIP	GA	CF, MD
Chen (1998)	Min INTER + OC + MHC + RCC	SO, DP, MPP, IP	DBH	CF, RCP
Chen (2001)	Min INTER + OC + IC + SC	SO, MINLP	SMH	CF, IL, PL
Baykasoglu et al. (2001)	Min DS + CLV + ECR	MO, NLP	SA	CF, PRP
Mak and Wang (2002)	Min TDT	SO, MINLP	GA	CF, PS
Muruganandam et al. (2005)	Min CLV + MM	CO, NLP	MA	CF
Tavakkoli-Moghaddam et al. (2005)	Min RCC + MAC + OC + INTER	SO, DP, NLP	GA, SA, TS	CF, RCP
Das et al. (2006)	Min 1.FRM	MO, MIP	SA	CF, PR
	2.VCM + INTER + NUC			
Chan et al. (2006)	Min VO + EE, TDT	SO, NLP, QAP, 2SP	GA	CF, MSE
Saidi-Mehrabad and Safaei (2007)	Min MAC + INTER + OC + RCC	SO, DP, MPP, NLP	NA	CF, CS, PR, RCP, MD
Jabal-Ameli and Arkat (2007)	Min INTER + MBC	SO, IP	B&B	CF, PR
Wu et al. (2007a)	Min 1.MM, 2.EE	MO, IP	HGA	CF, GL
Wu et al. (2007b)	Min CM	SO, NLP	HGA	CF, GL, PS
Das et al. (2007)	Min 1.ISR	MO, NLP	HIA	CF, OPMI
	2.VCM + INTER + NUC			
Mahdavi et al. (2007)	Min EE + VO	SO, NLP	B&B	CF
Schaller (2007)	Min PC + MAC + RCC	SO, DP, IP	CBP, TS	CF, RCP
Safaei et al. (2008)	Min INTER + INTRA + CCM + VCM + RCC	SO, DP, MPP, MIP	MFA-SA	CF, RCP
Bajestani et al. (2009)	Min 1.CLV	MO, DP, MINLP	MOSS	CF, RCP
	2.INTER + RCC + PUC			
Kioon et al. (2009)	Min RCC + PM + OC + SCC + HC + PC + INTER + INTRA	SO, DP, MINLP	B&C	CF, PP, PR, RCP, IL
Safaei and Tavakkoli-Moghaddam (2009)	Min CCM + VCM + IT + INTER + INTRA + RCC + SCC	SO, DP, MINLP	B&B	CF, RCP, BL, IL
Saidi-Mehrabad and Ghezavati (2009)	Min ICM + SCC + NUC	SO, STP, MINLP	B&B	CF
Mahdavi et al. (2010)	Min VO + EE	SO, NLP	GA	CF
(Tavakkoli-Moghaddam et al. 2010)	Min INTRAT + TT + CSC + CM	MC, MINLP	SS	CF, PS
Mahdavi et al. (2010)	Min HC + BC + INTER	SO, DP, MPP, NLP	B&B	CF, RCP, WA, IL
	MC + RCC + SAC + HIC + FIC			PL, BL, MD
Rafiee et al. (2011)	Min INTER + INTRA	SO, DP, MIP	PSO	CF, RCP, PR, PP, IL, SBQ
	OC + PUC + RCC + PRS + SCC + HC + CRC + PM			
Ariafar et al. (2011)	Min INTER + INTRA	SO, STP, MINLP	B&B	GL

researches are under dynamic conditions (Kioon et al. 2009; Mahdavi et al. 2010; Schaller 2007).

Inputs of classical cellular manufacturing system problems are certain, but in real problems, some input parameters such as costs, demands, processing times, and setup times are uncertain so that this uncertainty can affect on the results. Some approaches such as stochastic programming, queuing theory, and robust optimization can be applied for uncertain models (Saidi-Mehrabad and Ghezavati 2009). Ariafar et al. (2011) proposed a mathematical model for layout problem in cellular manufacturing systems that the demand of each product was described by uniform distribution. The objective function of their model minimizes total cost of inter-cell and intra-cell material handling, concurrently. It was solved by the Lingo software. Saidi-Mehrabad and Ghezavati (2009) presented a stochastic CMS problem applying queuing theory approach. They assumed parts as customers and machines as servers, in which the arrival rate of parts and service rate of machines were identified by the exponential distribution. Their objective function components are the following: non-utilization cost, machine idleness cost, and sub-contracting cost. They computed the utilization factor for each machine which shows the probability that each machine is busy.

Table 1 illustrates the summarization of more reviewed literatures consisting objectives, model definitions, solution methods, and outputs. This helps to compare our work with previous researches. Table 2 describes the abbreviations used in Table 1.

In this paper, we will develop a CMS problem as a queue system considering the average waiting time of parts behind each machine and holding cost of parts in cells. The goal of this model is to minimize summation of four cost types: (1) the idleness costs for machines, (2) total cost of sub-contracting, (3) non-utilization cost of machines in cells, and (4) holding cost of parts in cells. The rest of the paper is organized as follows: The proposed model is described in the 'Mathematical modeling' section. In the 'Experimental results' section, the experimental results and sensitivity analysis are presented. The last section is the 'Conclusions and future directions.'

Mathematical modeling
Problem description
In this section, a stochastic cellular manufacturing system problem will be formulated as a queue system, which considers parts and machines as customers and servers with the arrival and service rate of λ and μ, respectively. Also, we consider that at each time, only one part can be processed by a machine; thus, when a machine is processing a part, the others should wait, and a queue will be created behind the machine. The

population of this queue as a new part arrives (birth) can be increased, or it can be decreased (death) by service completion.

In a steady system, to avoid infinite growth of queue, the service rate must be greater than the arrival rate, so the utilization factor (probability of being busy) of each machine ($\rho = \lambda / \mu$) will be less than 1. Also, as each machine processes different parts with different arrival rates, according to this property, the minimum of some independent exponential random variables with the arrival rate of $\lambda_1, \lambda_2, ..., \lambda_m$ is also exponential with the arrival rate of $\lambda = \sum_{i=1}^{m} \lambda_i$. Hence, the utilization factor is $\rho = \sum \lambda_i / \mu$.

Our queue system is formulated as $M / M / 1$, where the arrival and processing time of parts are uncertain and described by exponential distribution, and as it was mentioned earlier, each machine can process at most one part at a time. In this problem, the decision maker needs to allocate parts and machines to cells in order to minimize objective function value. In the previous work of Saidi-Mehrabad and Ghezavati (2009), only the impact of utilization factor in designing CMS is considered, but our study shows the impact of utilization factor and maximum waiting time of parts simultaneously. Due to the uncertainty of arrival and service time of parts, the time that each part spends in the cell is uncertain, and as the time passes, the parts will be broken. Thus, in order to avoid long waiting time, a chance constraint is considered to show that the probability that the average waiting time of a part behind each machine exceeds the critical time is less than the service level (α). By knowing that, this probability affects on the utilization factor.

Notation
Indexes
The following are the indexes:

- i Part index, $i = 1, ..., p$
- j Machine index, $j = 1, ..., m$
- k Cell index, $k = 1, ..., c$

- $a_{ij} = \begin{cases} 1 & \text{if machine } j \text{ processes part } i \\ 0 & \text{otherwise} \end{cases}$

- C_i Sub-contracting cost per unit for part i
- u_j Idleness cost of machine j
- M_m Maximum number of machines permitted in a cell
- α Maximum allowed probability that the waiting time behind each machine can be more than the critical time
- t_{qj} Average time parts spend behind machine j
- t Critical waiting time
- np_i Total number of part i

Table 2 Abbreviations used in Table 1

Objective function	Abbreviation	Problem definition	Abbreviation	Solution and outputs	Abbreviation
Inter-cell material handling cost	INTER	Single objective	SO	Branch and bound	B&B
Intra-cell material handling cost	INTRA	Multi-objective	MO	Branch and cut	B&C
Machine operating cost	OC	Combined objective	CO	Genetic algorithm	GA
Material moves	MM	Multi-criteria	MC	Decomposition base heuristic	DBH
Backorder cost	BC	Multi-period planning	MPP	Silver meal heuristic	SMH
Production cost	PC	Stochastic problem	STP	Tabu search	TS
Inventory cost	IC	Dynamic programming	DP	Hierarchical genetic algorithm	AGA
Reconfiguration cost	RCC	Integer program	IP	Mean field annealing and simulated annealing	MFA-SA
Capital investment	CI	Mixed integer program	MIP	Multi-objective scatter search	MOSS
Machine holding cost	MHC	Non-linear program	NLP	Hierarchical approach	HIA
System setup cost	SC	Mixed integer non-linear program	MINLP	CB procedure	CBP
Dissimilarity between parts	DS	Quadratic assignment problem	QAP	Particle swarm optimization	PSO
Cell load variation	CLV	Two-stage scheduling problem	2SP	Scatter search	SS
Extra capacity requirement	ECR			Neural approach	NA
Total distance traveled	TDT			Cell formation	CF
Cell setup cost	CSC			Inventory level	IL
Holding cost	HC			Production level	PL
Machine cost	MC			Backorder level	BL
Machine amortization cost	MAC			Reconfiguration plan	RCP
Machine breakdown cost	MBC			Machine sequence	MSE
Failure rate of machine	FRM			Cell size	CS
Machine variable cost	VCM			Group layout	GL
Machine constant cost	CCM			Process route	PR
Machine non-utilization cost	NUC			Machine duplication	MD
Total number of voids	VO			Production plan	PP
Total exceptional elements	EE			Production scheduling	PS
Completion time	CM			Processing part requirement	PRP
Inverse of system reliability	ISR			Optimal preventive maintenance interval	OPMI
Purchase cost of machine	PUC			Subcontracted quantity	SBQ
Preventive maintenance cost	PM			Worker assignment	WA
Sub-contracting cost	SCC				
Inventory transportation	IT				
Intra-cell move time	INTRAT				
Tardiness time	TT				
Salary cost	SAC				
Firing cost	FIC				

Table 2 Abbreviations used in Table 1 *(Continued)*

Hiring cost	HIC
Process routes setup cost	PRS
Corrective repair cost	CRC
Machine idleness cost	ICM

- C_u Maximum number of cells
- λ_i Mean arrival rate for part i
- μ_j Mean service rate for machine j
- U_{ij} Cost of part i not utilizing machine j
- h_i Holding cost per unit for part i

Decision variables

The following are the decision variables:

- $X_{ik} = \begin{cases} 1 \text{ if part } i \text{ is assigned to cell } k \\ 0 \text{ otherwise} \end{cases}$

- $Y_{jk} = \begin{cases} 1 \text{ if machine } j \text{ is assigned to cell } k \\ 0 \text{ otherwise} \end{cases}$

- ρ_j Utilization factor for machine j (the value of ρ indicates the probability in which machine j is busy)

Mathematical model

In this section, details of the mathematical formulation which we are interested will be described. For this purpose, the following formulations are presented:

$$\text{Min } Z = \sum_j \left(1-\rho_j\right) u_j + \sum_k \sum_j \sum_i C_i a_{ij} X_{ik} \left(1-Y_{jk}\right)$$
$$+ \sum_k \sum_j \sum_i U_{ij} X_{ik} Y_{jk} \left(1-a_{ij}\right)$$
$$+ \sum_k \sum_j \sum_i \frac{h_i a_{ij} X_{ik} Y_{jk}}{np_i}$$

subject to   (1)

$$\sum_k X_{ik} = 1 \qquad i = 1, 2, ..., p, \tag{2}$$

$$\sum_k Y_{jk} = 1 \qquad j = 1, 2, ..., m, \tag{3}$$

$$\rho_j - \sum_k \frac{\sum_i \lambda_i a_{ij} X_{ik} Y_{jk}}{\mu_j} = 0 \qquad j = 1, ..., m, \tag{4}$$

$$\sum_j Y_{jk} \leq M_m \qquad k = 1, 2, ..., c, \tag{5}$$

$$P\left(t_{qj} > t\right) \leq \alpha, \tag{6}$$

$$\rho_j \leq 1 \; j = 1, 2, ..., m, \tag{7}$$

$$X_{ik}, Y_{jk} \in \{0, 1\}, \rho_j \geq 0. \tag{8}$$

Equation 1 indicates the objective function which can compute the total idleness cost for machines in cells,

sub-contracting cost, resource underutilization cost, and holding cost of parts in cells. Set constraint (2) restricts that each part is allocated to only one cell. Set constraint (3) ensures that each machine is allocated to only one cell. Set constraint (4) computes the utilization factor. Set constraint (5) guarantees that the number of machines in each cell will not exceed the maximum number. Equation (6) is a chance constraint and ensures that the probability that the average waiting time of parts behind each machine exceeds the critical time is less than the service level (α). Set constraint (7) says that the proportion of time that the machine is processing a part must be less than or equal to 1. Set constraint (8) specifies binary and non-negative variables.

Linearization of the model

As the objective function has non-linear terms, we can change it to a mixed integer linear programming. For this purpose, we replace a new binary variable Z_{ijk} instead of the multiple of X_{ik} and Y_{jk}. We reformulate the model, and the three auxiliary constraints (10), (11), and (12) are added to the model to guarantee the correctness of the replacement. Constraint (6) is equal to $e^{-\mu_j(1-\rho_j)t} \leq \alpha$ and contains a non-linear equation; therefore, Equation 14 denotes the linear form of this constraint.

$$\text{Min } Z = \sum_j \left(1-\rho_j\right) u_j + \sum_k \sum_j \sum_i C_i a_{ij} X_{ik}$$
$$- \sum_k \sum_j \sum_i C_i a_{ij} Z_{ijk}$$
$$+ \sum_k \sum_j \sum_i U_{ij} \left(1-a_{ij}\right) Z_{ijk}$$
$$+ \sum_k \sum_j \sum_i \frac{a_{ij} Z_{ijk}}{np_i}. \tag{9}$$

Subject to constraints (2), (3), (5), (7), and (8),

$$Z_{ijk} \leq X_{ik} \forall i, j, k, \tag{10}$$
$$Z_{ijk} \leq Y_{jk} \forall i, j, k, \tag{11}$$

$$X_{ik} + Y_{jk} - Z_{ijk} \leq 1 \forall i, j, k. \tag{12}$$

Constraints (4) and (6) are changed as follows:

$$\rho_j - \sum_k \frac{\sum_i \lambda_i a_{ij} Z_{ijk}}{\mu_j} = 0 \qquad \forall j \tag{13}$$

$$-\mu_j \left(1-p_j\right) t \leq \ln \alpha \forall j. \tag{14}$$

Table 3 Effectiveness of queuing approach in a CMS problem

Problem information					
Problem number	Number of parts × number of machines × number of cells	Maximum number of machine allowed in each cell	Idleness rate cost	Average utilization factor (%)	Number of inter-cellular movement
P1	30 × 10 × 3	4	300	24.27	43
P2			460	31.27	40
P3			654	34.66	38
P4			850	44.52	32
P5			905	48.36	30
P6			1,025	56.83	25
P7			1,350	59.90	21
P8			1,600	62.05	20
P9			2,105	65.12	16
P10			2,678	69.97	13
P11	38 × 10 × 3	4	300	46.91	52
P12			460	47.56	51
P13			654	48.55	50
P14			850	48.64	50
P15			905	48.71	49
P16			1,025	48.93	49
P17			1,350	49.63	49
P18			1,600	49.80	48
P19			2,105	49.96	48
P20			2,678	50.25	48

Experimental results

Consider a manufacturing system consisting of ten machines to process parts, wherein the decision maker should allocate machines and parts to three cells. Also, the maximum number of machines permitted to be located is four. In this section, we present some numerical examples which have been generated randomly, to illustrate the effect of changing the main parameters (α, t, u_j), on the number of sub-contracting movements and utilization factors. The proposed mixed integer model

was performed by GAMS and Cplex solver and was run on a processor Intel Core 2 Duo CPU running at 2 GHz with 2-GB RAM.

In Table 3, the results associated to solve two sets of examples for ten times for each is shown, where only idleness cost is not fix, and the impact of its changes on the average utilization factor and the number of sub-contracting movements are illustrated. As the utilization factor of the machines is directly related to the idleness cost, it can be found that the more the average of

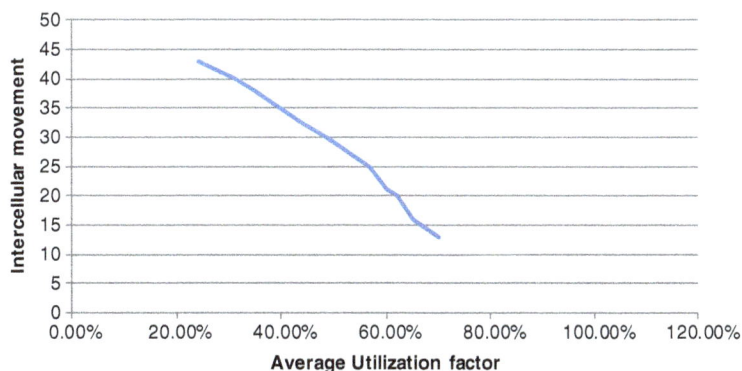

Figure 1 Relationship curve between average utilization factor and number of sub-contracting movements of set problem 1.

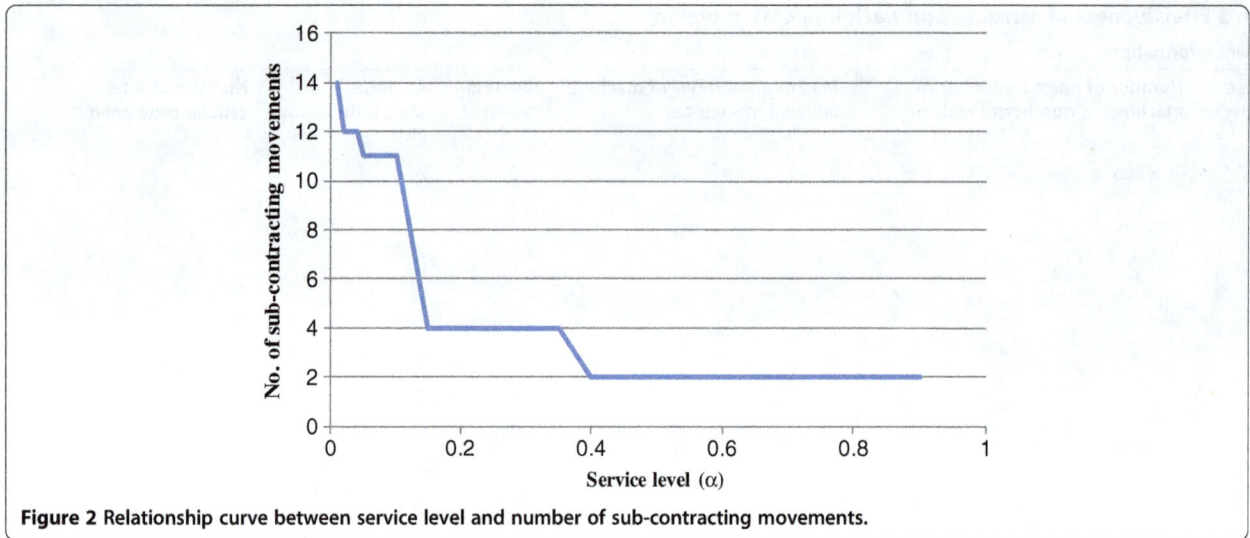

Figure 2 Relationship curve between service level and number of sub-contracting movements.

idleness cost leads to the better design of cellular manufacturing system. It means more costs lead to a higher utilization factor. Also, more utilization factor may lead to less number of sub-contracting movements. Results of both sets show the same changes.

Figure 1 indicates the relation between the average utilization factor and number of sub-contracting movements of set problem 1. It can be found that by increasing idleness cost, the average utilization factor will be increased, too. As the term $\Sigma(1 - \rho_j)u_j$ indicates, the direct relation between idleness cost and utilization factor is established. Therefore, in order to minimize this term, by increasing idleness cost, idleness rate of machine must be decreased. This means that the probability that each machine is busy increases. Therefore, the total number of sub-contracting movements must be decreased in order to decrease idleness of each machine. The less total number of sub-contracting movements

makes the queue system be more populated. Therefore, the total objective function value will be minimized.

The effect of service level's changes (the maximum allowed probability that the waiting time behind each machine can be more than the critical time) on the sub-contracting movements is shown in Figure 2. This figure illustrates that for a fixed critical waiting time, if the service level (α) increases, the upper bound of utilization factor (ρ_j) will increase, where this growth may cause decreasing in the total number of sub-contracting movements.

Figure 3 demonstrates the relation between changes of critical waiting time and number of sub-contracting movements. If we assume all the parameters to be fixed except the service level (α), increasing the critical waiting time may lead to the reduction of the number of sub-contracting movements, which is due to the increase of the upper bound of the utilization factor.

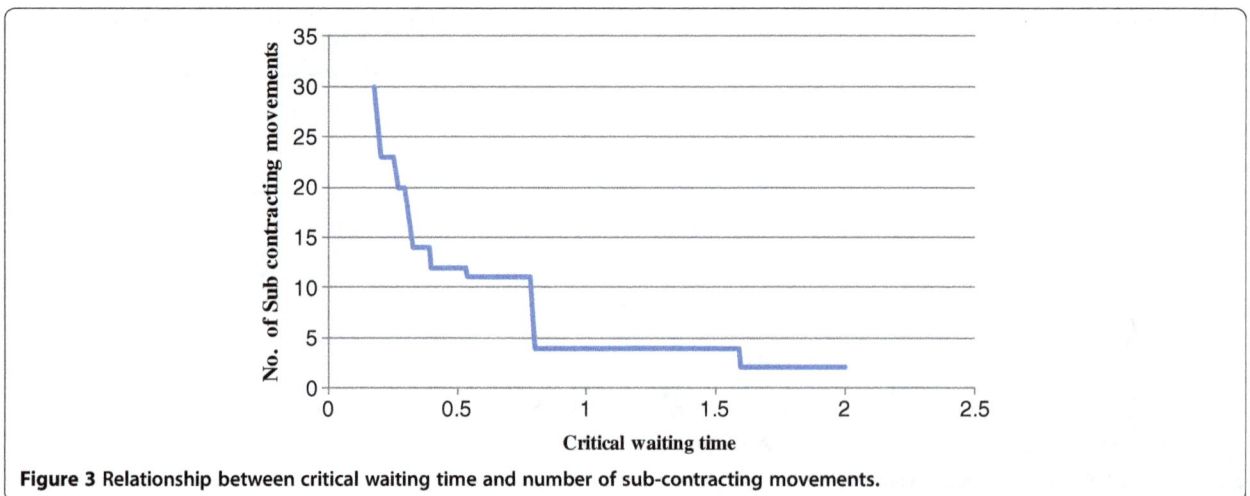

Figure 3 Relationship between critical waiting time and number of sub-contracting movements.

Conclusions and future directions

We have developed a stochastic CMS model that considers the arrival rate of parts into cells and machine service rate as uncertain parameters. The proposed non-linear mixed integer programming model was linearized using auxiliary variables. Then, the linearized model was solved using the Cplex solver of GAMS. As the CMS problem is NP-hard, by increasing the size of the problem, GAMS stops solving it; due to the increase of computational time, the branch-and-cut algorithm is unable to give good solutions. Therefore, it is necessary to present a heuristic or meta-heuristic approach to solve this model for large-scale problems. Also, the following directions can be applied for further considerations:

1. Developing the proposed model under new stochastic parameters such as capacities, lead times, and machine failures.
2. Analyzing the defined problem under scenario-based planning approach and robust optimization theory.
3. Incorporating our objectives with productivity and production planning aspects in uncertain situations.

Competing interests

The authors declare that they have no competing interests.

Authors' contributions

VG carried out the modeling, drafted the manuscript, involved in revising the manuscript, and gave final approval of the version to be published. FF was involved in performing the model and analyzing the results, participated in the sequence alignment, and helped draft the manuscript. AZ carried out the literature study and participated in the design of the study and the sequence alignment. All authors read and approved the final manuscript.

Author details

[1]Islamic Azad University, South Tehran Branch, Tehran, Iran. [2]Faculty of Industrial Engineering, Islamic Azad University, South Tehran Branch, Tehran, Iran.

References

Ariafar SH, Ismail N, Tang SH, Ariffin MKAM, Firoozi Z (2011) A stochastic facility layout model in cellular manufacturing systems. International Journal of the Physical Sciences 6(15):3666–3670

Bajestani AM, Rahimi-Vahed AR, Khoshkhou BG (2009) A multi-objective scatter search for a dynamic cell formation problem. Comp Oper Res 36(3):777–794

Baykasoglu A, Gindy NNZ, Cobb RC (2001) Capability based formulation and solution of multiple objective cell formation problems using simulated annealing. Integr Manuf Syst 12(4):258–274

Cesani VI, Steudel HJ (2005) A study of labor assignment flexibility in cellular manufacturing systems. Comp Indust Eng 48(3):571–591

Chan FFS, Lan KW, Chan PLY, Choy KLL (2006) Two stage approach for machine part grouping and cell layout problem. Robot Comp Integr Manuf 22(3):217–238

Chen M (1998) A mathematical programming model for system reconfiguration in a dynamic cellular manufacturing environment. Ann Oper Res 77:109–128

Chen M (2001) A model for integrated production planning in cellular manufacturing systems. Integr Manuf Syst 12(4):275–284

Das K, Lashkari RS, Sengupta S (2006) Reliability considerations in the design of cellular manufacturing systems: a simulated annealing based approach. Int J Qual Reliability Manage 23(7):880–904

Das K, Lashkari RS, Sengupta S (2007) Machine reliability and preventive maintenance planning for cellular manufacturing system. Eur J Oper Res 183(1):162–180

Jabal-Ameli MS, Arkat J (2007) Cell formation with alternative process routings and machine reliability consideration. Int J Adv Manuf Technol 35(7):761–768

Kioon SA, Bulgak AA, Bektas T (2009) Integrated cellular manufacturing system design with production planning and dynamic system reconfiguration. European Journal of Operation Research 192(2):414–428

Logendran R (1993) Methodology for converting a functional manufacturing system into a cellular manufacturing system. Int J Prod Econ 29(1):27–41

Mahdavi I, Mahadevan B (2008) CLASS: an algorithm for cellular manufacturing system and layout design using sequence data. Robot Comp Integr Manuf 24(3):488–497

Mahdavi I, Javadi B, Fallah-Alipour K, Slomp J (2007) Designing a new mathematical model for cellular manufacturing system based on cell utilization. App Math Comput 190(1):662–670

Mahdavi I, Alaei A, Paydar MM, Solimanpour M (2010) Designing a mathematical model for dynamic cellular manufacturing systems considering production planning and worker assignment. Comp Math Appl 60(4):1014–1025

Mak KL, Wang XX (2002) Production scheduling and cell formation for virtual cellular manufacturing system. Int J Adv Manuf Technol 20(2):144–152

Muruganandam A, Prabharan G, Asokan P, Baskaran V (2005) A memetic algorithm approach to the cell formation problem. Int J Adv Manuf Technol 25:988–997

Olorunniwo F, Udo G (2002) The impact of management and employees on cellular manufacturing implementation. Int J Prod Econ 76(1):27–38

Rafiee K, Rabbani M, Rafiei H, Rahimi-Vahed A (2011) A new approach towards integrated cell formation and inventory lot sizing in an unreliable cellular manufacturing system. Appl Math Modeling 35(4):1810–1819

Safaei N, Tavakkoli-Moghaddam R (2009) Integrated multi-period cell formation and sub-contracting production planning in dynamic cellular manufacturing system. Int J Prod Econ 120(2):301–314

Safaei N, Saidi-Mehrabad M, Jabal-Ameli MS (2008) A hybrid simulated annealing for solving an extended model for dynamic cellular manufacturing system. Eur J Oper Res 185(2):563–592

Saidi-Mehrabad M, Ghezavati VR (2009) Designing cellular manufacturing system under uncertainty. J Uncert Syst 3(4):315–320

Saidi-Mehrabad M, Safaei N (2007) A new model of dynamic cell formation by a neural approach. Int J Adv Manuf Technol 33:1001–1009

Sarker BR, Li K (1997) Simultaneous route selection and cell formation: a mixed-integer programming time–cost model. Integr Manuf Syst 8(6):374–377

Schaller J (2007) Designing and redesigning cellular manufacturing system to handle demand change. Comput Ind Eng 53(3):478–490

Tavakkoli-Moghaddam R, Aryanezhad MB, Safaei N, Azaran A (2005) Solving a dynamic cell formation problem using meta-heuristics. Appl Mathe Comp 170(2):761–780

Tavakkoli-Moghaddam R, Javadian N, Javadi B, Safaei N (2007) Design of facility layout problem in cellular manufacturing system with stochastic demand. Appl Math Comput 184(2):721–728

Tavakkoli-Moghaddam R, Javadian N, Khorrami A, Gholipor-Kanani Y (2010) Design of a scatter search method for a novel multi-criteria group scheduling problem in a cellular manufacturing system. Expert Syst Appl 37(3):2661–2669

Wemmerlov U, Hyer NL (1989) Cellular manufacturing in the U.S. industry: a survey of users. Int J Product Res 27(9):1511–1530

Wicks EM, Reasor RJ (1999) Designing cellular manufacturing systems with dynamic part populations. Inst Indust Eng Trans 31(1):11–20

Wu X, Chu CH, Wang Y, Yan W (2007a) A genetic algorithm for cellular manufacturing design and layout. European J Oper Res 181(1):156–167

Wu X, Chu CH, Wang Y, Yue D (2007b) Genetic algorithm for integrating cell formation with machine layout and scheduling. Comp Indust Eng 53(2):277–289

Yasuda K, Hu L, Yin Y (2005) A grouping genetic algorithm for the multi-objective cell formation problem. Int J Prod Res 43(4):829–853

4

A three-stage assembly flow shop scheduling problem with blocking and sequence-dependent set up times

Aref Maleki-Darounkolaei[1], Mahmoud Modiri[1*], Reza Tavakkoli-Moghaddam[2] and Iman Seyyedi[3]

Abstract

This paper considers a three-stage assembly flowshop scheduling problem with sequence-dependent setup times at the first stage and blocking times between each stage in such a way that the weighted mean completion time and makespan are minimized. Obtaining an optimal solution for this type of complex, large-sized problem in reasonable computational time using traditional approaches or optimization tools is extremely difficult. Thus, this paper proposes a meta-heuristic method based on simulated annealing (SA) in order to solve the given problem. Finally, the computational results are shown and compared in order to show the efficiency of our proposed SA.

Keywords: Assembly flowshop scheduling, Sequence-dependent setup times, Blocking times, Weighted mean completion time, Makespan, Simulated annealing

Background

Because of strong competition and limitation of resources in our environment, scheduling is a very important decision-making process in production and service industries. In common flowshop scheduling, we have two main elements, namely a group of M machines and a set of N jobs to be processed on this group of machine [1]. Assembly flowshop scheduling is a type of flowshop that at first each of n jobs has to be processed at the first stage consisting of m different parallel machines and then assembled at the second stage including only one assembly machine [2]. Assembly-type production systems have evolved partially as an answer to the market pressure for larger product variety [3]. Most of studies considered a two-stage assembly flowshop scheduling problem (AFSP) defined as follows. M machines are available in the first stage and only one machine is available in the assembly stage. There are n jobs, which should be scheduled and each of them includes $m + 1$ operations. The first m operations of a job are performed at the first stage in parallel by m

machines and the final operation is conducted at the second stage. Each of m operations of a job at the first stage is performed by a different machine, and the assembly operation on the machine at the second stage starts when all m operations at the first stage are completed. Each machine works just on one job at a time. It should be noted that when there is only one machine at the first stage [4]. In the two-stage AFSP, assumed collecting and transferring time of components from the first stage to assemble is negligible. This is unrealistic especially when a two-stage assembly problem is used to simulate production systems with a multi-facilities plant and a final assembly plant. But to have more realistic environments of a production system, it is required that the intermediate operation is devoted to collect and transport the manufactured parts from the various production areas to the assembly line. This stage is important especially when parts are manufactured in multiple production sites. The three-stage AFSP is the extended model of two-stage assembly flowshop that the collecting and transferring actions are regarded as the second stage, and assembly machine is in the third stage [3].

Suppose there are n jobs for scheduling, in which each job includes m components. At the first stage, there are m parallel and independent machines, in which each

* Correspondence: mahmoud.modiri@gmail.com
[1]Department of Management and Accounting, South Tehran Branch, Islamic Azad University, Tehran, Iran
Full list of author information is available at the end of the article

machine can process just one component. When all of m components of each job are processed on the first stage machines, they will be collected and transferred to the assembly machine (i.e., third stage) by passing the second stage (i.e., transportation stage). Then the machine at the third stage assembles m components of job that are transferred from the first stage together for completing a job. Koulamas and Kyparisis [3] proposed this type of an assembly line problem with the objective of minimizing the makespan. Hatami et al., [5] developed this model with sequence-dependent setup time for first stage machines.

In this paper, we consider a three-stage AFSP with blocking times and sequence dependent setup times. To make this type of assembly flowshop more realistic our research added the blocking times limitation (buffer = 0) to the model presented in [5]. Sequence-dependent setup time says that setup time of a job in position i on machine j depends on the current job and the previous job on this machine. Once its processing is completed on a processor in the first or second stage, a product is transferred directly to either an available processor in the next stage (or another downstream stage depending on the product processing route), or a buffer ahead of that stage when such an intermediate buffer is available. However, when an intermediate buffer is unavailable, the product remains blocking the processor until a downstream processor becomes available [6]. In general, blocking scheduling problems arise in modern manufacturing environments with limited intermediate buffers between processors, such as just-in-time production systems or flexible assembly lines, and those without intermediate buffers, such as surface mount technology (SMT) lines in the electronics industry for assembling printed circuit boards, which includes three different stages in the following sequence: solder printing, component placement and solder reflow [7].

Yokoyama and Santos [8] presented a branch-and-bound method for three-stage flowshop scheduling with assembly operations to minimize the weighted sum of product completion times where there is only one machine in each stage. Koulamas and Kyparisis [3] analyzed a three-stage assembly scheduling problem by minimizing the makespan and analyzed the worst-case ratio bound for several heuristics for this problem. Hatami et al., [5] extended the three-stage assembly flowshop model presented in [3] with sequence-dependent setup time by minimizing the mean flow time and maximum tardiness and they proposed two meta-heuristics, namely simulated annealing (SA) and tabu search (TS). Allahverdi and Al-Anzi [9] addressed a two-stage AFSP with setup time by minimizing the total completion time and they proposed a dominance

relation and three heuristics, such as Ntabu, SDE and NSDE.

Lee et al., [10] studied a two-stage AFSP with considering two machines at the first stage. Al-Anzi and Allahverdi [4] considered a two-stage AFSP with the objective of minimizing the weighted sum of makespan and maximum lateness and presented heuristics namely TS, PSO, and SDE. Cheng et al., [11] studied two-stage differentiation flowshop consisting of a common critical machine in stage one and two independent dedicated machines in stage two by minimizing the weighted sum of machine completion times. Ng et al., [1] proposed a branch-and-bound algorithm for solving a two-machine flow shop problem with deteriorating jobs. Ruiz and Allahverdi [12] minimized the bi-criteria of makespan and maximum tardiness with an upper bound on maximum tardiness of the flowshop scheduling problem. Sun et al., [13] addressed powerful heuristics to minimize makespan in fixed, 3-machine, assembly-type flowshop scheduling.

In some environments, there are limited buffers or zero buffers between stages. Hall and Sriskandarajah [7] reviewed machine scheduling problems with blocking and no wait in process. Qian et al., [14] presented an effective hybrid algorithm based on deferential evolution (DE) for multi-objective flow shop scheduling with limited buffers. Liu et al., [15] solved flow shop scheduling with limited buffers with an effective hybrid PSO-based algorithm to minimize the maximum completion time. Wang et al., [16] introduced a hybrid genetic algorithm (GA) for flowshop scheduling with limited buffers with the objective to minimize the total completion time. Grabowski and Pempera [17] developed a fast tabu search (TS) algorithm to minimize the makespan in a flow shop problem with blocking.

Ronconi [18] analyzed the minimization of the makespan criterion for the flowshop problem with blocking by proposing constructive heuristics, namely MM, MME and PFE. Tavakkoli-Moghaddam et al., [6] presented an efficient memetic algorithm (MA) combined with (NVNS) to solve the flexible flow line with blocking (FFLB). Sawik [19] addressed a new mixed integer programming for the FFLB. Norman [20] explored a flowshop scheduling problem with finite buffer and sequence-dependent setup times and proposed a TS method. Ronconi and Henriques [21] introduced a GRASP-based heuristic method for a scheduling problem with blocking to minimizing the total tardiness. Tozkapan et al., [22] developed a lower bounding procedure and a dominance criterion incorporated into a branch-and-bound procedure for the two-stage AFSP to minimize the total weighted flowtime. Yagmahan and Yenisey [23] offered a multi-objective ant colony algorithm for flowshop scheduling to minimizing the makespan and total flow time. Sung and Kim [24] developed a branch-and-bound algorithm for two stage multiple assembly flowshop to minimize the sum of completion times. Yokoyama [25] considered flowshop scheduling with setup and assembly operation and to solve used pseudo-dynamic programming and a branch-and-bound. Liu and Kozan [26] studied scheduling flowshop with combined buffer condition considering blocking, no-wait and limited-buffer. Lee et al., [27] brought the concept of blocking into the deteriorating job scheduling problem on the two-machine flow- shop. They proposed A branch-and-bound algorithm incorporating with several dominance rules and a lower bound as well as several heuristic algorithms. Gong et al., [28] studied two-stage flow shop scheduling problem on a batching machine and a discrete machine with blocking and shared setup times. Wang et al., [29] proposed a HDDE algorithm for solving a flowshop scheduling with blocking to minimize the makespan.

Since blocking has been never considered in three-stage assembly flowshop so we add blocking as a constraint to the Hatami's problem [5]. Thus according to the new objective functions (i.e., weighted mean completion time and makespan) and blocking, we present a new mathematical model for this case.

The rest of this paper is come up as follows. In the next section, we explain the new mathematical model. In Section 3, we propose a meta-heuristic method based on SA to solve the given problem. Section 4 discusses the computational results and finally, the conclusion is presented in Section 5.

Problem Description

The problem considered in this paper is a three-stage AFSP with sequence-dependent setup times at the first stage and blocking time between stages minimizing the weighted mean completion time. In this problem, there are n jobs available at zero time. Job preemption is not allowed and each job includes m parts or components. We have m independent parallel machines at the first stage and every part of a job should be processed on just one machine at the first stage and each machine can process only one part. Setup time of a job on machines depend on job and previous job. After completing all m parts of job at the first stage, if the next stage machine is available, they are collected and transferred by an automatic transportation system. We assume that transfer is done in the second stage. At the second stage and third stage, there is only one machine so each job needs to $m + 2$ operations to be completed. There is no buffer storage between stages so if the downstream machine is not available the job stays on the current machine and blocks, it will be available until the next stage.

Objective functions

By considering a scalarizing method, the bi-criteria problem is converted to a single objective problem. So, the objective function is computed by:

$$OF = \alpha \left(\frac{\sum_{j=1}^{n} \sum_{i=1}^{n} wjC_i}{W} \right) + (1 - \alpha)C_{max}$$

where, $0 < \alpha < 1$. Note that when α is equal to 0 or 1, the problem is reduced to the single criterion of C_{max} or the weighted mean weighted completion time, respectively. The objective is to find a schedule which yields a minimum objective function value (OFV). In addition, we use the following notations in the presented model.

n	Number of jobs
m	Number of machines
$e_{i,h}$	Starting time of job in position i at stage h
$D_{i,h}$	Departure time of job in position i from stage h
$t_{j,k}$	Processing time of job j on machine k at first stage
$Si-1,i,k$	Set up time on machine k from job in position i-1 to job i at the first stage
T_i	Time of collecting and transferring job in position i to third stage
At_i	Assembly time of job in position i
w_j	Assigned weight to job j
$C_{i,1}$	Completion time of job j in position i at the end of first stage
$C_{i,2}$	Completion time of job j in position i at the end of second stage
C_i	Completion time of job j in position i at the end of third stage
$[C_{max}] = Cn$	Completion time of job in last sequence
$x_{i,j}$	If job j is in position i of sequence
$x_{i,j} = 1$	otherwise, it is 0.

Mathematical model

$$min : Z = \alpha \left(\frac{\sum_{j=1}^{n} \sum_{i=1}^{n} wjC_i}{W} \right) + (1 - \alpha)C_{max} \qquad (1)$$

$$s.t. \sum_{i=1}^{n} x_{i,j} = 1; \forall j \qquad (2)$$

$$\sum_{j=1}^{n} x_{i,j} = 1; \forall i \qquad (3)$$

$$e_{i,h} \geq D_{i,h-1}; \forall i \quad h = 1, 2, 3 \qquad (4)$$

$$e_{i,h} \geq D_{i-1,h}; \forall i \quad h = 1, 2, 3 \qquad (5)$$

$$D_{i,h-1} \geq D_{i-1,h}; \forall i \quad h = 1, 2, 3 \tag{6}$$

$$B_{i,h} = \begin{cases} D_{i,h} - C_{i,h}, D_{i-1,h+1} > C_{i,h}; \\ 0, D_{i-1,h+1} > C_{i,h} \\ \quad ; \forall i \quad h = 1, 2, 3 \end{cases} \tag{7}$$

$$D_{i,h} = C_{i,h} + B_{i,h}; \forall i \quad h = 1, 2, 3 \tag{8}$$

$$\begin{aligned} C_{i,1} &= e_{i,1} \\ &+ \left\{ max_{k=1,...,m} \left(\sum_{j=1}^{n} \left(t_{j,k} + S_{j-1,j,k} \right) \right) \right\} \\ &\times x_{i,j} \cdot \forall i \end{aligned} \tag{9}$$

$$C_{i,2} = e_{i,2} + \sum_{j=1}^{n} \left(T_i \times x_{i,j} \right); \forall i \tag{10}$$

$$Ci = e_{i,3} + \sum_{j=1}^{n} \left(Ati \times xi, j \right); \forall i \tag{11}$$

$$C_{max} = Cn \tag{12}$$

$$S_{j,j,k} = 0 \tag{13}$$

$$C_o = 0 \tag{14}$$

$$D_{i,0} = 0 \tag{15}$$

$$D_{0,h} = 0 \tag{16}$$

$$B_{1,h} = 0 \tag{17}$$

$$B_{i,3} = 0 \tag{18}$$

$$W = \sum_{j=1}^{n} w_j \tag{19}$$

$$0 < w_j < 1 \tag{20}$$

$$x_{i,j} \in \{0, 1\} \tag{21}$$

Eq. (1) presents the objective function. Eqs. (2) and (3) show that each job can only be placed in one position and only one job can be placed in each position, respectively. Eqs. (4) and (5) express that processing of job in position i and will start at stage h when it is left the previous stage and the job in position i-1 has left stage h, respectively. Eq. (6) shows that the departure time of a job in position i at stage h-1 is after departure of the job in the previous sequence from stage h. Eq. (7) calculates blocking time of job in position i at stage h. Eq. (8) shows the departure time of a job in position i from stage h. Eqs. (9), to (11) calculate the completion time of a job in position i at the first, second and the last stages (i.e., completion time of the job in position i). Eq. (12) calculates the completion time of the job in the last sequence (i.e., makespan). Eqs. (17) and (18) show that a job in the first sequence has no blocking time and job in position i has no blocking time at the third stage, respectively. Eq. (20) indicates w_j takes value only between

0 and 1. Eq. (21) shows that x_{ij} can only take 0 or 1 value.

Meta-heuristic Method

From [3], we know (AF $(m, 1, 1)//$) is an NP-hard problem, so by adding sequence-dependent setup time and blocking times to this model it is strongly NP-hard too. Because of this to solve the problem, we propose a meta-heuristic algorithm, namely simulated annealing (SA). This algorithm was originally proposed by Metropolis et al. [30] to simulate the annealing process. This algorithm starts with a high temperature. After generating an initial solution, it attempts to move from the current solution to one of its neighborhood solutions. The changes in the objective function values ΔE are computed. If the new solution results in a better objective value, it is accepted. However, if the new solution yields the worse value, it can be accepted according to the probability function considering the Boltzmann's constant and the initial (or current) temperature. By accepting worse solutions, SA can avoid being trapped on local optima. It repeats this process L times at each temperature to reach the thermal equilibrium, where L is a control parameter, called the Markov chain length. The temperature T is gradually decreased by a cooling function as SA proceeds until the stopping condition is met.

Setting the parameters for the proposed SA algorithm is essential in achieving a good performance. An initial estimation for the best value of a given parameter is obtained by changing the values of that parameter while keeping all other parameters as constant. We use the following values as initial estimates of the parameters; (initial temperature $T_1 = 0.5$, 0.1, and 0.01), (cooling factor = 0.99, 0.98, 0.97, 0.96, and 0.95), and (final temperature = 0.1, 0.125, 0.15, 0.175). Once these initial values are determined, then, the method of factorial experimental design (three values for each parameter including the initial best value of that parameter, one value above and one value below that value) is used to fine tune the values of the parameters. After these experimentations, the parameters for the SA algorithm are set as follows. The initial temperature, cooling factor, final temperature and number of iterations per fixed temperature are set to 0.1, 0.98, 0.001, and 50, respectively. Following is a general steps of the SA procedure.

- Initialization
- Set cooling schedule
- Define neighborhood solution
- Generate initial solution (s_0)
- Evaluate the initial solution and set $s^* = s_0$
- Generate a new neighborhood solution (s_0)

- Set $s^* = s_0$ if $f(s_0) < f(s^*)$ or set $s^* = s_0$ if $p > r$
- Stop the algorithm if the stopping criteria is met.

Computational Results

To compare the related results, a number of test problems are solved by the use of SA and Lingo. Although GAMS and CPLEX are useful to solve the mixed-integer model, we solve a small case of this problem and the same solutions are obtained. Therefore because of our experience and knowledge, we solve all cases with Lingo 8, which is well-known optimization software using a branch-and-bound algorithm. The proposed SA is implemented in Delphi 10. The computer used in this research is a Laptop with Core 2Duo CPU processors of 1.5 GHz running under the Windows 7 operating system with 2 GB of RAM. To measure the efficiency of the proposed SA, we compare its performance against Lingo. In this paper, the processing times of the first and third stages are integer values that are randomly generated from the uniform distribution (1,100) on all m first stage machines and single third stage machine, second stage processing times are integers that are randomly generated from the uniform distribution (1, 10) on single second stage machine [3]. Setup times are integers and randomly generated from uniform distribution (1, 20) on all m machines [31]. The problem data are generated for different numbers of jobs, say 6, 7, 8, 9, 20, 40, 60, and

Table 1 Computational results for $n = 6$ and $n = 7$

n	m	α	Lingo		SA	
			OFV	CPU time	OFV	CPU time (Sec.)
6	2	0	60.132	>4 h	60.132	61
	2	0.3	118.892	>4 h	118.892	61
	2	0.7	197.239	>4 h	197.239	60
	2	1	256	>4 h	256	60
	4	0	76.688	>4 h	76.688	63
	4	0.3	145.482	>4 h	145.482	58
	4	0.7	237.206	>4 h	237.206	59
	4	1	306	>4 h	306	59
	6	0	97.053	>4 h	97.053	61
	6	0.3	200.837	>4 h	200.837	61
	6	0.7	339.216	>4 h	339.216	62
	6	1	434	>4 h	434	62
	8	0	95.049	>4 h	95.049	63
	8	0.3	179.934	>4 h	179.934	62
	8	0.7	293.114	>4 h	293.114	64
	8	1	377	>4 h	377	69
7	2	0	69.444	>4 h	69.444	61
	2	0.3	142.211	>4 h	142.211	64
	2	0.7	239.233	>4 h	239.233	64
	2	1	312	>4 h	312	65
	4	0	90.457	>4 h	90.457	58
	4	0.3	171.32	>4 h	171.32	59
	4	0.7	279.137	>4 h	279.137	60
	4	1	360	>4 h	360	61
	6	0	120.86	>4 h	120.86	62
	6	0.3	243.007	>4 h	243.007	63
	6	0.7	405.860	>4 h	405.860	62
	6	1	520	>4 h	520	61
	8	0	98.505	>4 h	98.505	61
	8	0.3	187.753	>4 h	187.753	61
	8	0.7	306.751	>4 h	306.751	65
	8	1	389	>4 h	389	65

Table 2 Computational results for $n = 8$ and $n = 9$

n	m	α	Lingo		SA	
			OFV	CPU time	OFV	CPU time (Sec.)
8	2	0	74.955	>5 h	74.955	61
	2	0.3	161.968	>5 h	161.968	69
	2	0.7	277.986	>5 h	277.986	61
	2	1	365	>5 h	365	65
	4	0	104.216	>5 h	104.216	60
	4	0.3	207.051	>5 h	207.051	60
	4	0.7	344.165	>5 h	344.165	60
	4	1	447	>5 h	447	60
	6	0	132.839	>5 h	132.839	64
	6	0.3	274.487	>5 h	274.487	64
	6	0.7	463.351	>5 h	463.351	61
	6	1	597	>5 h	597	61
	8	0	105.240	>5 h	105.240	65
	8	0.3	219.768	>5 h	219.768	62
	8	0.7	372.472	>5 h	372.472	64
	8	1	480	>5 h	480	69
9	2	0	-	>4 h	99.156	61
	2	0.3	-	>4 h	206.509	62
	2	0.7	-	>4 h	349.646	60
	2	1	-	>4 h	457	63
	4	0	-	>4 h	113.592	61
	4	0.3	-	>4 h	224.114	60
	4	0.7	-	>4 h	371.477	62
	4	1	-	>4 h	482	61
	6	0	-	>4 h	139.329	62
	6	0.3	-	>4 h	288.930	63
	6	0.7	-	>4 h	488.398	62
	6	1	-	>4 h	630	61
	8	0	-	>4 h	129.774	65
	8	0.3	-	>4 h	260.341	65
	8	0.7	-	>4 h	434.432	66
	8	1	-	>4 h	562	66

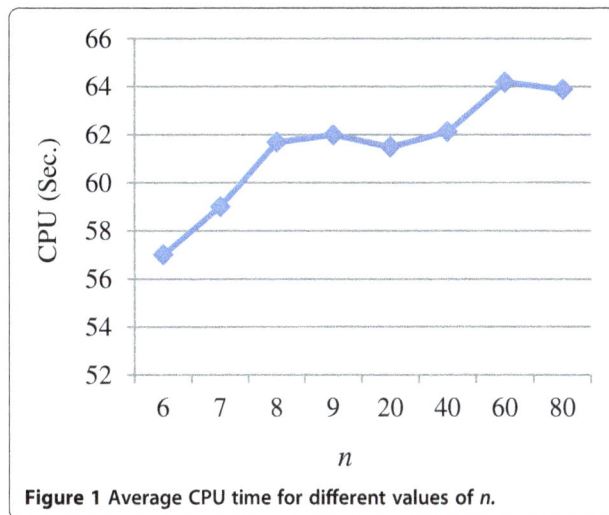

Figure 1 Average CPU time for different values of *n*.

80. The experimentation is conducted for the number of machines at the first stage being 2, 4, 6 or 8. We choose the values of the weight α to be 0, 0.3, 0.7 or 1.

The weight values of more than 0.5 give a more weight to the weighted mean completion time criterion, whereas the values of less than 0.5 give a more weight to the C_{max} criterion. The obtained results are shown in Tables 1 and 2. Some of the problems cannot be solved in reasonable time using Lingo. As shown in Table 2, because of an overmuch number of variables and constraints, it shows SA better time of solution than Lingo. In Figure 1, we show the average CPU time for the SA algorithm for jobs 6, 7, 8, 9, 20, 40, 60 and 80. It indicates that SA has reasonable computational time to obtain the objective function value (OFV).

Conclusions

In this paper, we have investigated a new 3-stage assembly flow shop scheduling problem with sequence-dependent setup time with blocking time that minimizes the weighted mean completion time and makespan. We have also proposed a new mathematical model solved by the proposed simulated annealing (SA) algorithm and Lingo 8. The obtained results have shown that our proposed SA was able to solve a number of test problems in a more reasonable time in comparison with Lingo. For future research, the case can be considered with machine breakdown, fuzzy data input. Also some other meta-heuristics methods can be used for soling the model.

Competing interests
The author(s) declare that they have no competing interests.

Authors' contributions
AMD surveyed the literature review and presented the model. He also drafted the manuscript. MM was responsible for revising and improving the quality of the manuscript. RTM participated in the design of problem solving

approach. IS analyzed the data and the prepared the figures and tables. All authors read and approved the final manuscript.

Author details
[1]Department of Management and Accounting, South Tehran Branch, Islamic Azad University, Tehran, Iran. [2]Department of Industrial Engineering, College of Engineering, University of Tehran, Tehran, Iran. [3]Department of Industrial Engineering, Payame Noor University, Tehran, Iran.

References
1. Ng CT, Wang JB, Cheng TCE, Liu LL (2010) A branch-and-bound algorithm for solving a two-machine flow shop problem with deteriorating jobs. Comput Oper Res 37(1):83–90
2. Allahverdi A, Al-Anzi FS (2007) The two-stage assembly flowshop scheduling problem with bicriteria of makespan and mean completion time. Int J Adv Manuf Technol 37(1):166–177
3. Koulamas S, Kyparisis G (2001) The three-stage assembly flowshop scheduling problem. Comput Oper Res 28(7):689–704
4. Al-Anzi FS, Allahverdi A (2009) Heuristics for a two-stage assembly flowshop with bicriteria of maximum lateness and makespan. Comput Oper Res 36(9):2682–2689
5. Hatami S, Ebrahimnejad S, Tavakkoli-Moghaddam R, Maboudian Y (2010) Two meta-heuristics for three-stage assembly flowshop scheduling with sequence-dependent setup times. Int J Adv Manuf Technol 50:1153–1164
6. Tavakkoli-Moghaddam R, Safaei N, Sassani F (2009) A memetic algorithm for the flexible flow line scheduling problem with processor blocking. Comput Oper Res 36(2):402–414
7. Hall NG, Sriskandarajah C (1996) A survey of machine scheduling problems with blocking and no wait in process. Oper Res 44(3):510–525
8. Yokoyama M, Santos DL (2005) Three-stage flow-shop scheduling with assembly operations to minimize the weighted sum of product completion times. Eur J Oper Res 161(3):754–770
9. Allahverdi A, Al-Anzi FS (2009) The two-stage assembly scheduling problem to minimize total completion time with setup times. Comput Oper Res 36 (10):2740–2747
10. Lee CY, Cheng TCE, Lin BMT (1993) Minimizing the makespan in the 3-machine assembly type flowshop scheduling problem. Manag Sci 39(5):616–625
11. Cheng TCE, Lin BMT, Tian Y (2009) Scheduling of a two-stage differentiation flowshop to minimize weighted sum of machine completion times. Comput Oper Res 36(11):3031–3040
12. Ruiz R, Allaherdi A (2009) Minimizing the bicriteria of makespan and maximum tardiness with an upper bound on maximum tardiness. Comput Oper Res 36(4):1268–1283
13. Sun X, Morizawa K, Nagasawa H (2003) Powerful heuristics to minimize makespan in fixed, 3-machine, assembly-type flowshop scheduling. Eur J Oper Res 146(3):498–516
14. Qian B, Wang L, Huang DX, Wang W, Wang X (2009) An effective hybrid DE-based algorithm for multi-objective flow shop scheduling with limited buffers. Comput Oper Res 36(1):209–233
15. Liu B, Wang L, Jin YH (2008) An effective hybrid PSO-based algorithm for flow shop scheduling with limited buffers. Comput Oper Res 35(9):2791–2806
16. Wang L, Zhang L, Zheng DZ (2006) An effective hybrid genetic algorithm for flow shop scheduling with limited buffers. Comput Oper Res 33(10):2960–2971
17. Grabowski J, Pempera J (2007) The permutation flowshop problem with blocking .A tabu search approach. Omega 35(3):302–311
18. Ronconi DP (2004) A note on constructive heuristics for the flowshop problem with blocking. Int J Prod Econ 87(1):39–48
19. Sawik T (2000) Mixed integer programming for scheduling flexible flow lines with limited intermediate buffers. Math Comput Model 31(13):39–52
20. Norman AB (1999) scheduling flowshops with finite buffers and sequence-dependent setup times. Comput Ind Eng 36(1):163–177
21. Ronconi DP, Henriques LRS (2009) Some heuristic algorithms for total tardiness minimization in a flowshop with blocking. Omega 37(2):272–281

22. Tozkapan A, Kirca O, Chung CS (2003) A branch and bound algorithm to minimize the total weighted flowtime for the two-stage assembly scheduling problem. Comput Oper Res 30(2):309–320

23. Yagmahan B, Yenisey MM (2010) A multi-objective ant colony system algorithm for flow shop scheduling problem. Expert Syst Appl 37(2):1361–1368

24. Sung CS, Kim HK (2008) A two-stage multiple-machine assembly scheduling problem for minimizing sum of completion times. Int J Prod Econ 113(2):1038–1048

25. Yokoyama Y (2008) Flow-shop scheduling with setup and assembly operations. Eur J Oper Res 161(3):754–770

26. Liu SQ, Kozan E (2009) scheduling a flow shop with combined buffer conditions. Int J Prod Econ 117(2):371–380

27. Lee WC, Shiuan YR, Chen SK, Wu CC (2010) A two-machine flowshop scheduling problem with deteriorating jobs and blocking. Int J Prod Econ 124(1):188–197

28. Gong H, Tang L, Duin CW (2010) A two-stage flow shop scheduling problem on a batching machine and a discrete machine with blocking and shared setup times. Comput Oper Res 37(5):960–969

29. Wang L, Pan QK, Suganthan PN, Wang WH, Wang YM (2010) A novel hybrid discrete differential evolution algorithm for blocking flow shop scheduling problems. Comput Oper Res 37(3):509–520

30. Metropolis N, Rosenbluth AW, Teller AH (1953) Equation of state calculations by fast computing machines. J Chem Phys :1087–1092

31. Lin SW, Ying KC, Lee ZJ (2009) Meta-heuristics for scheduling a non-permutation flow line manufacturing cell with sequence dependent family setup times. Comput Oper Res 36(4):1110–1121

5

Modeling and development of a decision support system for supplier selection in the process industry

Pandian Pitchipoo[1*], Ponnusamy Venkumar[2†] and Sivaprakasam Rajakarunakaran[2†]

Abstract

This paper presents the development of a model based decision support system with a case study on solving the supplier selection problem in a chemical processing industry. For the evaluation and selection of supplier, the analytical hierarchy process (AHP) and grey relational analysis (GRA) were used. The intention of the study is to propose an appropriate platform for process industries in selecting suppliers, which was tested with an electroplating industry during the course of development. The sensitivity analysis was performed in order to improve the robustness of the results with regard to the relative importance of the evaluation criteria and the parameters of the evaluation process. Finally, a practical implementation study was carried out to reveal the procedure of the proposed system and identify the suitable supplier with detailed discussions about the benefits and limitations.

Keywords: Supplier evaluation; Supplier selection; Model base; Decision support system; Analytical hierarchy process; Grey relational analysis

Background

Supplier selection is one of a critical decision in supply chain management in both manufacturing and process industries. In these industries the purchase department often plays an important role in reducing the purchasing cost and selecting appropriate suppliers. Improper evaluation and selection of potential supplier can reduce the organization's supply chain performance. Besides, supplier selection process is a multi-criteria decision making problem, in which both qualitative and quantitative factors are included. Because of the complexity and importance of supplier selection decisions, decision support systems (DSS) are generally used for decision-making. An effective DSS has the following components such as inputs [factors, numbers, and characteristics to analyze], user knowledge and expertise [inputs requiring manual analysis by the user], outputs [transformed data from which DSS "decisions" are generated] and decisions [results generated by the DSS based on user criteria] (Turban et al., 2004). In this paper the development of a model based DSS is discussed with a case study.

The rest of the paper is organized as follows: Literature review section illustrates the review of the relevant literatures in this area and also describes the problem context, Decision support system section, the steps and details of the proposed decision support system integrating AHP, GRA and Hybrid model for supplier selection. In Case study section, an actual application of the proposed method is presented. Systems integration and implementation section, System integration and implementation are explained. Finally, Conclusion section concludes the study and outlines some future research directions.

Literature review

The literature review is categorized into outline of supplier evaluation and selection, overview of models and decision support system applications.

Outline of supplier evaluation and selection

The identification of influencing criteria for the evaluation and selection of suppliers was focused by many researchers.

* Correspondence: drpitchipoo@gmail.com
†Equal contributors
[1]Department of Mechanical Engineering, P.S.R. Engineering College, Sivakasi–626140, Tamil Nadu, India
Full list of author information is available at the end of the article

Dickson (1966) carried out a study by the help of a survey which was conducted in 300 business organizations. The purchasing managers of those organizations were requested to identify the factors that were influencing the supplier selection. As an outcome of the survey, totally 23 factors were identified as important factors for the supplier selection decision problem. Weber et al. (1991) reviewed a total of 74 research papers on supplier selection and identified net price (cost), delivery, quality, production capability, geographical location, technical capability, reputation, financial position, performance history and warranty are the most contributed criteria for supplier selection. Boer et al. (2001) presented a review of decision methods based on an extensive search in the academic literature for the supplier selection process. Ho et al. (2010) reviewed the literatures related to the multi-criteria decision making approaches for supplier evaluation and selection appearing in the international journals from the year 2000 to 2008. For the supplier selection various techniques such as mathematical programming (linear programming, integer programming and goal programming), data envelopment analysis, AHP, analytical networking process, fuzzy set theory, and genetic algorithm were found in literatures. In most of the literatures, quality, delivery and price/cost were considered as the most influencing criteria for supplier evaluation and selection.

Overview of models
In this section, the relevant literatures to the proposed models such as GRA, AHP and hybrid AHP-GRA were reviewed.

Grey relational analysis (GRA) applications
Deng (1989) introduced the Grey theory that provides an effective means to solve problems containing uncertainty and indetermination. It was suitable to the decision-making under more uncertain environments. Tsai et al. (2003) have developed a supplier selection model for a garment industry using grey relational analysis. For this work the quality of the product, price, delivery date, quantity and services were considered to evaluate the suppliers. Hsu et al. (2008) evaluated the competencies of various suppliers of a centrifugal pump manufacturer in Taiwan using GRA. Li et al. (2008) proposed the grey rough set theory approach for supplier selection. In that approach the linguistic variable assigned to the alternatives were converted into grey number. The most suitable supplier was determined by GRA based on grey number. Tseng (2010) proposed GRA approach to deal with supplier evaluation of environmental knowledge management capacities with uncertainty and lack of information.

Analytical hierarchy process (AHP) applications
Narasimahn (1983) suggested the use of the AHP approach for vendor selection problems. Kumar and Parashar

(Kumar et al. 2009) applied analytical hierarchy process (AHP) for vendor selection process with the evaluation of multiple criteria and various constraints associated with small, medium and large scale industries. Chakraborty et al. (2005) evaluated the performance of existing die casting vendors of a light engineering industry using AHP tool. Cebi and Bayraktar (2005) solved the supplier selection problem for a food processing industry considering the qualitative and quantitative factors by using AHP and lexicographic goal programming (LGP) integrated model. Water and Peet (2006) presented a decision making model for taking make or buy decision using AHP for a shipyard in the Netherlands. Chan et al. (2007) developed an AHP based decision making approach to solve the supplier selection problem. Potential suppliers were evaluated and a sensitivity analysis using expert choice was performed to examine the responses of various alternatives. Tahriri et al. (2008) developed an AHP-based supplier selection model and applied to a real case study for a steel manufacturing company in Malaysia. Pitchipoo et al. (2011) developed an AHP model for the evaluation and selection of supplier for an electroplating industry in India.

Hybrid approach applications
Yang and Chen (2006) proposed an integrated model by combining AHP and GRA to evaluate and select the supplier for a notebook computer manufacturer. In this study an integrated model was formulated to examine the feasibility in selecting a best supplier and the effectiveness of the integrated model was demonstrated. Haq and Kannan (2006) proposed an integrated model comprising of AHP and GRA for the evaluation and selection of vendor in a forward supply chain. Pitchipoo et al. (2012) constructed an Excel based DSS for supplier selection process by integrating the AHP and GRA. First the highly influencing criteria were selected using Shannon mutual information based feature selection method and then the weights of the selected criteria were calculated using AHP. Finally the best supplier was selected using GRA.

Decision support system (DSS) applications
Son (1991) developed a decision support system for the automation of the processes in a factory. The cost effectiveness of the DSS was justified with the help of a case study conducted in a manufacturing firm. Chan (2003) suggested an interactive supplier selection model using AHP. The DSS was developed with the help of the multi-criterion decision making software called Expert Choice. Snijders et al. (2003) developed an electronic DSS to help the procurement managers and also compare the human decision with the decision made by the DSS. Moynihan (2006) developed a decision support system for procurement operations, which focuses primarily on procurement operations within a manufacturing environment. Hou and Su (2007)

developed an AHP based decision support system for the supplier selection problem in a mass customization environment. Gerardo (2007) proposed an alternative decision support system using Visual Interactive Goal Programming (VIG) for supplier selection problem.

Kumar and Rajender Singh (2008) developed a rule-based expert system for selection of piloting to assist die designers and process planners working in stamping industries. Montazer et al. (2009) developed an expert decision support system (EDSS) using a fuzzy version of ELECTRE III method for ranking the alternatives based on the experts' knowledge. This EDSS was applied to a vendor selection process in an Iranian oil industry. Razmi et al. (2009) developed a fuzzy analytic network process (ANP) model based DSS to evaluate the potential suppliers and select the best. Elanchezhian et al. (2010) proposed AHP for the selection of supplier for the glass product manufacturing industry. This software package was developed to meet the requirement of the purchase manager in the purchase of raw material. Kumar et al. (2011) developed a DSS for supplier selection based on fuzzy decision making techniques. The DSS was developed for a metal fabrication industry. Miah and Huth (2011) demonstrated a decision support system for supplier selection using multi attribute utility approach. The application of the approach was demonstrated using a subset of sample data from a real-world project.

Decision support system

Developing a decision support system for supplier selection is to assist the decision makers to obtain quick and accurate actions. In an organization, the decision support system was built to reduce the risk faced by procurement department because the procurement is a profit-contributing activity instead of a routine order-placing function (Moynihan, 2006). However, the absence of an adequate support system to assist the decision maker which results in inefficient and ineffective procurement may affect the profitability of the entire organization.

Proposed DSS framework

DSS is an interactive computer based system which helps the decision makers to utilize the data and models to solve large and complex decision problems. The framework of the proposed DSS is depicted in Figure 1. This system aims to develop appropriate decision strategies for supplier evaluation and selection in a process industry.

The proposed DSS contains of the following three components:

- Database system: It is enhancing the availability and access of various data necessary for model development in an accurate and precise manner.
- Model base system: This subsystem contains three supplier evaluation and selection models namely AHP model, GRA model and hybrid model with a sensitivity analysis model for testing the robustness of results.
- User interface system: Interact directly with a decision maker or a participant in the decision making process in such a way that the user has a flexible choice and sequence of selection strategies.

The proposed DSS can be used by the decision makers for the selection of suitable suppliers with appropriate

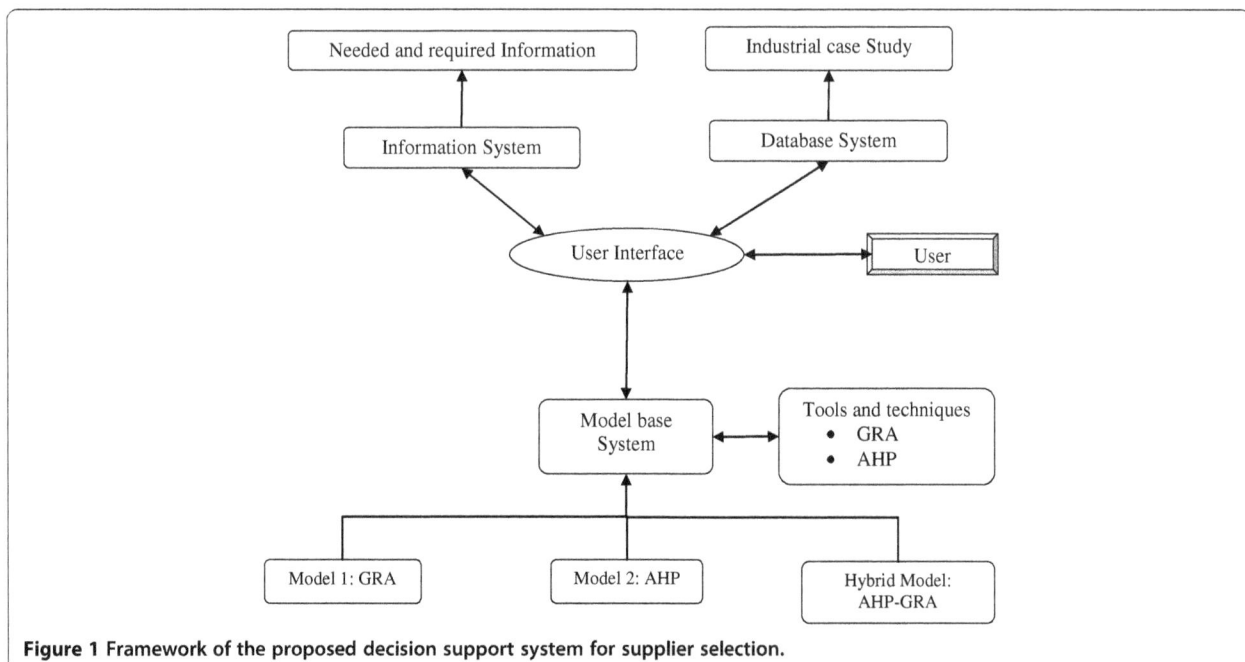

Figure 1 Framework of the proposed decision support system for supplier selection.

evaluation strategies. It will help them to make strategic decisions in the purchase activity where the risks are involved in the selection of suitable suppliers to be tackled and in deciding which procurement limitations are to be resolved. The proposed DSS will be tested and validated with sample data collected from electroplating industry in Case study section.

Database management system (DBMS)

A database is an integrated collection of data records, files, and other data related to the decision problem. It allows organizations to conveniently develop databases for supplier selection, which typically supports the needed query language and dedicated database language. The proposed DBMS provides facilities for creation, controlling data access, enforcing data integrity, recovering the database after failures and restoring it from backup files, maintaining database security with updation and report generation options (Lee et al. 2006).

To enhance the data quality and avoid data redundancy, it is necessary to develop DBMS with data extracted from the internal sources such as finance, production and procurement departments and external sources including journals, reference books, websites and materials from similar industries. The data categories in this study that are relevant to supplier evaluation and selection can be defined as quantitative criteria (cost of the material and delivery lead time) and qualitative criteria (quality of the supplied materials, warranty given on the material by the supplier and production capacity of the suppliers). The theoretical information related to the organization such as organization structure, raw materials, products, processes and the details about the tools used for the evaluation of the suppliers were included as help menu of DSS. It helps the users to know about the organization and tools which are used for this study.

Model base management system (MBMS)

The model base in a DSS gives decision makers access to a variety of models which are developed for this specific application to assist the decision makers in the decision making process. These models mathematically represent

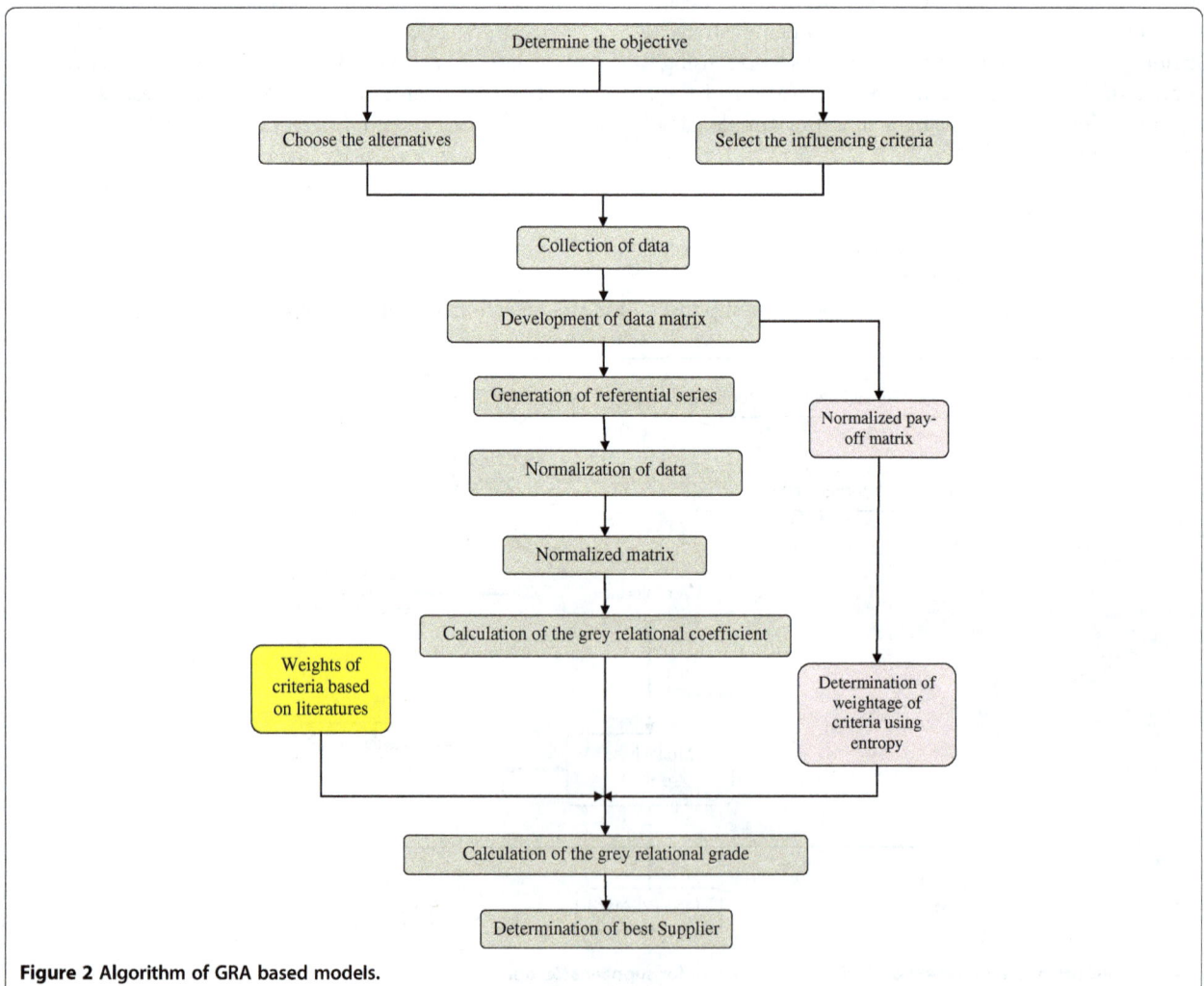

Figure 2 Algorithm of GRA based models.

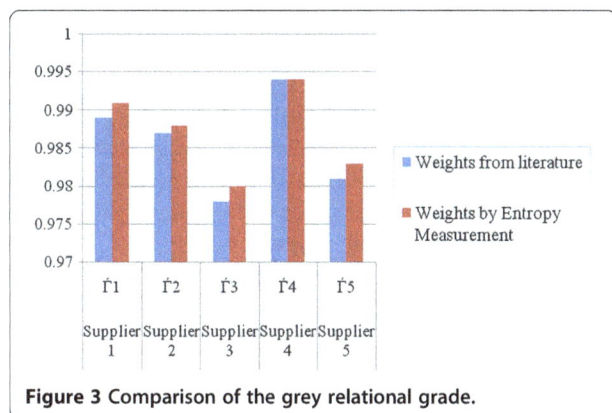

Figure 3 Comparison of the grey relational grade.

Table 2 Grey relational coefficient

Suppliers	Grey relational coefficient	Cost	Delivery	Capacity	Warranty
Supplier 1	γ_1	0.99995027	0.966	0.998	1.000
Supplier 2	γ_2	0.99990381	0.966	0.999	0.988
Supplier 3	γ_3	0.99985134	0.933	1.000	0.988
Supplier 4	γ_4	0.99971035	1.000	0.999	0.976
Supplier 5	γ_5	1.00000000	0.949	1.000	0.983

wide variety of different types of models in an integrated manner.

the various decision making activities and provide necessary decision support based on built in analytical tools which are used in this proposed DSS namely, AHP and GRA. Generally the model formulation process plays a major role in the result of the decision-making process. The purpose of the MBMS is to transform data from the DBMS into information that is useful in decision making. The developed model base management system has the facilities such as model base, model directory, model execution, integration of models and command processor. The model directory for this specific application consists of the following:

- GRA model in which the evaluation was done based on quantitative criteria.
- AHP model which is used to evaluate the suppliers based on both qualitative and quantitative criteria.
- Hybrid model which is developed to use the advantages of both AHP and GRA.

The important functions of the MBMS allow the users to manipulate models so that they can conduct experiments and sensitivity analysis from what – if to goal seeking (Ma et al. 2010). It also stores, retrieves and manages a

Table 1 Criteria matrix and normalized matrix for GRA model

	Suppliers	Cost (Rs)	Delivery (Days)	Capacity (Units)	Warranty (days)
Criteria matrix	Supplier 1	2439	12	170	28
	Supplier 2	2567	12	260	21
	Supplier 3	2711	14	280	21
	Supplier 4	3099	10	260	14
	Supplier 5	2302	13	290	18
Normalize matrix	Supplier 1	0.828	0.500	0.000	1.000
	Supplier 2	0.668	0.500	0.750	0.500
	Supplier 3	0.487	0.000	0.917	0.500
	Supplier 4	0.000	1.000	0.750	0.000
	Supplier 5	1.000	0.250	1.000	0.286

User system interface and hardware

The hardware needed for this DSS includes input data entry devices, central processing unit, data storage files and output devices. Input data entry devices allow entering, verifying, and updating the required data. The central processing unit (CPU) is the core part to control the other computer system components. Data storage files are the saved useful information, and this part also helps the decision maker to search past history easily. Output devices provide a visual or permanent record for the decision maker to store or read. This output device refers to the visual output device such as monitor or printer.

A DSS needs to be efficient to retrieve relevant data for decision makers, so the user system interface is very important. It is crucial that the interface must fit the decision maker's decision-making style (O'Brien and Marakas, 2004). If the decision maker is not comfortable with the user system interface the purpose of DSS will not be attained. The interfaces available in this developed DSS are scheduled reports, questions/answers and menu driven input/output.

Results and Discussion

The developed DSS is tested with the case study in an Electroplating industry to select the best supplier for its raw materials. The nature of electroplating industry selected for this study is involved in nickel coating and chrome plating for auto components. Moreover this type of industries is highly hazardous in nature and these industries are operating as small factories and workshops. At present the supplier selection is done based on bidding technique. In this current practice, the quality and supply lead time were not considered. But

Table 3 Relative weights of criteria

	Cost	Delivery	Capacity	Warranty
Weights of criteria from literature (Cheraghi, 2004)	0.2337	0.2896	0.2352	0.2415
Weights of criteria by using entropy measurement	0.2445	0.2450	0.2518	0.2587

Table 4 Grey relational grade

Suppliers		Grey relational grade	
		Weights of criteria from literature	Weights of criteria by using entropy measurement
Supplier 1	$\acute{r}_1$	0.989	0.991
Supplier 2	$\acute{r}_2$	0.987	0.988
Supplier 3	$\acute{r}_3$	0.978	0.980
Supplier 4	$\acute{r}_4$	0.994	0.994
Supplier 5	$\acute{r}_5$	0.981	0.983

these industries have to ensure that their product must meet the international standards and quality requirements to remain competent in the market. To achieve this, it is a necessary requirement that the supply of raw material or any other kind of necessary inputs should be selected appropriately. For this study the evaluation of suppliers is carried out based on the criteria such as performance assessment criteria, manufacturing criteria, quality system assessment criteria and business factors.

Model 1: GRA base models

Grey system theory was proposed by Deng (1989) and it is a mathematical theory that was born by the concept of grey set. It is one of the effective methods that are used to solve uncertainty problems under discrete data and partial information. The grey relation is the relation with incomplete information. GRA is an important approach of grey system theory in the application of estimating a set of alternatives in terms of decision attributes. The major advantage of grey theory is that it is suitable to handle both incomplete information and unclear problems. For the development of GRA based model, the following criteria such as performance assessment criteria (delivery lead time and cost of the material), manufacturing criteria (production capacity of the suppliers) and quality system assessment criteria (warranty given on the material by the supplier) are considered. The algorithm used for the

evaluation and selection process for GRA based models involve the following steps which are shown in Figure 2.

- Step 1: generation of referential series: The evaluation criteria for the available alternatives were tabulated and this was known as criteria matrix (Table 1). The referential series is the optimal values for each criterion from the criteria matrix.
- Step 2: normalization of data set: The criteria matrix was normalized using two approaches: For criteria if the larger value is better (warranty and capacity), the matrix can be normalized using larger the better concept using equation (1), while for a criteria if the smaller value is better (cost and delivery), the matrix can be normalized using smaller the better concept using equation (2). The normalized matrix is shown in Table 1.

$$X_i^*(j) = \frac{x_i(j) - min \ x_i(j)}{max \ x_i(j) - min \ x_i(j)} \tag{1}$$

$$X_i^*(j) = \frac{max \ x_i(j) - x_i(j)}{max \ x_i(j) - min \ x_i(j)} \tag{2}$$

where, i= 1, 2, ... m (Alternatives); j=1, 2 ... n (Criteria).
Then the absolute difference was calculated between the normalized cell value and the corresponding referential series value by using equation (3).
Absolute difference

$$\Delta_i(j) = abs \ (X_o^*(j) - X_i^*(j)) \tag{3}$$

where $X_o^*(j)$ = referential series value of jth criteria; $X_i^*(j)$ = normalized cell value of jth criteria
- Step 3: calculation of the grey relational coefficient: Grey relational coefficient is calculated to express

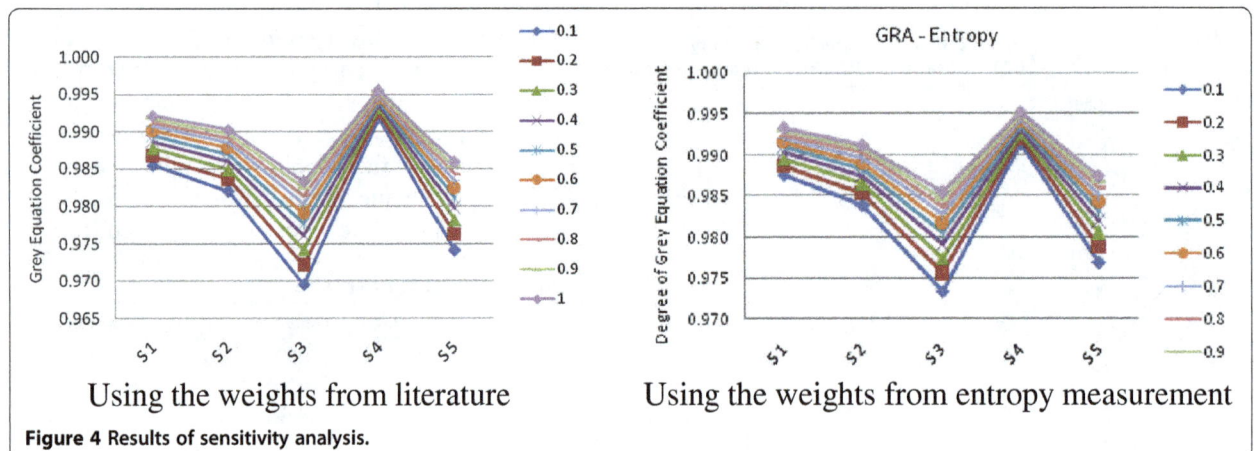

Using the weights from literature   Using the weights from entropy measurement

Figure 4 Results of sensitivity analysis.

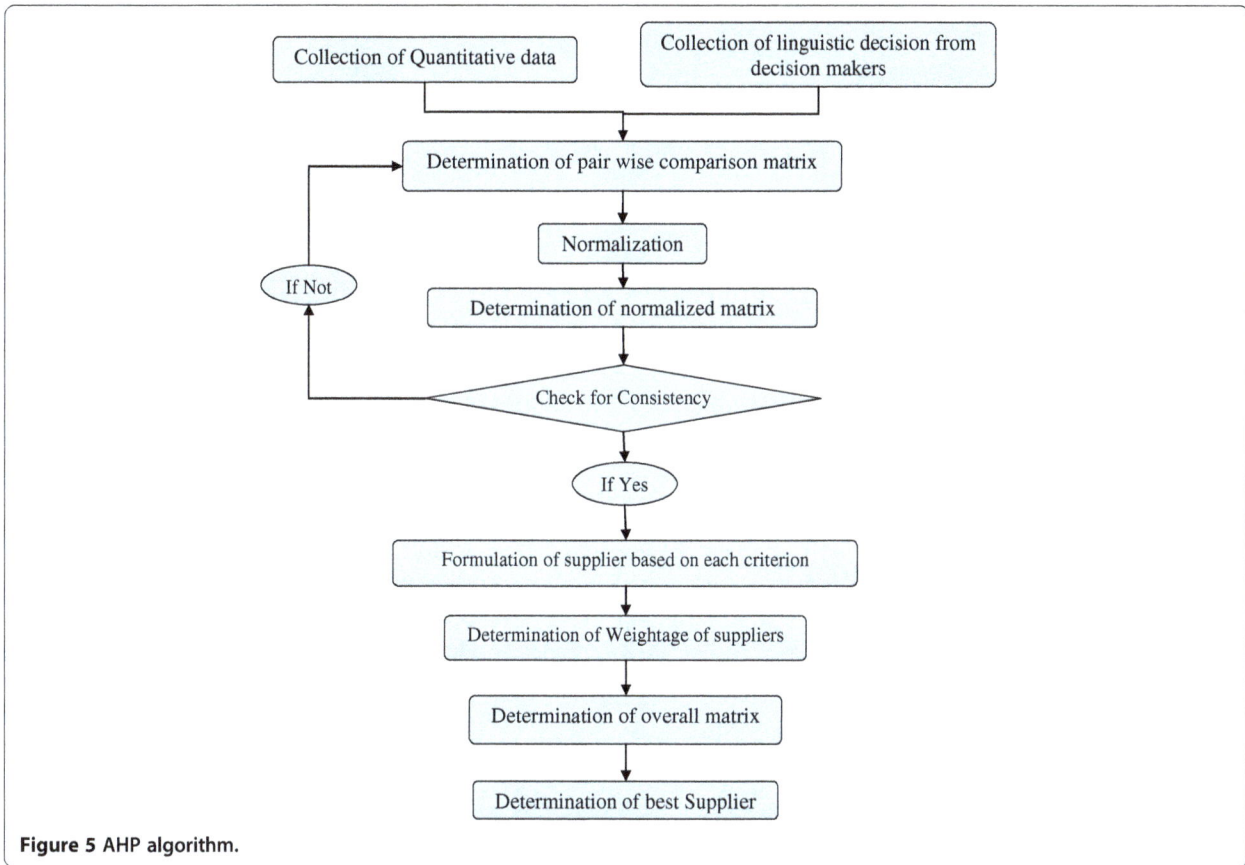

Figure 5 AHP algorithm.

the relationship between the best and the actual results, by using the following equation (4). The grey relational coefficient is given in Table 2.

$$\gamma_i(j) = \frac{\Delta \min + \xi \Delta \max}{\Delta_i(j) + \xi \Delta \max} \qquad (4)$$

where, $\Delta \min = \min_i \min_j \Delta_i(j)$; $\Delta \max = \max_i \max_j \Delta_i(j)$ and ξ = distinguished coefficient $\xi \varepsilon$ [0, 1]

- Step 4: calculation of the grey relational grade (GRG): The grey relational grade / grey relational

grade is calculated using the equation (5).

$$\hat{\Gamma}_i = \sum_{j=1}^{3} [W_i(j) \times \gamma_i(j)] \qquad (5)$$

where $W_i(j)$ = weightage of criteria j.

In this work the weights of the criteria were taken from the previous literature (Cheraghi, 2004) and by using entropy measurement approach. The relative weights of criteria are shown in Table 3. In the entropy measurement approach, the weights of criteria were determined using the following steps (Aomar, 2002):

Table 5 Criteria matrix – original matrix

	Cost	Quality	Delivery	Warranty	Capacity	Reputation	Fin. Posi.
Cost	1	0.333	3	5	4	6	5
Quality	3	1	5	6	7	6	7
Delivery	0.333	0.2	1	3	4	5	4
Warranty	0.2	0.167	0.333	1	2	3	4
Capacity	0.25	0.143	0.25	0.5	1	3	3
Reputation	0.167	0.167	0.2	0.333	0.333	1	2
Fin. Posi	0.2	0.143	0.25	0.25	0.333	0.5	1
Total	5.15	2.153	10.033	16.083	18.666	24.5	26

Table 6 Measurement scale for pairwise comparison

Verbal judgment	Numerical rating
Extremely preferred	9
Very strongly to extremely preferred	8
Very strongly preferred	7
Strongly to very strongly preferred	6
Strongly preferred	5
Moderately to strongly preferred	4
Moderately preferred	3
Equally to moderately preferred	2
Equally preferred	1

Table 8 Random indices

m	1	2	3	4	5	6	7	8	9	10
RI	0	0	0.58	0.90	1.12	1.24	1.32	1.41	1.45	1.49

$$D_j = 1 - E_j \qquad (8)$$

- Determination of weights of criteria: Finally the weights of each criterion were calculated by the following equation (9).

$$W_i(j) = \frac{D_j}{\sum_{j=1}^{n} D_j} \qquad (9)$$

From the grey relational grade, the supplier with larger coefficient was selected as the best supplier. The priority of the five suppliers was supplier 4 > supplier 1 > supplier 2 > supplier 5 > supplier 3. From GRA model application, supplier 4 will be declared as the best supplier. Figure 3 depicts the comparison of the grey relational grade. Table 4 shows the grey relational grade computed by using two different weights.

- Formulation of normalized pay-off matrix (P_{ij}) : The criteria matrix (Table 1) was normalized using the equation (6). This is called normalized pay-off matrix (P_{ij}).

$$P_{ij} = \begin{bmatrix} N_{11} & N_{12} & \cdots\cdots\cdots N_{1n} \\ N_{21} & N_{22} & \cdots\cdots\cdots N_{2n} \\ \cdots\cdots\cdots\cdots\cdots\cdots \\ \cdots\cdots\cdots\cdots\cdots\cdots \\ . \\ N_{m1} & N_{m2} & \cdots\cdots\cdots N_{mn} \end{bmatrix} \; where \qquad (6)$$

$$N_{ij} = \frac{X_{ij}}{\sum_{i=1}^{m} X_{ij}}$$

where, $x_i(j)$ = j^{th} criteria value for i^{th} alternative

- Determination of entropy: The entropy E_j of the set of alternatives for criterion j from the normalized pay-off matrix (P_{ij}) is determined by using the equation (7).

$$E_j = \frac{1}{\ln(m)} \sum_{i=1}^{m} p_{ij} \ln \left(p_{ij} \right) \qquad (7)$$

where, 'm' is the number of alternatives.

- Determination of degree of diversification: Next the degree of diversification of the information provided by the outcomes of the criterion j is determined using the equation (8).

Sensitivity analysis

Sensitivity analysis is a technique used to determine how different values of an independent variable will impact a particular dependent variable under a given set of assumptions. It is a method to predict the outcome of a decision if a situation turns out to be different compared to the key predictions. In this sub division, sensitivity analysis for the above said GRA models have been carried out to observe whether the optimal setting is sensitive to the individual response weightages.

That means, if there is any change in optimal setting due to change in relative weightages of the responses; it can be concluded that the optimal setting is sensitive to the individual weightage values. Different weightages [0.1 to 1.0] have been assigned to distinguished coefficient and the results of sensitivity analysis have been presented in Figure 4. Based on the outcome of the

Table 7 Adjusted matrix (normalized matrix)

	Cost	Quality	Delivery	Warranty	Capacity	Reputation	Fin. posi	Weights
Cost	0.194	0.155	0.299	0.311	0.214	0.245	0.192	0.230
Quality	0.583	0.464	0.498	0.373	0.375	0.245	0.269	0.401
Delivery	0.065	0.093	0.100	0.187	0.214	0.204	0.154	0.145
Warranty	0.039	0.078	0.033	0.062	0.107	0.122	0.154	0.085
Capacity	0.049	0.066	0.025	0.031	0.054	0.122	0.115	0.066
Reputation	0.032	0.078	0.020	0.021	0.018	0.041	0.077	0.041
Fin. Posi	0.039	0.066	0.025	0.016	0.018	0.020	0.038	0.032

Table 9 Original supplier matrix

Criteria	Suppliers	Supplier 1	Supplier 2	Supplier 3	Supplier 4	Supplier 5
Cost	1	1.000	5.000	7.000	9.000	0.111
	2	0.200	1.000	3.000	7.000	0.200
	3	0.143	0.333	1.000	5.000	0.200
	4	0.111	0.143	0.200	1.000	0.111
	5	9.000	5.000	5.000	9.000	1.000
	Net	10.454	11.476	16.200	31.000	1.622
Quality	1	1.000	7.000	9.000	5.000	3.000
	2	0.143	1.000	5.000	0.333	0.200
	3	0.111	0.200	1.000	0.200	0.143
	4	0.200	3.000	5.000	1.000	0.333
	5	0.333	5.000	7.000	3.000	1.000
	Net	1.787	16.200	27.000	9.533	4.676
Delivery	1	1.000	0.143	0.200	5.000	0.333
	2	7.000	1.000	3.000	7.000	5.000
	3	5.000	0.333	1.000	7.000	3.000
	4	0.200	0.143	0.143	1.000	0.143
	5	3.000	0.200	0.333	7.000	1.000
	Net	16.200	1.819	4.676	27.000	9.476
Warranty	1	1.000	8.000	8.000	6.000	7.000
	2	0.125	1.000	1.000	7.000	8.000
	3	0.125	1.000	1.000	7.000	8.000
	4	0.167	0.143	0.143	1.000	0.167
	5	0.143	0.125	0.125	6.000	1.000
	Net	1.560	10.268	10.268	27.000	24.167
Capacity	1	1.000	0.167	0.143	0.167	0.111
	2	6.000	1.000	0.200	1.000	0.167
	3	7.000	5.000	1.000	5.000	0.200
	4	6.000	1.000	0.200	1.000	0.167
	5	9.000	6.000	5.000	6.000	1.000
	Net	29.000	13.167	6.543	13.167	1.645
Reputation	1	1.000	1.000	5.000	1.000	5.000
	2	1.000	1.000	5.000	1.000	5.000
	3	0.200	0.200	1.000	0.200	1.000
	4	1.000	1.000	5.000	1.000	5.000
	5	0.200	0.200	1.000	0.200	1.000
	Net	3.400	3.400	17.000	3.400	17.000
Payment terms	1	1.000	0.200	7.000	6.000	4.000
	2	5.000	1.000	9.000	7.000	5.000
	3	0.143	0.111	1.000	0.200	0.167
	4	0.167	0.143	5.000	1.000	0.333
	5	0.250	0.200	6.000	3.000	1.000
	Net	6.560	1.654	28.000	17.200	10.500

Table 10 Adjusted supplier matrix

Criteria	Suppliers	Supplier 1	Supplier 2	Supplier 3	Supplier 4	Supplier 5	Weights
Cost	1	0.096	0.436	0.432	0.290	0.068	0.264
	2	0.019	0.087	0.185	0.226	0.123	0.128
	3	0.014	0.029	0.062	0.161	0.123	0.078
	4	0.011	0.012	0.012	0.032	0.068	0.027
	5	0.861	0.436	0.309	0.290	0.617	0.502
Quality	1	0.560	0.432	0.333	0.524	0.642	0.498
	2	0.080	0.062	0.185	0.035	0.043	0.081
	3	0.062	0.012	0.037	0.021	0.031	0.033
	4	0.112	0.185	0.185	0.105	0.071	0.132
	5	0.186	0.309	0.259	0.315	0.214	0.257
Delivery	1	0.062	0.079	0.043	0.185	0.035	0.081
	2	0.432	0.550	0.642	0.259	0.528	0.482
	3	0.309	0.183	0.214	0.259	0.317	0.256
	4	0.012	0.079	0.031	0.037	0.015	0.035
	5	0.185	0.110	0.071	0.259	0.106	0.146
Warranty	1	0.641	0.779	0.779	0.222	0.290	0.542
	2	0.080	0.097	0.097	0.259	0.331	0.173
	3	0.080	0.097	0.097	0.259	0.331	0.173
	4	0.107	0.014	0.014	0.037	0.007	0.036
	5	0.092	0.012	0.012	0.222	0.041	0.076
Capacity	1	0.034	0.013	0.022	0.013	0.067	0.030
	2	0.207	0.076	0.031	0.076	0.102	0.098
	3	0.241	0.380	0.153	0.380	0.122	0.255
	4	0.207	0.076	0.031	0.076	0.102	0.098
	5	0.310	0.456	0.764	0.456	0.608	0.519
Reputation	1	0.294	0.294	0.294	0.294	0.294	0.294
	2	0.294	0.294	0.294	0.294	0.294	0.294
	3	0.059	0.059	0.059	0.059	0.059	0.059
	4	0.294	0.294	0.294	0.294	0.294	0.294
	5	0.059	0.059	0.059	0.059	0.059	0.059
Payment terms	1	0.152	0.121	0.250	0.349	0.381	0.251
	2	0.762	0.605	0.321	0.407	0.476	0.514
	3	0.022	0.067	0.036	0.012	0.016	0.030
	4	0.025	0.086	0.179	0.058	0.032	0.076
	5	0.038	0.121	0.214	0.174	0.095	0.129

Table 11 Overall AHP score

Suppliers	Cost	Quality	Delivery	Warranty	Capacity	Reputation	Fin.Pos	Score
Supplier 1	0.061	0.200	0.012	0.046	0.002	0.012	0.008	0.340
Supplier 2	0.029	0.032	0.070	0.015	0.006	0.012	0.016	0.181
Supplier 3	0.018	0.013	0.037	0.015	0.017	0.002	0.001	0.103
Supplier 4	0.006	0.053	0.005	0.003	0.006	0.012	0.002	0.088
Supplier 5	0.116	0.103	0.021	0.006	0.034	0.002	0.004	0.287

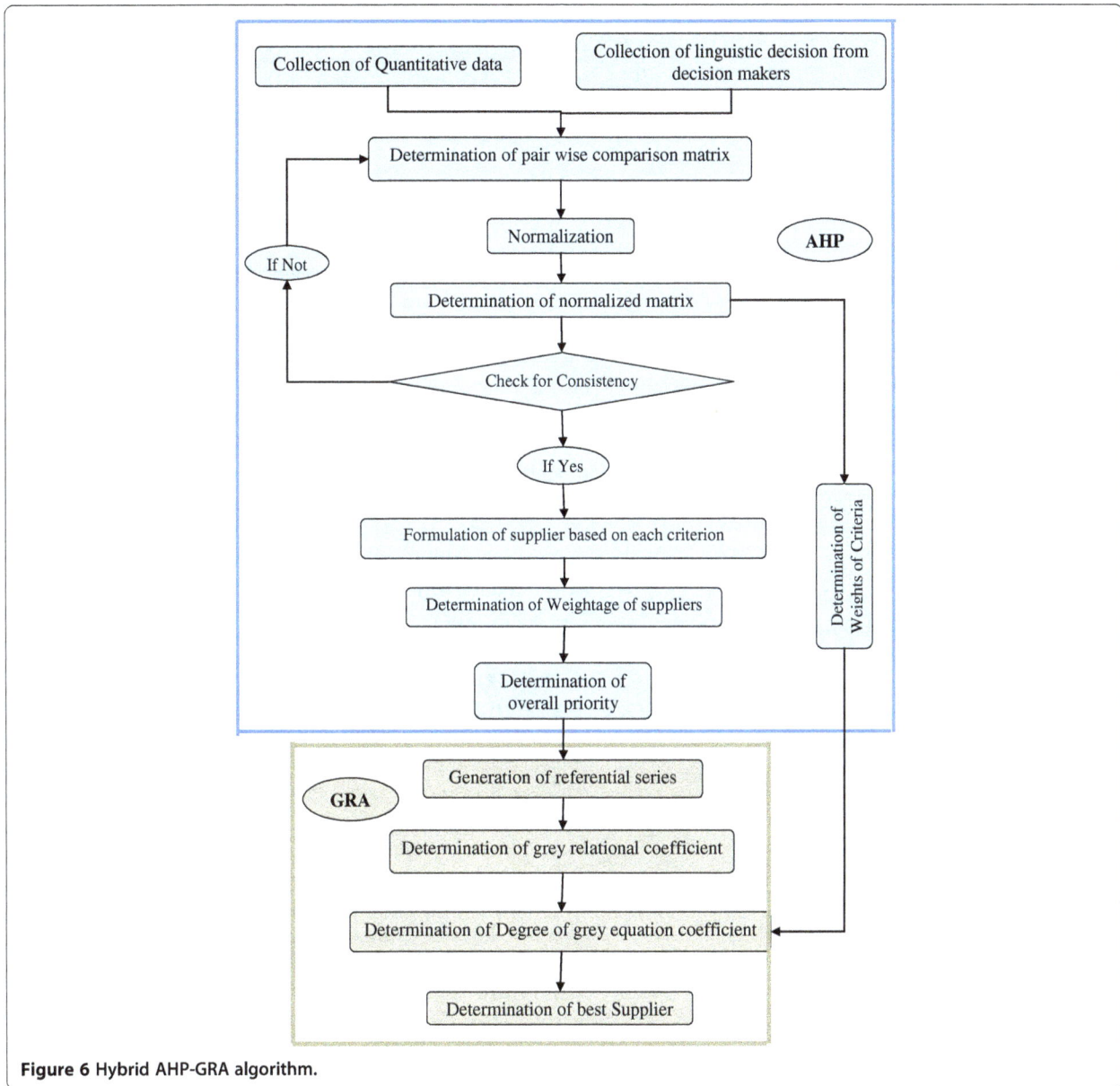

Figure 6 Hybrid AHP-GRA algorithm.

analysis the robustness of an optimal solution was tested and the sensitive variables were identified.

Model 2: analytical hierarchy process (AHP)

The Analytic Hierarchy Process (Thomas L. Saaty, 1980) is a powerful tool used to make decisions in situations when multiple and conflicting objectives/criteria are present. AHP helps capture both subjective and objective evaluation measures and provides a useful mechanism for checking the consistency of the evaluation measures and alternatives suggested by the team thus reducing bias in decision making. For the development of AHP models,

Table 12 Overall weights of criteria: hybrid AHP-GRA model

Suppliers	Cost	Quality	Delivery	Warranty	Capacity	Reputation	Fin. position
Supplier 1	0.061	0.200	0.012	0.046	0.002	0.012	0.008
Supplier 2	0.029	0.032	0.070	0.015	0.006	0.012	0.016
Supplier 3	0.018	0.013	0.037	0.015	0.017	0.002	0.001
Supplier 4	0.006	0.053	0.005	0.003	0.006	0.012	0.002
Supplier 5	0.116	0.103	0.021	0.006	0.034	0.002	0.004

along with the criteria which were considered for the GRA based models, the following criteria such as performance assessment criteria (quality of the supplied materials), and business factors (reputation or brand image of the supplier and financial position of the supplier) were used. The advantage of the AHP model over the GRA based models is that the AHP model can be suitable to accommodate the qualitative criteria also. The AHP algorithm is shown in Figure 5.

In the AHP model, first a set of pair wise comparisons of the criteria which is known as criteria matrix (Table 5) was developed to prioritize the criteria based on a measurement scale (Table 6) defined by Saaty (1990, 2008). The criteria matrix was normalized and from the normalized matrix which is shown in Table 7, the weightage of each criterion was calculated.

Next the consistency of the proposed comparison matrix was checked. To check the consistency, the consistency ratio (CR) was calculated using the equation (10).

$$CR = CI/RI \qquad (10)$$

where CI = Consistency Index and RI = Random indices.

$$CI = \frac{\lambda_{max} - m}{m - 1} \qquad (11)$$

Where λ_{max} = Max of B or m;

$$B = \left(\frac{\frac{A_1}{w_1} + \frac{A_2}{w_2} + \frac{A_3}{w_3} + \dots + \frac{A_m}{w_m}}{m} \right) \qquad (12)$$

where m=Number of criteria and A_1, A_2 . A_m are calculated using the equation (13).

$$[Xatt][Watt] = [A] \qquad (13)$$

where X_{att} = criteria matrix;

$$W_{att} = \text{Weights matrix} = \begin{bmatrix} W1 \\ W2 \\ W3 \\ :::: \\ :::::: \\ Wn \end{bmatrix} \text{ and } [A] = \begin{bmatrix} A1 \\ A2 \\ A3 \\ :::: \\ :::::: \\ Am \end{bmatrix}$$

Random indexes (RI) for various matrix sizes, m, have been approximated by Saaty (1990, 2008) as shown in Table 8.

If the CR < 0.10 the decision maker's pairwise comparison matrix is acceptable (Saaty, 1990). For the proposed AHP model, the consistency ratio was calculated as 0.079. Since this is less than 0.1, this model is acceptable.

Then the weights of the suppliers based on each criterion were calculated by using the same procedure. The original supplier matrix and the normalized/adjusted supplier matrix are shown in Table 9 and Table 10. After that the overall matrix is calculated by multiplying the weightage

Table 13 Grey relational grade

Suppliers	Grey equation coefficient	Value
Supplier 1	$\Gamma1$	0.800
Supplier 2	$\Gamma2$	0.485
Supplier 3	$\Gamma3$	0.487
Supplier 4	$\Gamma4$	0.636
Supplier 5	$\Gamma5$	0.492

of each criterion with the weightage of the supplier for that criterion. From the overall matrix the higher priority will be selected as best alternative. Based on the total weight calculated the best supplier was calculated. Based on the outcomes of model application, the overall AHP score obtained for each supplier is given in Table 11.

The final score will be calculated by summing the weightage of all criteria for a particular supplier. From this Table 11, the supplier with higher score (Supplier 1) will be selected as best supplier.

Hybrid model: AHP-GRA

The major shortcoming of the GRA model is that it is difficult to incorporate the linguistic variable and sometimes there is not much difference between the degrees of the grey equation coefficient. Consequently it is the very critical situation for a decision maker to take a concrete decision. The limitations of the AHP are that it only works if the matrices are in the same mathematical form and other apparent drawback is the formation and usage of the scaling factors. To overcome these drawbacks, hybridization approach is proposed. In this hybrid approach AHP and GRA are combined together for the evaluation and selection of the best supplier. The algorithm for the hybrid AHP-GRA algorithm is shown in Figure 6. The solution methodology of this hybrid approach consists of two stages:

Stage 1: Determination of the weights of the attributes by using AHP.

Stage 2: Selection of supplier is done by grey relational grading using GRA.

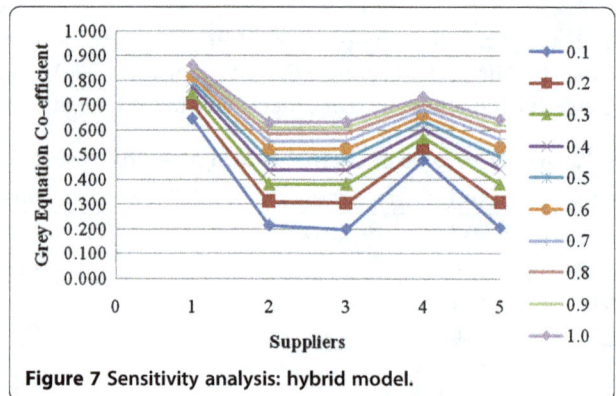

Figure 7 Sensitivity analysis: hybrid model.

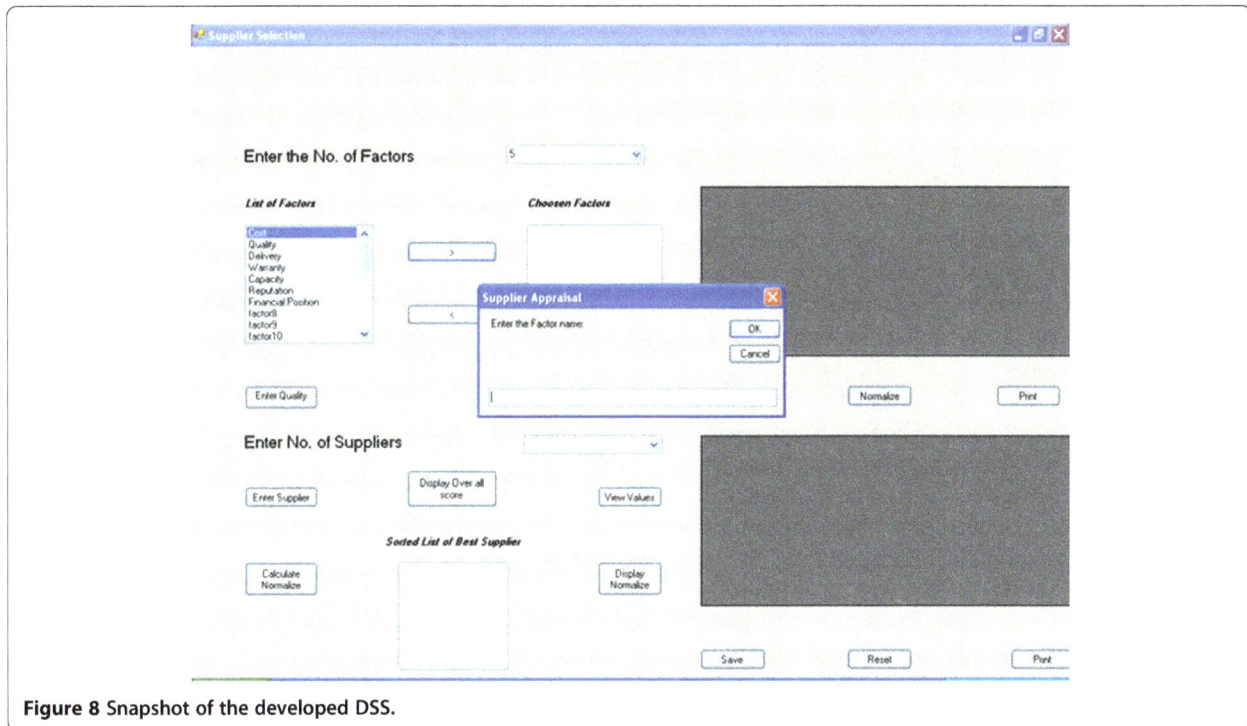

Figure 8 Snapshot of the developed DSS.

The weights for all criteria and the weights for alternatives were determined by AHP. The overall weights of criteria for the suppliers are shown in Table 12.

After the determination of weights using AHP, the ranking of suppliers is determined by GRA. The grey relational grade is determined and shown in Table 13. Finally the supplier with the higher grey relational grade (supplier 1) is selected as the best supplier.

The sensitivity of the grey relational grade was determined with the different distinguished coefficient (ξ) which varies from 0.1 to 1.0, which is shown in Figure 7. From the Figure 7, it is understood that the ranking of supplier is highly sensitive with the change in coefficient.

Systems integration and implementation

Integration architecture configured all the subsystems design to assure easy and secure data sharing across subsystems. User integration enabled a system user to concentrate on the tasks to be accomplished and not on the specific details of the technological system being integrated. The integration can be done based on vertical, star and horizontal integration (Holsapple and Whinston, 1996). Vertical integration is the process of integrating subsystems according to their functionality. Star integration is a process of integration of the systems where each system is interconnected to each of the remaining subsystems. Horizontal integration is an integration method in which a specialized subsystem is dedicated to communication between other subsystems. In this proposed DSS all subsystems should be interconnected to perform the decision making process. Hence star integration technique was used in the DSS.

To prove the system's efficiency and practical application, the model based DSS for supplier selection and evaluation has been built. During the implementation, confirmation testing and validation have been properly carried out. Confirmation testing is the process of checking that the DSS has been developed according to the specifications, and that it is perfect and error-free. It evaluates the DSS for its completeness and accuracy. Set of confirmation tests such as interface integrity test, information content test and performance test were employed to determine the integrity of the individual sub system functions of a DSS (Turban et al., 2004). Interface integrity test first ensured the internal and external interfaces and then the integration of each module was incorporated into the DSS structure. Information content test was designed to reveal the errors associated with data base. The performance test was carried out to verify performance limits for the number of suppliers and number of criteria which were established during DSS design.

Validation is the determination of the correctness of the developed DSS with respect to the user's needs. Validation is generally accomplished by verifying each stage of the DSS by face validation and predictive validation. Face validation is the technique by which a system is evaluated by the feedback from several experts. Predictive validation makes use of case studies from either the literature or real world situations. In this work the developed DSS was validated by the feedback from the decision makers in the organization where the case study was conducted, the industrial experts who are in similar kind of industries and the subject experts from academic institutions.

In the implementation, the first entry of the DSS is the login menu. The users of the proposed DSS in the organization where the case study was performed are the purchase manager, purchase department staff specialists, staff assistant and system analyst. The duties and responsibilities of the staff specialists and staff assistant are collection and updation of supplier related data and the system analyst is maintaining the DSS. They can access the system through the intranet and can look through useful information stored in the database. Users must pass the authentication stage of security management. Without registration the users are not allowed to enter the system. Figure 8 shows the snapshot of the developed DSS.

The developed DSS has the flexibility of incorporate 23 criteria which were identified through literatures (Dickson, 1966). By using the DSS maximum upto 25 suppliers can be evaluated. The performance of the DSS is based on the number of criteria and the number of suppliers. When the matrix size is more DSS consume more time to evaluate which are not able to evaluate manually.

Conclusion

This paper presented a case study on solving the supplier selection problem in the Process industry through a model based decision support system that employs the analytical hierarchy process and grey relational analysis, a multicriteria methodology aiming towards providing comprehensive support to decision makers in a process industry for supplier selection. GRA model is an effective model where limited data are available. It can be used to evaluate the quantitative data. AHP model has the ability to handle problems which cannot be handled by mathematical models. It also used to solve the problem with qualitative and quantitative criteria in the same decision framework. AHP model can handle more number of criteria. The advantages of both GRA and AHP models are achieved in the hybrid model. The methodology has been implemented in an integrated DSS and provides the users with enhanced database management capabilities, several analysis options and reporting tools. The sensitivity analysis was also performed in order to improve the robustness of the results.

The proposed decision support system provides appropriate strategies to procurement operations that would meet the desired requirements and speed up the process of decision making in supplier selection process. This support system replaces the conventional methods employed by human decision makers with a systematic, consistent model based approach through the use of operations research models. The future scope will be the upgradation and expansion with knowledge based models which are developed by using computational intelligent techniques.

Competing interests
The authors declare that they have no competing interests.

Authors' contributions
Pitchipoo et al. developed a decision support system for the supplier selection process. It helps the decision makers to understand the specific information regarding the criteria used for the supplier selection process and various methodologies to be used by organizations for the evaluation and selection of suppliers. The proposed DSS will enable the purchasing officials to make the right sourcing decisions. All authors read and approved the final manuscript.

Acknowledgements
The authors wish to thank the Department of Mechanical Engineering of P.S. R Engineering College and Kalasalingam University for their kind permission to carry out the research.

Author details
[1]Department of Mechanical Engineering, P.S.R. Engineering College, Sivakasi–626140, Tamil Nadu, India. [2]Department of Mechanical Engineering, Kalasalingam University, Krishnankoil - 626126, Tamil Nadu, India.

References
Cebi F, Bayraktar D (2005) An integrated approach for supplier selection. Logist Inform Manage 16(16):395–400

Chakraborty PS, Majumder G, Sarkar B (2005) Performance evaluation of existing vendors using AHP. J Sci Ind Res 64(9):648–652

Chan FTS (2003) Interactive selection model for supplier selection process: an analytical hierarchy process approach. Int J Prod Res 41(15):3549–3579

Chan FTS, Chan HK, Ip RWL, Lau HCW (2007) A decision support system for supplier selection in the airline industry. P I Mech Eng B-J Eng 221(5):741–758

de Boer L, Labro E, Morlacchi P (2001) A review of methods supporting supplier selection. Eur J Purch Supply Manage 7(2):75–89

Deng JL (1989) Introduction to grey system. J Grey Sys 1(1):1–24

Dickson GW (1966) An analysis of vendor selection systems and decisions. J Purch 2(1):5–17

Efraim T, Aronson JE, Liang T-P (2004) Decision support systems and intelligent system. Prentice-Hall, NJ

Elanchezhian C, Vijaya Ramnath B, Kesavan R (2010) Vendor selection using analytical hierarchy process in supply chain management. J Engg Res Studies 1(1):118–127

Farzad Tahriri M, Osman R, Ali A, Yusuff RM, Esfandiary A (2008) An analytic hierarchy process approach for supplier evaluation and selection in a steel manufacturing company. J Ind Engg Manage 1(2):54–76

Henny van de Water, Heleen P van Peet (2006) A decision support model based on the analytic hierarchy process for the make or buy decision in manufacturing. J Purch Supply Manage 12(5):258–271

Holsapple CW, Whinston AB (1996) Decision support systems: A knowledge based approach. Thomson Learning Publishing, UK

Hou J, Daizhong S (2007) EJB-MVC oriented supplier selection system for mass customization. J Manuf Technol Manage 18(1):54–71

Hsu L-H, Ken M-L, Lein C-F (2008) The evaluation of the supplier's competencies for product innovation based on grey relational analysis -A Case for centrifugal pumps. J Grey Syst 11(1):1–10

Kumar S, Singh R (2008) An expert system for selection of piloting for sheet metal work on progressive die. J Sci Ind Res 67(10):774–779

Kumar S, Parashar N, Haleem A (2009) Analytical hierarchy process applied to vendor selection problem: Small scale, medium scale and large scale industries. Busi Intel J 2(2):355–362

Kumar V, Srinivasan S, Das S (2011) A multi-agent system for management of supplier selection process in a fuzzy supply chain. I J Comput Appl 23(6):31–37

Lee D, Lee T, Lee S-k, Jeong O-r, Eom H, Lee S-g (2006) Bestchoice: a decision support system for supplier selection in e-marketplaces. Lect Notes Comput Sci 4055:198–208

Li G-D, Yamaguchi D, Nagai M (2008) A grey-based rough decision-making approach to supplier selection. I J Adv Manuf Technol 36(9–10):1032–1040

Ma J, Jie L, Zhang G (2010) Decider: A fuzzy multi-criteria group decision support system. Knowledge Base Syst 23(1):23–31

Martinez-Martinez JG (2007) Use of an alternative decision support system in vendor selection decisions. Int Met Buss J 3(2):1–14

Miah SJ, Huth M (2011) Cross-functional decision support systems for a supplier selection problem. I J Manage Decis Mak 11(3/4):217–230

Montazer GA, Saremi HQ, Ramezani M (2009) Design a new mixed expert decision aiding system using fuzzy ELECTRE III method for vendor selection. Expert Syst Appl 36(8):10837–10847

Moynihan GP, Puneet S, Fonseca DJ (2006) Development of a decision support system for procurement operations. I J Logist Syst Manage 2(1):1–18

Narasimahn R (1983) An analytical approach to supplier selection. J Purch Mat Manage 19(4):27–32

Noorul Haq A, Kannan G (2006) An integrated approach for selecting a vendor using grey relational analysis. I J Inform Techol Decis Mak 5(2):277–295

O'Brien JA, Marakas GM (2004) Management information systems: Managing information technology in the business enterprise. McGraw-Hill/ Irwin, Boston

Pitchipoo P, Venkumar P, Muthugurupackiam K, Rajakarunakaran S (2011) Modelling and development of decision model for supplier selection in process industry. I J Comput Aid Engg Technol 3(5/6):504–516

Pitchipoo P, Venkumar P, Rajakarunakaran S (2012) A distinct model for evaluation and selection of supplier for an electro plating industry. Int J Prod Res 50(16):4635–4648

Razmi J, Rafiei H, Hashemi M (2009) Designing a decision support system to evaluate and select suppliers using fuzzy analytic network process. Comput Ind Engg 57(4):1282–1290

Saaty TL (1980) The analytic hierarchy process: planning, priority setting, resource allocation. McGraw-Hill, New York

Saaty TL (1990) How to make a decision: The analytic hierarchy process. European J Oper Res 48(1):9–26

Saaty TL (2008) Decision making with the analytic hierarchy process. I J Ser Sci 1(1):83–98

Snijders C, Tazelaar F, Batenburg R (2003) Electronic decision support for procurement management: evidence on whether computers can make better procurement decisions. J Purch Supply Manage 9(5–6):191–198

Son YK (1991) A decision support system for factory automation: A case study. Int J Prod Res 29(7):1461–1473

Tsai C-H, Chang C-L, Chen L (2003) Applying grey relational analysis to the vendor evaluation model. Int J Comp Internet Manage 11(3):45–53

Tseng ML (2010) Using linguistic preferences and grey relational analysis to evaluate the environmental knowledge management capacity. Expert Syst Appl 37(1):70–81

Weber CA, Current JR, Benton WC (1991) Vendor selection criteria and methods. Eur J Oper Res 50(1):2–18

William H, Xiaowei X, Dey PK (2010) Multi-criteria decision making approaches for supplier evaluation and selection: A literature review. Eur J Oper Res 202 (1):16–24

Yang C-C, Chen B-S (2006) Supplier selection using combined analytical hierarchy process and grey relational analysis. J Manuf Technol Manage 17(7):926–941

6

An employee transporting problem

Ümit Yüceer

Abstract

An employee transporting problem is described and a set partitioning model is developed. An investigation of the model leads to a knapsack problem as a surrogate problem. Finding a partition corresponding to the knapsack problem provides a solution to the problem. An exact algorithm is proposed to obtain a partition (subset-vehicle combination) corresponding to the knapsack solution. It requires testing and matching too many alternatives to obtain a partition. The sweep algorithm is implemented in obtaining a partition (subset-vehicle combination) in an efficient manner. Illustrations are provided to show how the algorithms obtain solutions.

Keywords: Employee transportation; Set partitioning; Knapsack problem; Sweep algorithm

Background

A number of employees will be picked up from various places (bus-stops) within a city and brought to the plant. At the end of the work day, they will be returned back to where they were picked up. A similar situation holds for school students. The students are picked up from their homes and returned back to their homes after school. Transporting the passengers needs to be accomplished within a given time period. It is obvious that work at a plant starts at a specified time (8:30 in the morning for instance), and the first class starting time at some schools is 8:40 in the morning. The workers or the students must reach the plant or the school on time. We call this problem an *employee transporting* problem.

A set of different types of vehicles is used during the transportation of those passengers. Vehicles may differ in capacity, operating cost, and speed. Cruising within a city has speed limits, and transporting passengers within a city requires strict obedience to these speed limits. Subsequently, all the vehicles are assumed to have the same speed while traveling in the city streets.

Even though the vehicles may have different operating costs, they are chartered according to their capacities and paid accordingly. Every vehicle will be assigned to a subset of bus stops according to its capacity and the distance (time) required for traveling to this subset. Clearly, the number of passengers picked up or delivered cannot exceed the capacity of the vehicle. Further, the distance

(time) required to travel through this subset of destinations (bus stops) via a route must remain less than a predetermined distance (time) amount. Any route is acceptable as long as the traveling distance (time) does not exceed the predetermined time, and thus, it is not necessarily the traveling salesman (TSP) route.

The vehicle routing problem (VRP) or vehicle scheduling problems concern with the distribution of goods between the center (depot) and the customers (destinations-final users). The objective is to determine an optimal set of routes by a fleet of identical vehicles to serve these customers. The total demand of the customers on a route cannot exceed the vehicle capacity. This type of problem is called capacitated vehicle routing problem (CVRP). The fleet of the vehicles may differ in size, capacity, and the operating costs. This gives rise to heterogeneous vehicle routing problem (HVRP) or fleet mix problem in the literature. All these problems are the extensions and/or generalizations of the vehicle routing problems. Toth and Vigo (2002) provide an extensive discussion on VRP and solution methods with some applications. There are also numerous excellent articles on those problems and solution methods in the literature. A general approach to modeling this class of problems is based on the integer formulation of Miller et al. (1960). Various researchers have modified and/or extended their formulation to handle different situations.

The latest developments and the challenges in terms of modeling and the solution techniques for VRP and its

Correspondence: umit.yuceer@toros.edu.tr
Department of Industrial Engineering, Toros University, Mersin 33140, Turkey

variants are presented in Bruce et al. (2008). Yaman (2006) presents different formulations for HVRP and one formulation with flow variables and also develops strong bounds. Baldacci et al. (2011a) developed an exact method for solving VRP after modeling it based on a partitioning problem. Again in another attempt, Baldacci et al. (2011b) describe an effective exact method for solving CVRP based on the set partitioning formulation.

One possible approach for dealing with the existence of binary and integer variables in those models is to rely on the heuristic solution methods for CVRP and HVRP. Desrochers and Verhoog (1991) propose a new savings heuristic based on successive route fusion. Gheysens et al. (1986) developed another heuristic based on a lower-bound procedure. Numerous others have developed various heuristic methods to solve the problem of determining fleet size and the mix vehicle routing problems in the literature. There have also been attempts to apply metaheuristic methods too. Önder (2007) attempts using metaheuristics to solve HRVP without an investigation of the structure of the model but fails unfortunately. Evidently, the application of heuristics and even the metaheuristics requires a thorough investigation of the structure of the problem and the model.

The sweep algorithm described in Toth and Vigo (2002) is a heuristic for a possible solution method for VRP and some other problems. Nurchanyo et al. (2002) investigates the capability of the sweep algorithm in solving the VRP for public transport. After trying various approaches, they conclude that it is capable of solving VRP for public transport under certain conditions. Renaud and Boctor (2002) apply the sweep algorithm for the fleet size and the mix vehicle routing problem. Two kinds of decisions are considered: selecting a mix of vehicles and the routing of the selected fleet. Their algorithm generates a large number of routes to be serviced by one or two vehicles. Then the selection of the vehicles is accomplished by solving a set partitioning problem with a special structure.

The employee transporting problem is a variant of mix vehicle problem without routing. This article takes a different view from the literature and presents a set partitioning model for the employee transporting problem. The thrust of this article lies in the thorough investigation of the structure of the model. Subsequently, a surrogate problem is developed from this model which turns out to be a knapsack problem. The assignments of the vehicles are accomplished by matching subsets. For large-size problems, the sweep algorithm is proposed to obtain the assignments of the vehicles and consequently a partition in an efficient manner. The next section presents a set partitioning model, then the section 'A surrogate constraint' develops the surrogate problem. The solution methods are described in the section 'A solution method,' and some illustrations are presented in the section 'An illustration'. Finally, conclusions and findings of this article are summarized in the 'Conclusions' section.

A set partitioning model for employee transporting problem

Let A be the set of n destinations (bus stops) in an area, for instance, vicinity of a city. Each destination has a demand: the number of employees to be picked up and/or delivered. Let $s(a)$ denote the number of employees at destination $a \in A$. Naturally, $s(a) \geq 1$ and an integer. Let V be the set of the vehicles available with different capacities. Then $A_v \subset A$ represents a number of destinations to be serviced by the vehicle v of capacity Q_v. If m vehicles are required for servicing those n destinations, then each vehicle v will serve a subset A_v, and none of the subsets have any intersection. More specifically, a partition of the set A is obtained in the assignment of m vehicles. Let Ω be the class of all partitions of the set A. A partition $P = (A_1, A_2 \ldots, A_m)$ is a collection of subsets of A where $1 \leq m \leq n$, $A_i \cap A_j = \emptyset$, for $i \neq j$, and $\cup_{j=1}^{m} A_j = A$.

The cost per trip of assigning a vehicle $v \in V$ with capacity Q_v to a subset $A_j \in P \subset \Omega$ is c_v. This cost depends on the capacity of the vehicle only, and it is independent of the subset assigned. However, the total cost of the assignment depends on the partition $P \in \Omega$. Consequently, the decision of assigning a vehicle to a subset is not independent of the partition $P \in \Omega$. Therefore, the decision variable of assigning a vehicle to a subset is defined as follows:

$$y_{vj} = \begin{cases} 1 \text{ if the vehicle } v \in V \text{ is assigned to the subset } A_j \in P \\ 0 \text{ otherwise} \end{cases}$$

$$\tag{1}$$

Thus, the employee transporting model is given by the following:

$$\min_{P \in \Omega} \sum_{j \in J^P} \sum_{v \in V} c_v y_{vj} \tag{2}$$

subject to

$$y_{vj} \sum_{a \in A_j} s(a) \leq Q_v \qquad \text{for all } v \in V, A_j \in P, P \in \Omega \tag{3}$$

$$y_{vj} \delta(A_j) \leq T_{\max} \quad \text{for all } v \in V, A_j \in P, P \in \Omega \tag{4}$$

$$y_{vj} \in \{0, 1\} \tag{5}$$

where J^P is the index set of the subsets in the partition $P \in \Omega$, $\delta(A_j)$ is the traveling distance (time) to the destinations in a subset A_j via some route for $j \in J^P$, and $T_{\max}$ is the maximum allowed total traveling distance (time) for any subset in any partition. The sum $s(A_j) = \sum_{a \in A_j} s(a)$ represents the total number of passengers to be transported in a subset A_j, and it reduces constraint 3 to $y_{vj}S(A_j) \leq Q_v$. This is a linear constraint. It simply expresses the fact that the total number of passengers to be transported from any subset cannot exceed the capacity of the vehicle.

The set of constraint 4 states that the traveling distance (time) to a subset $A_j \in P$ for any partition $P \in \Omega$ cannot exceed the maximum allowed distance (time). It is tacitly assumed that the vehicles start from the center and return back to the center after visiting every destination in their routes. Traveling distance (time) to a subset is independent of the vehicle; therefore, $\delta(A_j) \leq T_{max}$ for all $A_j \in P$, $P \in \Omega$. This is not a linear constraint. The total distance (time) needs to be calculated for each subset A_j. Any route to the subset A_j is acceptable as long as the distance (time) does not exceed T_{max}. This means that it is not necessary to solve the TSP to determine the least-distance (time) route for each subset.

The set of constraint 4 can also be treated as a set of soft constraints. The distance (time) traveled to any subset can be increased by a small amount ($\epsilon > 0$) to make the right-hand side of the constraint $T_{max} + \epsilon$ at an added cost of opportunity loss of being late to work (school). Consequently, the objective function will be increased by $\eta(\delta(A_j))$ if the vehicle v traverses the subset A_j in $T_{max} + \epsilon$. In this expression, $\eta(.) > 0$ is the cost of opportunity loss of being late by an amount of ϵ for the subset A_j. For a given value of ϵ, this problem is solved by setting the right-hand side of the set of constraint 4 to $T_{max} + \epsilon$.

The objective function is then the total cost of all assignments of vehicles to all subsets of any partition. The problem is then to find a partition $P \in \Omega$ and a corresponding set of assignments of the vehicles to those subsets of this partition such that the total cost of such partition-vehicle assignments is minimum. This expression clearly defines the employee transporting problem as a heterogenous fleet mix problem without routing.

A surrogate constraint

Multiplying the set of constraint 3 by y_{vj} then summing the capacity constraints over $v \in V$ and $j \in J^P$ yields $\sum_{j \in J^P} \sum_{v \in V} y_{vj}^2 s(A_j) \leq \sum_{j \in J^P} \sum_{v \in V} y_{vj} Q_v$. The variables y_{vj} are binary variables, and thus, $y_{vj}^2 = y_{vj}$. On the other hand, every subset $A_j \in P$ must be assigned to a vehicle $v \in V$, but not every vehicle $v \in V$ may be assigned to a subset. Consequently, $\sum_{v \in V} y_{vj} = 1$ for all $j \in J^P$, and $\sum_{j \in J^P} y_{vj} \leq 1$ for all $v \in V$. In fact, a new variable can be defined now representing whether a vehicle is assigned or not as follows:

$$y_v = \sum_{j \in J^P} y_{vj} = \begin{cases} 1 & \text{if the vehicle } v \in V \text{ is assigned} \\ 0 & \text{otherwise} \end{cases}$$

(6)

Then a sequence of algebraic manipulations yields the right-hand side of the inequality 3 to be the total number of passengers to be transported from all the destinations.

$$\sum_{j \in J^P} \sum_{v \in V} y_{vj}^2 s(A_j) = \sum_{j \in J^P} \sum_{v \in V} y_{vj} s(A_j) \tag{7}$$

$$= \sum_{j \in J^P} s(A_j) \sum_{v \in V} y_{vj} \tag{8}$$

$$= \sum_{j \in J^P} s(A_j) = s(A) \tag{9}$$

On the other side of the inequality, $\sum_{v \in V} Q_v \sum_{j \in J^P} y_{vj} = \sum_{v \in V} Q_v y_v$. Then constraint 3 becomes $\sum_{v \in V} Q_v y_v \geq S(A)$. A simple modification in the objective function gives $\sum_{v \in V} \sum_{j \in J^P} c_v y_{vj} = \sum_{v \in V} c_v \sum_{j \in J^P} y_{vj} = \sum_{v \in V} c_v y_v$. Consequently, a knapsack problem is obtained for a partition $P \in \Omega$ as follows:

$$\min \sum_{v \in V} c_v y_v \tag{10}$$

subject to

$$\sum_{v \in V} Q_v y_v \geq s(A) \tag{11}$$

where $y_v \in \{0, 1\}$ for all $v \in V$ and the additional set of constraints $\delta(A_j) \leq T_{max}$ for all $j \in J^P$, $P \in \Omega$.

For any partition $P \in \Omega$, this surrogate problem needs to be solved. The Stirling number of the second kind ($\mathcal{S}_n^{(k)}$) gives the number of partitions with exactly k nonempty subsets of n elements. For instance, the number of partitions of 15 elements with exactly six nonempty subsets is $\mathcal{S}_{15}^{(6)} = 420,693,273$. Consequently, the total number of all partitions of a set of n elements is given by $\sum_{k=1}^{n} \mathcal{S}_n^{(k)}$. In fact, the total number of all partitions of a set of 15 elements is $\sum_{k=1}^{n} \mathcal{S}_{15}^{(k)} = 1,382,858,545$.

Let $P^* \in \Omega$ be an optimal partition with a corresponding vehicle assignment ($V^* = \{1, 2, \ldots, v_p\}$). Hence, $\sum_{v \in V^*} c_v y_v$ is the minimum cost and the total transporting capacity is given by $\sum_{v \in V^*} Q_v y_v = b \geq s(A)$. The right-hand side of the constraint can be greater than or equal to the total number of passengers ($s(A)$) to be carried due to the discreteness of the number of passengers at different locations and the traveling time constraint. For each subset $A_j \in P^*$, the distance (time) constraint $\delta(A_j) \leq T_{max}$ holds.

The following knapsack problem is independent of any partition. If a partition (vehicle-subset assignment satisfying the capacity and the distance (time) constraints) can be obtained for this solution, a solution to the original problem is obtained.

$$\min \sum_{v \in V} c_v y_v \tag{12}$$

subject to

$$\sum_{v \in V} Q_v y_v \geq r \tag{13}$$

$$y_v \in \{0, 1\} \qquad \text{for all } v \in V \tag{14}$$

Theorem 1. *Let $Y^*(r = b)$ be an optimal solution to the knapsack problem above. If there exists a partition (satisfying the capacity and the distance constraints) corresponding to $Y^*(b)$, then this solution is optimal to the problems (2), (3), and (4).*

Proof. Let $P* \in \Omega$ be an optimal partition and V^* be the corresponding vehicle assignment. This solution satisfies the surrogate problem necessarily. Clearly, $\sum_{v \in V*} c_v y_v$ is the minimum cost solution for the optimal partition $P*$, and further, $\sum_{v \in V*} c_v y_v \leq \sum_{v \in V} c_v y_v$ for any set V and any partition $P \in \Omega$. The knapsack problem with the right-hand side $s(A) \leq r < b$ can have a solution in $Y(r)$, but there cannot be any corresponding vehicle-subset assignments and/or partitions available satisfying the distance (time) constraint. If this were possible, then there would be a partition P' and V' such that $\sum_{v \in V'} c_v y_v < \sum_{v \in V*} c_v y_v$. This is a contradiction to the hypothesis. Therefore, a knapsack problem with $r = b$ needs to be considered.

Let $Y(r = b) = (y_{v1}, y_{v2}, \ldots, y_{vm})$ be a solution to this knapsack problem for some $1 \leq m \leq n$ together with $V(r) = \{v_1, v_2, \ldots, v_m\}$ corresponding to a vehicle-subset assignment $(A_1, A_2, \ldots, A_m)$ satisfying the distance (time) constraints. Then $\sum_{v \in V(r)} c_v y_v$ is the minimum cost satisfying the constraint $\sum_{v \in V(r)} Q_v y_v \geq b > s(A)$. Subsequently, $\sum_{v \in V(r)} c_v y_v \leq \sum_{v \in V*} c_v y_v$ implies that they must be equal. The sets V^* and $V(r)$ may not be equal because of the existence of alternating optimal vehicle-partition combinations. Thus, an optimal partition (vehicle-subset combination) exists for the solution $Y(r)$. $\qquad\square$

The argument until now assumes that there is only one type of each vehicle; hence, y_v is a binary variable, and the surrogate problem is a (0,1) knapsack problem. However, if the number of vehicles of each capacity is more than one, then the surrogate problem becomes a bounded integer knapsack problem. All the arguments up to now apply equally well to the case of bounded integer knapsack problem. In this form, the employee transporting problem is reduced to the common sense problem of determining the number of vehicles of each capacity to transport all the passengers at a minimum cost provided that the distance (time) of any route to a subset of destinations does not exceed a predetermined limit without any consideration of routing and vehicle-subset assignments.

A solution method

The knapsack problem can be solved by any method in the literature, see for instance Kellerer et al. (2004). It contains $|V|$ variables, binary or integer. Fortunately, in this class of problems, $|V|$ is a rather small number possibly not exceeding 10. This fact simplifies the numerical computations somehow while solving this knapsack problem. After obtaining a solution to the knapsack problem, a partition of the set A is sought, satisfying the capacity and the distance constraints. If such a partition is obtained, then an optimal solution to the original problem is found. Otherwise, the right-hand side value of the knapsack problem is increased to the next range, and the procedure is repeated. Before describing the algorithms, first, how to determine the range on the right-hand side of the knapsack problem is presented. The knapsack problem

$$\min z = \sum_{j=1}^{m} c_j y_j$$

subject to

$$\sum_{j=1}^{m} a_j y_j \geq b$$

where $y_j \geq 0$, and the integer for $j = 1, 2, \ldots, m$ can be converted to a maximization knapsack problem by setting $y_j = w_j - x_j$ where $w_j = \lceil \frac{b}{a_j} \rceil$ is the smallest integer greater than or equal to b/a_j for $j = 1, 2, \ldots, m$. The new knapsack problem is given as follows:

$$\max z' = \sum_{j=1}^{m} c_j x_j - \sum_{j=1}^{m} c_j w_j$$

subject to

$$\sum_{j=1}^{m} a_j x_j \leq \sum_{j=1}^{m} a_j w_j - b$$

where $0 \leq x_j \leq w_j$, and the integer for $j = 1, 2, \ldots, m$. Let $(x_1^*, x_2^*, \ldots, x_m^*)$ be an optimal solution to this knapsack problem. Then $\sum_{j=1}^{m} a_j x_j^* = b_* \leq \sum_{j=1}^{m} a_j w_j - b$. Consequently, the upper bound on r is $r_u = \sum_{j=1}^{m} a_j(w_j - x_j^*)$ and the lower bound on r is $r_\ell = b$.

Intuitively, having a sufficient number of vehicles will provide a partition (vehicle-subset assignment). Obviously, the algorithm will converge to a solution after a finite number iterations. A main algorithm to solve the knapsack problem consecutively and a subalgorithm to obtain a partition will be described next. Let $Y^t(r) = (y_{v1}^t, y_{v2}^t, \ldots, y_{vm}^t)$ represent the number of each type of vehicle in the knapsack solution in the iteration t.

The main algorithm is as follows:

Step 0. Initialize: Set $t = 1$, $r = s(A)$, go to Step 1.
Step 1. Solve the knapsack problem to obtain a solution $Y^t(r)$, and go to Step 2.
Step 2. Use the subalgorithm to obtain a partition corresponding to the solution $Y^t(r)$. If such a partition is obtained, go to Step 4, otherwise go to Step 3.
Step 3. Increase the right-hand side range to the next level for which r is the lower bound. Set $t := t + 1$ and go to Step 1.
Step 4. Terminate: An optimal partition (vehicle-subset assignment) is obtained.

The subalgorithm to find a partition corresponding to $Y^t(r)$ will determine all the possible subsets for each vehicle in the solution $Y^t(r)$, and subsequently, it will check whether a match is possible. The total capacity of the vehicles used in the solution $Y^t(r)$ is $\sum_{v \in V(r)} Q_v = r \geq s(A)$. Let $s_0 = s(A) - r$ be the slack representing the unused capacity of all the vehicles initially. Let R_v represent the set of not-yet-assigned destinations for the vehicle v. Initially, $R_1 = A$. P_v represents the incomplete partition for the vehicles $1, 2, \ldots, v$, and $P_0 = \emptyset$.

An exact algorithm to find a partition is as follows:

Step 0. Initialize: Set $v = 1$, $R_1 = A$, and $P_0 = \emptyset$. Go to Step 1.
Step 1. $L_1 = \{A_x | A_x \subset A, Q_1 - s_0 \leq s(A_x) \leq Q_1, \delta(A_x) \leq T_{max}\}$. If $L_1 = \emptyset$, go to Step 5, otherwise go to Step 2.
Step 2. Let $A_v \in L_v$ be the first subset in this list. Set $P_v := P_{v-1} + A_v$, $s_v := s_{v-1} - (Q_v - s(A_v))$, and $v := v + 1$. Go to Step 3.
Step 3. If ($v > v_m$), go to Step 5. Otherwise, $R_v := R_{v-1} - A_v$, and $L_v = \{A_x \subset R_v \mid Q_v - s_{v-1} \leq s(A_x) \leq Q_v\}$. If $L_v = \emptyset$, go to Step 4, otherwise go to Step 2.
Step 4. $v := v - 1$, $P_{v-1} := P_v - A_v$, $L_v := L_v - A_v$, $s_{v-1} := s_v + (Q_v - s(A_v))$, and $R_v := R_v + A_v$. If $v = 1$, go to Step 1, otherwise go to Step 2.
Step 5. Terminate by declaring that either a feasible partition is obtained if $R_v = \emptyset$ or there are no feasible partitions if $R_v \neq \emptyset$ or $L_1 = \emptyset$.

This algorithm finds a partition for the solution $Y^t(r)$ if it exists by checking all the possibilities. An example is provided to illustrate how this algorithm works. Unfortunately, it requires checking so many alternatives; it is not practical for solving large-size problems. Consequently, an adaptation of the sweep method (Toth and Vigo 2002) is proposed as an easier and efficient approach for finding the subset-vehicle assignments. This sweep algorithm is also a subalgorithm to

be called by the main algorithm in Step 2 to obtain a partition.

The sweep algorithm is as follows:

Step 0. Initialize: Convert the Cartesian coordinates to the polar coordinates for each destination $a_j(x, y) \rightarrow a_j(r, \theta)$ for $j = 1, 2, \ldots, n$. The destinations are called nodes from now on. Set $v := 1$. Start with the vehicle v in the solution $Y^t(r)$ with the capacity Q_v. Go to Step 1.
Step 1. Set $B := \emptyset$, $A_v := \emptyset$, $s(A_v) = 0$, $k = 0$.
Step 1a. Start rotating from the x-axis and find a node a_{vk} such that $\min_\theta a_j(r, \theta) = a_{vk}(r, \theta)$ where $\theta_k \leq \theta_j$ for all $a_j \in A$. If a_{vk} exists, then go to Step 1b, otherwise go to Step 3.
Step 1b. If $s(A_v) + s(a_{vk}) \leq Q_v$, then go to Step 1c, otherwise go to Step 1e.
Step 1c. Calculate $\delta(A_v + \{a_{vk}\}) = \delta$. If $\delta \leq T_{max}$, then go to Step 1d, otherwise go to Step 1e.
Step 1d. $A_v := A_v + \{a_{vk}\}$, and $A := A - \{a_{vk}\}$. Set $k := k + 1$ and go to Step 1a.
Step 1e. Set $B := B + \{a_{vk}\}$ and $A := A - \{a_{vk}\}$. If $A = \emptyset$, go to Step 2, otherwise go to Step 1a.
Step 2. Set $A := A \cup B$, and $v := v + 1$. If $v \leq v_m$, then go to Step 1, otherwise go to Step 3.
Step 3. Terminate by declaring either that a partition is obtained if $A = \emptyset$ or a partition is not obtained if $A \neq \emptyset$.

An illustration

An implementation of the main algorithm and the sub-algorithms is essential to observe how the employee transporting problem can be solved to optimality in an efficient manner. A small example with ten destinations and 91 passengers to be transported is used in testing the knapsack solution and the exact algorithm. A larger-size example with 79 destinations and 694 passengers is used to implement the sweep method. Both examples are randomly generated.

A small town with ten destinations is considered. The coordinates and the number of passengers to be transported from/to each destination is given in Table 1. The distance matrix is given in Table 2. There are three types of vehicles available for transportation: mini-busses with 15 seats, midi-busses with 30 seats, and coaches with 50 seats with costs of 35, 55, and 105 TL per trip, respectively. There are altogether 91 passengers to be transported. How many vehicles of each type are needed and to which subsets will they be assigned to accomplish the task at a minimum cost?

The initial knapsack problem to be solved is given as follows:

$$\min z = 35y_1 + 55y_2 + 105y_3$$

Table 1 The coordinates and the number of passengers

a	x_a	y_a	r_a	θ_a	$s(a)$
1	0.90	7.10	7.16	0.13	10
2	0.70	10.20	10.22	0.07	8
3	1.50	12.20	12.29	0.12	13
4	0.20	10.90	10.90	0.02	6
5	0.60	14.30	14.31	0.04	6
6	2.20	14.50	14.67	0.15	4
7	0.10	14.60	14.60	0.01	10
8	0.50	16.50	16.51	0.03	11
9	1.80	18.00	18.09	0.10	8
10	2.50	15.80	16.00	0.16	15

subject to

$$15y_1 + 30y_2 + 50y_3 \geq 91$$

where $y_i \geq 0$, and the integer for $i = 1, 2, 3$. In this problem, y_1 is the number of mini-busses, y_2 is the number of midi-busses, and y_3 is the number of coaches to employ. The solution to this knapsack problem yields $Y^1 = (1, 1, 1)$ with a cost of 195 TL. The following argument shows that this solution is valid for $91 \leq r \leq 95$. A simple computation shows $w_1 = 7$, $w_2 = 4$, and $w_3 = 2$. The transformation $y_j = w_j - x_j$ for $j = 1, 2, 3$ gives the following knapsack problem:

$$\max z' = 35x_1 + 55x_2 + 105x_3 - 675$$

subject to

$$15x_1 + 30x_2 + 50x_3 \leq 325 - 91 = 234$$

where $0 \leq x_1 \leq 7$, $0 \leq x_2 \leq 4$, and $0 \leq x_3 \leq 2$ are all integers. The optimal solution to this knapsack is $X = (6, 3, 1)$ and $Y = (1, 1, 1)$ with $z' = 480 - 675 = -195 = -z$ and $15x_1 + 30x_2 + 50x_3 = 230$. Therefore, $r_u =$

Table 2 The distances between the destinations

a	0	1	2	3	4	5	6	7	8	9	10
0	-	75	103	126	111	144	152	146	167	185	166
10	166	91	67	42	68	40	14	47	40	26	
9	185	110	82	59	79	45	35	47	30		
8	167	94	64	48	57	23	40	20			
7	146	75	45	35	38	8	40				
6	152	78	53	27	55	32					
5	144	71	41	28	35						
4	111	40	10	30							
3	126	52	26								
2	103	30									
1	75										

$325 - 230 = 95$ and $r_\ell = 91$. Obviously, if the right-hand side of this knapsack is less than 230, a new solution is required.

The next range is $96 \leq r \leq 105$ with an optimal solution $Y^2 = (1, 3, 0)$ with a cost of 200 TL. Further, the next range is $106 \leq r \leq 110$ with an optimal solution of $Y^3 = (0, 2, 1)$ with a cost of 215 TL. $T_{\max} = 370$ distance (time) units for this town. First, the exact algorithm is implemented for finding a partition corresponding to the solution $Y^t(r)$. There are no partitions for the solution $Y^1(91) = (1, 1, 1)$ with $T_{\max} = 370$. Then the next best solution (next range) is attempted. $Y^2(96) = (1, 3, 0)$. The total capacity of the vehicles now is 105. Therefore, $s_0 = 105 - 91 = 14$. The list L_1 of all subsets satisfying the condition $30 - 14 \leq s(A_x) \leq 30$ and the distance constraint has 118 subsets of A. First, the subset $A_1 = \{4, 7, 8\}$ with $s(A_1) = 27$ and $\delta(A_1) = 336$ is chosen. The set of remaining destinations is $\{1, 2, 3, 5, 6, 9, 10\}$ for the vehicle $v = 2$ with $Q_2 = 30$. Now $s_1 = 14 - (30 - 27) = 11$. The list L_2 of all subsets satisfying the condition $30 - 11 = 19 \leq s(A_x) \leq 30$ contains 22 subsets. First one in this list, $A_2 = \{6, 10\}$ with $s(A_2) = 19$ and $\delta(A_2) = 332$ yields $s_2 = 11 - (30 - 19) = 0$, and there are no subsets for $v = 3$ satisfying the condition $30 \leq s(A_x) \leq 30$. Subsequently, the subset $A_2 = \{6, 10\}$ is deleted from L_2. Next, the subset $A_2 = \{3, 9\}$ is tested. $s(A_2) = 21$ and $\delta(A_2) = 370$. Now $s_2 = 11 - (30 - 21) = 2$. The set of remaining destinations now is $\{1, 2, 5, 6, 10\}$. The set L_3 of all subsets satisfying the condition $30 - 2 \leq s(A_x) \leq 30$ and $\delta(A_x) \leq 370$ for the vehicle $v = 3$ with $Q_3 = 30$ contains three elements only. Taking $A_3 = \{2, 5, 10\}$ with $s(A_3) = 29$ and $\delta(A_3) = 350$ yields $L_4 = A_4 = \{1, 6\}$ with $s(A_4) = 14$ and $\delta(A_4) = 305$ for the vehicle $v = 4$ with $Q_v = 15$. Hence, an optimal partition is obtained. There are also alternating optimal partitions: $A_3 = \{1, 2, 5, 6\}$ with $s(A_3) = 28$ and $\delta(A_3) = 330$ for $Q_3 = 30$ and $A_4 = \{10\}$ with $s(A_4) = 15$ and $\delta(A_4) = 332$ for $Q_4 = 15$ with a cost of 200 TL.

Solving the same problem by the sweep algorithm yields $A_1 = \{2, 4, 5, 7, 8\}$ with $s(A_1) = 41$, $\delta(A_1) = 346$ for the vehicle of the capacity $Q_1 = 50$; $A_2 = \{3, 9\}$ with $s(A_2) = 21$, $\delta(A_2) = 370$ for the vehicle of the capacity $Q_2 = 30$; and $A_3 = \{1, 6, 10\}$ with $s(A_3) = 29$, $\delta(A_3) = 333$ for the vehicle of capacity $Q_3 = 30$ corresponding to the knapsack solution $Y^3 = (0, 2, 1)$ with a cost of 215 TL and $T_{\max} = 370$. This is a good example that the sweep algorithm may miss the optimal partition.

Treating the constraint $\delta(A_x) \leq T_{\max}$ as a soft constraint is also considered. $T_{\max} = 370$ distance time units in this example. Now it is increased by 1% to 373.7 at an additional cost of opportunity loss of 2% of the total cost of the solution. Then $Y^1 = (1, 1, 1)$ yields the partition $A_1 = \{1, 4, 5, 6, 7, 8\}$ with $s(A_1) = 47$, $\delta(A_1) = 370$; $A_2 = \{2, 3, 9\}$ with $s(A_2) = 29$, $\delta(A_2) = 373$; and $A_3 =$

Table 3 The locations and the number of passengers to be transported

a	y_a	x_a	r_a	θ_a	$s(a)$	a	y_a	x_a	r_a	θ_a	$s(a)$
1	0.50	1.50	1.58	1.25	7	41	12.20	1.50	12.29	0.12	13
2	1.00	2.40	2.60	1.18	8	42	10.80	2.60	11.11	0.24	14
3	1.80	3.30	3.76	1.07	9	43	10.90	0.20	10.90	0.02	6
4	2.60	4.40	5.11	1.04	9	44	13.40	11.40	17.59	0.70	5
5	3.50	3.50	4.95	0.79	6	45	13.50	7.50	15.44	0.51	5
6	5.00	3.60	6.16	0.62	13	46	13.40	5.80	14.60	0.41	8
7	4.70	4.60	6.58	0.77	3	47	14.00	5.50	15.04	0.37	12
8	4.40	5.60	7.12	0.90	11	48	19.40	4.30	19.87	0.22	4
9	4.60	8.10	9.32	1.05	7	49	13.90	3.50	14.33	0.25	7
10	5.70	5.80	8.13	0.79	8	50	13.10	2.50	13.34	0.19	3
11	5.50	9.00	10.55	1.02	8	51	14.30	0.60	14.31	0.04	6
12	6.60	7.50	9.99	0.85	2	52	14.80	12.50	19.37	0.70	6
13	7.00	2.50	7.43	0.34	8	53	14.20	11.00	17.96	0.66	6
14	7.10	0.90	7.16	0.13	10	54	14.40	9.80	17.42	0.60	3
15	8.10	11.00	13.66	0.94	6	55	14.50	8.10	16.61	0.51	8
16	8.00	8.50	11.67	0.82	2	56	14.50	2.20	14.67	0.15	4
17	7.90	6.20	10.04	0.67	6	57	14.60	0.10	14.60	0.01	10
18	7.80	2.60	8.22	0.32	10	58	15.90	11.10	19.39	0.61	9
19	8.00	1.40	8.12	0.17	11	59	15.80	10.50	18.97	0.59	13
20	9.00	1.00	9.06	0.11	8	60	15.10	9.10	17.63	0.54	11
21	8.70	7.00	11.17	0.68	13	61	15.50	7.00	17.01	0.42	10
22	9.20	2.50	9.53	0.27	6	62	15.10	4.30	15.70	0.28	10
23	10.50	10.40	14.78	0.78	8	63	15.80	2.50	16.00	0.16	15
24	10.20	8.30	13.15	0.68	5	64	16.10	13.00	20.69	0.68	15
25	10.30	7.30	12.62	0.62	13	65	16.70	12.40	20.80	0.64	10
26	10.00	6.40	11.87	0.57	12	66	16.20	9.70	18.88	0.54	10
27	10.10	4.50	11.06	0.42	14	67	16.80	8.80	18.97	0.48	10
28	10.00	2.40	10.28	0.24	13	68	16.20	8.10	18.11	0.46	5
29	10.20	0.70	10.22	0.07	8	69	16.10	6.20	17.25	0.37	3
30	11.70	11.50	16.41	0.78	12	70	16.60	5.30	17.43	0.31	3
31	11.50	9.60	14.98	0.70	11	71	16.50	3.30	16.83	0.20	11
32	11.40	6.70	13.22	0.53	15	72	16.50	0.50	16.51	0.03	11
33	11.50	3.50	12.02	0.30	13	73	18.00	11.20	21.20	0.56	5
34	12.70	11.00	16.80	0.71	13	74	18.00	7.50	19.50	0.39	13
35	12.60	9.70	15.90	0.66	8	75	18.00	5.00	18.68	0.27	12
36	12.50	8.30	15.00	0.59	7	76	17.80	3.10	18.07	0.17	12
37	13.00	6.50	14.53	0.46	8	77	18.00	1.80	18.09	0.10	8
38	12.60	3.50	13.08	0.27	6	78	18.50	11.40	21.73	0.55	8
39	12.00	3.30	12.45	0.27	14	79	19.00	9.00	21.02	0.44	11
40	12.10	2.20	12.30	0.18	8		Total				694

{10} with $s(A_3) = 15$, $\delta(A_3) = 332$ at a total cost of $195 + 2\%(195) = 198.90$ TL. This is another alternative to consider for the fleet mix problem.

Now a larger-size problem shows how efficient the sweep algorithm is in obtaining a partition if it exists or not for the given fleet of vehicles. This example has 79 destinations located in a metropolitan area, and the total distance to be traveled should not exceed 50 km. Table 3 provides the locations in (x, y) and (r, θ) polar coordinates and the number of passengers to be transported from those 79 destinations to the center in the morning and back later in the evening in this metropolitan area. There are 694 passengers to be transported; therefore, the knapsack problem to be solved is given as follows:

$$\min \ z = 35y_1 + 55y_2 + 105y_3$$

subject to

$$15y_1 + 30y_2 + 50y_3 \geq 694$$

where $y_i \geq 0$ and the integer for $i = 1, 2, 3$. The solution is $Y^1 = (1, 21, 1)$ with a cost of 1,295 in the range $694 \leq r \leq 695$. The next best solution is $Y^2 = (1, 23, 0)$ with a cost of 1,300 in the range $696 \leq r \leq 710$. Further, the next best solution is $Y^3 = (0, 22, 1)$ with a cost of 1,315 in the range $711 \leq r \leq 720$. Finally, the next solution is $Y^4 = (2, 23, 0)$ with a cost of 1,335 in the range $721 \leq r \leq 725$. The sweep algorithm does not obtain any partitions for the solutions $Y^1 = (1, 21, 1)$, $Y^2 = (1, 23, 0)$, and $Y^3 = (0, 22, 1)$, but finds a partition for the solution $Y^4 = (2, 23, 0)$ given in Table 4. Therefore, the relative error is less than $(1,335 - 1,295)/1,295 = 3.09\%$. As an indication of the efficiency of the sweep algorithm, one should point out that a small laptop computer (netbook) with a processor chip of 1.60 GHz solves the last problem in 1.73 s of CPU time.

Conclusions

The employee transporting problem is actually a set partitioning problem for the assignment of the vehicles. The capacities of the vehicles vary in size. Therefore, it is a heterogenous fleet problem. These facts make the employee transporting problem belong to a different class of problems than the HVRP in the literature. Although in both problems the aim is to determine the fleet mix and composition, the employee transporting problem strives for this aim without routing. A set partitioning model is developed, deviating from the integer formulations in the literature. An important aspect of this article is the thorough investigation of the model structure. Consequently, this investigation leads to a knapsack problem from the model structure. This knapsack problem is the common sense problem of determining the number of vehicles of each capacity to transport all the passengers at a minimum cost by ignoring the subset-vehicle assignments.

The partition obtained for the knapsack solution is an optimal solution. There is an exact algorithm to find the partition for a given knapsack solution. However, it requires matching too many subsets; it is impractical for large-size problems. The sweep algorithm is suitable for large-size problems. It is a heuristic method which obtains a partition in a very efficient manner computationally

Table 4 Solution of the large-size problem for $Y = (2, 23, 0)$

	v	$s(A_v)$	Q_v	$\delta(A_v)$	A_v
Vehicle	1	30	30	41.98	Part: 50 57 43 72
Vehicle	2	30	30	42.78	Part: 51 56 77 48 29
Vehicle	3	30	30	29.35	Part: 41 49 14
Vehicle	4	30	30	38.02	Part: 63 76 70
Vehicle	5	30	30	33.66	Part: 19 40 71
Vehicle	6	30	30	34.77	Part: 28 42 69
Vehicle	7	29	30	39.96	Part: 22 39 38 54
Vehicle	8	30	30	37.40	Part: 13 62 75
Vehicle	9	30	30	48.37	Part: 16 18 33 68
Vehicle	10	30	30	41.10	Part: 47 45 74
Vehicle	11	30	30	29.94	Part: 37 46 27
Vehicle	12	29	30	43.15	Part: 55 61 79
Vehicle	13	30	30	42.86	Part: 67 73 32
Vehicle	14	29	30	45.97	Part: 66 60 78
Vehicle	15	30	30	43.31	Part: 12 26 36 58
Vehicle	16	29	30	38.13	Part: 7 25 59
Vehicle	17	29	30	41.63	Part: 6 53 65
Vehicle	18	27	30	34.06	Part: 21 17 35
Vehicle	19	26	30	41.46	Part: 52 64 24
Vehicle	20	29	30	35.22	Part: 31 34 44
Vehicle	21	28	30	32.81	Part: 20 23 30
Vehicle	22	25	30	16.68	Part: 5 8 10
Vehicle	23	30	30	27.69	Part: 4 9 15 11
Vehicle	24	9	15	7.52	Part: 3
Vehicle	25	15	15	5.21	Part: 2 1
Total		694	720		

but may miss the optimal partition. A few examples are provided to illustrate how the main algorithm and each subalgorithm works. It is worth mentioning at this point that other heuristics and/or metaheuristics can be implemented in obtaining the partition after deciding the mix of the fleet. The employee transporting problem differs in nature from HVRP by not considering the routing, thus a computational comparison with other methods in the literature will not serve any purpose and such a task was not undertaken.

The model for the employee transporting problem and its solution can provide useful insights to practitioners in order to decide the mix of vehicles in the fleet and subsequently the fleet size to transport the passengers from/to the destinations. Furthermore, vehicle drivers are human beings; hence, they may have preferences and/or restrictions where to go. The approach of this article allows including the preferences and/or restrictions of the drivers in the choice of destination subsets.

Competing interests
The author declares that he has no competing interests.

Acknowledgements
The author thanks the anonymous Referees for their constructive comments for an improved version.

References

Baldacci R, Mingozzi A, Roberti R (2011a) New route relaxation and pricing strategies for the vehicle routing problem. Oper Res 59: 1269–1283

Baldacci R, Bartolini E, Mingozzi A (2011b) An exact algorithm for the pickup and delivery problem with time windows. Oper Res 59: 414–426

Bruce GL, Raghavan S, Wasil EA (eds) (2008) The vehicle routing problem: latest advances and new challenges. Operations research/computer science interface series, vol. 43. Springer, Heidelberg

Desrochers M, Verhoog TW (1991) A new heuristic for the fleet size and the mix vehicle routing problem. Comput Oper Res 18: 263–274

Gheysens F, Golden B, Assad A (1986) A new heuristic for determining the fleet size and composition. Math Programming Study 26: 233–236

Kellerer H, Pferschy U, Pisinger D (2004) Knapsack problems. Springer, Heidelberg

Miller C, Tucker A, Zemlin R (1960) Integer programming formulations and traveling salesman problem. J ACM 7: 326–329

Nurchanyo GW, Alias A, Shamsuddin SMM, MNMd Sap (2002) Sweep algorithm in vehicle routing problem for public transport. J Antarabanga (Teknologi Maklumat) 2: 51–64

Önder I (2007) An employee transporting problem and its heuristic solution. M.S Thesis. Çankaya University, Ankara

Renaud J, Boctor FF (2002) A sweep based algorithm for the fleet size and the mix vehicle routing problem. Eur J Oper Res 140: 618–628

Toth P, Vigo D (2002) The vehicle routing problem. Monographs on discrete mathematics and applications, Vol. 9. SIAM, Philadelphia

Yaman H (2006) Formulations and valid inequalities for the heterogeneous vehicle routing problem. Math Programming A 106: 365–390

7

Finding efficient frontier of process parameters for plastic injection molding

Wu-Lin Chen[1*], Chin-Yin Huang[2] and Ching-Ya Huang[2]

Abstract

Product quality for plastic injection molding process is highly related with the settings for its process parameters. Additionally, the product quality is not simply based on a single quality index, but multiple interrelated quality indices. To find the settings for the process parameters such that the multiple quality indices can be simultaneously optimized is becoming a research issue and is now known as finding the efficient frontier of the process parameters. This study considers three quality indices in the plastic injection molding: war page, shrinkage, and volumetric shrinkage at ejection. A digital camera thin cover is taken as an investigation example to show the method of finding the efficient frontier. Solidworks and Moldflow are utilized to create the part's geometry and to simulate the injection molding process, respectively. Nine process parameters are considered in this research: injection time, injection pressure, packing time, packing pressure, cooling time, cooling temperature, mold open time, melt temperature, and mold temperature. Taguchi's orthogonal array L27 is applied to run the experiments, and analysis of variance is then used to find the significant process factors with the significant level 0.05. In the example case, four process factors are found significant. The four significant factors are further used to generate 3^4 experiments by complete experimental design. Each of the experiments is run in Moldflow. The collected experimental data with three quality indices and four process factors are further used to generate three multiple regression equations for the three quality indices, respectively. Then, the three multiple regression equations are applied to generate 1,225 theoretical datasets. Finally, data envelopment analysis is adopted to find the efficient frontier of the 1,225 theoretical datasets. The found datasets on the efficient frontier are with the optimal quality. The process parameters of the efficient frontier are further validated by Moldflow. This study demonstrates that the developed procedure has proved a useful optimization procedure that can be applied in practice to the injection molding process.

Keywords: Injection molding; Taguchi's orthogonal array; Mutiple regression analysis; Data envelopment analysis; Optimization

Introduction

Along with the rapid progress of production techniques for high-tech products, better and better quality of products is required for the survival in the current market. Besides providing various functions, the trend of the design for plastic products is light, thin, short, and small. Therefore, the setting of process parameters for plastic products has a remarkable influence on their quality (Huang and Tai 2001).

Injection molding is one of the most important techniques for polymer processing (to manufacture plastic products) because of its high speed for molding and its capability of manufacturing complex geometric shapes of products. Besides, injection molding is capable of mass production, so it is widely used for many products, especially for electronic products, such as computers and communication products. Injection molding is usually adopted to produce thin parts or thin covers for these products.

Currently, there are two categories for the setting of process parameters for injection molding: one is based on the technicians' previous experience and the other takes advantage of mold flow analysis softwares, such as Moldflow (used in this study) to find the initial values for process parameters (by running various simulations on these moldflow analyses). However, no method can quickly find the reasonable combination of process parameters. In addition, trial and error is required for both

* Correspondence: wlchen@pu.edu.tw
[1]Department of Computer Science and Information Management, Providence University, Taichung, Republic of China (Taiwan)
Full list of author information is available at the end of the article

Figure 1 The geometry.

methods, and the process of trial and error consumes a significant amount of time and cost. Therefore, both methods cannot meet the requirement of the current market.

Researchers have applied various kinds of methods, e.g., artificial neural network and/or fuzzy logic (Liao et al. 2004a, b; Kurtaran et al. 2005; Ozcelik and Erzurumlu 2006), genetic algorithm (Kurtaran et al. 2005; Ozcelik and Erzurumlu 2006), design of experiments (Huang and Tai, 2001; Liao et al. 2004a, b), and response surface method (Ozcelik and Erzurumlu 2005; Kurtaran and Erzurumlu 2006; Chen et al. 2010) to optimize the initial process parameter setting of plastic injection molding. However, most studies focus on the single optimal combination of process parameters by different optimization

techniques. It is well known that when multiple quality characteristics are considered, the trade-off relationships exist among these quality characteristics, and these relationships make the task of finding the optimal combination rather complicated if not impossible. Instead of finding a single optimal combination of process parameters, this research seeks the efficient frontier of process parameters by data envelopment analysis (DEA).

The remainder of this paper is organized as follows: the 'Literature review on the optimization of process parameters for injection molding' section will review related work in the literature. The properties of the material and product used in this paper will be addressed in the 'Materials and product' section. The 'Experimental design and methodology' section will discuss the experimental design and the procedure of finding the efficient frontier of process parameters. Finally, the summary and concluding remarks are provided in the 'Summary and conclusions' section.

Literature review on the optimization of process parameters for injection molding

The literature of optimization for injection molding is briefly addressed in this section. Kim and Lee (1997) discussed different geometries for plastic parts to improve the parts' warpage by Taguchi's orthogonal experiment design. To avoid producing flaws of silver streaks for automobile plastic bumpers, Taguchi's optimization method is utilized to decide the optimal values for the process parameters by Chen et al. (1997). The same

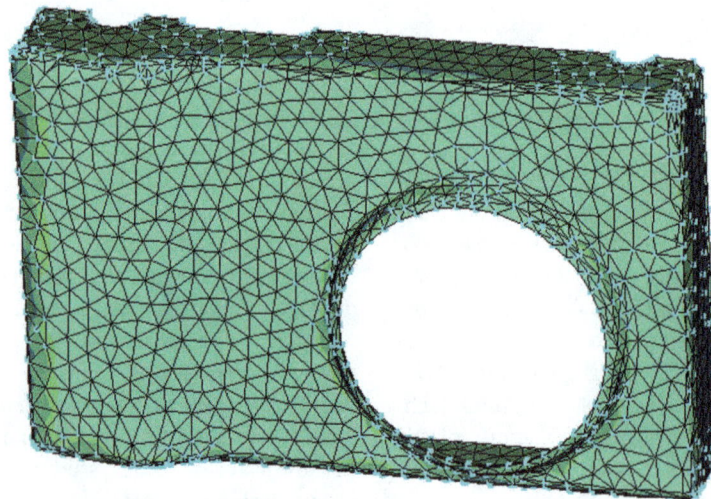

Figure 2 The finite element model. Courtesy of Moldflow Corporation.

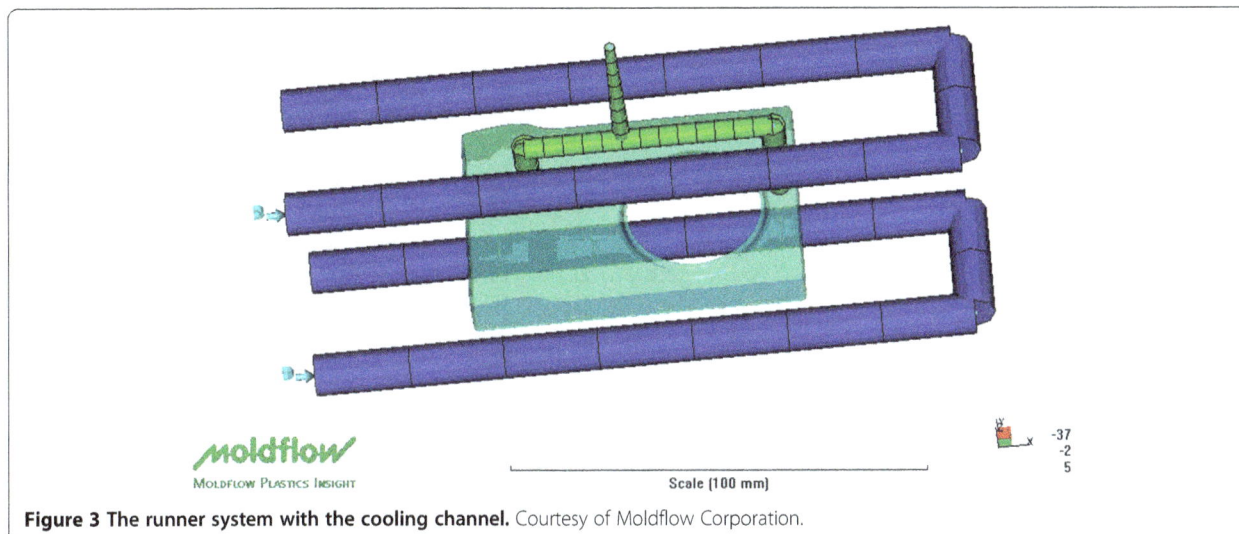

Figure 3 The runner system with the cooling channel. Courtesy of Moldflow Corporation.

optimization method was also used by several works (Huang and Tai 2001; Liu and Chen 2002; Liao et al. 2004a, b; Oktem et al. 2007) to find the optimal combinations of process parameters for different plastic products. In these works, warpages and shrinkages of plastic parts were usually considered as their quality indices. Moldflow, a mold flow analysis software, was used to simulate a real injection machine by Erzurumlu and Ozcelik (2006). Several techniques including the Taguchi's method, the neural networks, and the genetic algorithm were combined to optimize the process parameters.

The response surface methodology (RSM) is another popular method to optimize the process parameters in the literature. The complete model of RSM was first established by Box and Wilson (1951). To improve two quality indices, within-wafer non-uniformity and the removal rate, of the chemical–mechanical planarization process in semiconductor manufacturing, dual RSM was proposed by Fan (2000) to optimize five process parameters. In order to avoid the difficulty of minimizing both quality indices, one was treated as the primary response put in the objective function and the other was the secondary response placed in the constraint. Two works developed RSM combined with different optimization techniques (Ozcelik and Erzurumlu 2006; Chiang and Chang 2007). Chen et al. (2010) applied dual RSM to improve the quality of plastic injection molding. Warpage is the primary response (and is treated as the objective function), while shrinkage is the secondary response (and is then set as the constraint) in their work.

For multiple quality indices, the above two classes of optimization focus on searching the single optimal combination of process parameters. However, it is known that the trade-off relationships exist among multiple quality indices, so the searching task of the single optimum is not easy. Castro et al. (2007) used DEA technique to find the

efficient frontier when six quality indices (related with the part's geometry) were considered. This research proposes to combine several techniques including experimental design, analysis of variance (ANOVA), multiple regression analysis, and DEA to find the efficient frontier of the process parameters when three essential quality indices, warpage, shrinkage, and volumetric shrinkage at ejection, are under consideration simultaneously.

Materials and product

This study demonstrates how to determine the efficient frontier of process parameters for injection molding by taking an example of the thin cover of a digital camera. The CAD software, Solidworks, is the first to prepare the geometry of the product as shown in Figure 1.

Next, the commercial CAE simulation software tool, Moldflow, is utilized to create the finite element model and uses the finite element and finite difference method to solve pressure, flow, and temperature fields of injection molding (Walsh 1993; Mackerle 2005; Shoemaker 2006).

Table 1 Material properties of GE Cycoloy C2950 PC/ABS

Material property	Value
Recommended mold surface temperature (°C)	70
Recommended melt temperature (°C)	275
Melt density (g/cm^3)	0.97618
Solid density (g/cm^3)	1.1161
Eject temperature (°C)	113
Maximum shear stress (MPa)	0.4
Maximum shear rate (s^{-1})	40,000
Mold thermal conductivity (W/m °C)	29
Elastic module (MPa)	200,000
Poisson's ratio	0.33

Table 2 Key process parameters and their ranges of operation

Parameter	Variable	Initial value	Range of operation
Injection time	INT (x_1)	1 s	0.5 to 1.5 s
Injection pressure	INP (x_2)	120 MPa	100 to 140 MPa
Packing pressure	PP (x_3)	100 MPa	80 to 120 MPa
Packing time	PT (x_4)	10 s	7.5 to 12.5 s
Cooling time	COTI (x_5)	19 s	14 to 24 s
Coolant temperature	COTE (x_6)	25°C	20°C to 30°C
Mold open time	MOO (x_7)	5 s	4 to 6 s
Melt temperature	MET (x_8)	275°C	270°C to 280°C
Mold surface temperature	MOTE (x_9)	70°C	65°C to 75°C

The setting in Moldflow to simulate production of the parts is described below.

Figure 2 shows the finite element model of the plastic part, and Figure 3 is the runner system with the cooling channel. The material used in this paper is GE Cycoloy C2950 PC/ABS (Gardena, CA, USA), and the simulating molding machine used is Roboshot 330i (330 tons, 8.90 oz, 44 mm) high speed/pressure (Milacron, Cincinnati, OH, USA). Table 1 lists the main properties of the material.

This paper discusses nine key process parameters which are suggested in the literature (Chen and Turng 2005; Chen et al. 2010) because these parameters are most likely to affect products' quality. Table 2 shows nine process parameters and their ranges of operation suggested by Moldflow. Three indices of quality, warpage, shrinkage, and volumetric shrinkage at ejection, are considered in this paper.

The warpages are measured at four reference nodes, and the shrinkages are measured at four edges in this paper; these nodes and edges are depicted in Figure 4 of the *XY* horizontal plane. In Figure 4, four reference nodes N1991, N1962, N1955, and N2261 are shown, while four edges are formed by these four nodes and are denoted by E1, E2, E3, and E4, respectively.

Experimental design and methodology

This section presents how to design the experiment as well as the procedure to determine the efficient frontier of process parameters. In the 'Experimental design' subsection, the experimental design will be addressed. To effectively find the efficient frontier, ANOVA will be firstly executed to determine the significant process parameters out of the original nine process parameters in 'Finding significant process parameters by ANOVA' subsection. The complete design of experiment with four significant process parameters is again executed on Moldflow to have better accuracy of the following regression equations. Then, response regression model will be established in which only significant process parameters are considered in 'Setting up the regression response model to create the complete dataset'

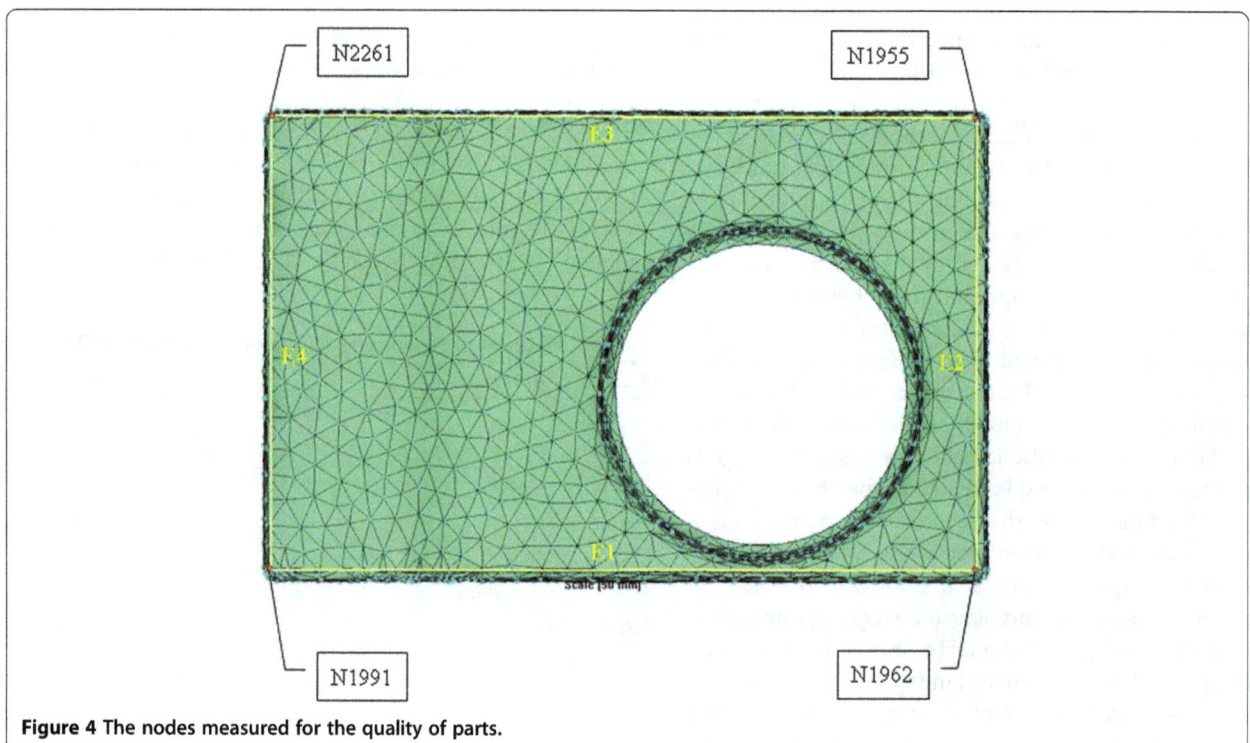

Figure 4 The nodes measured for the quality of parts.

Table 3 Results of ANOVA

	N1991	N1962	N1955	N2261	E1	E2	E3	E4	Volume	U
INT									◎	•
INP	◎	◎	◎	◎	◎	◎	◎	◎		•
PP	◎	◎	◎	◎	◎	◎	◎	◎		•
PT									◎	•
COTI										
COTE										
MOO										
MET										
MOTE										

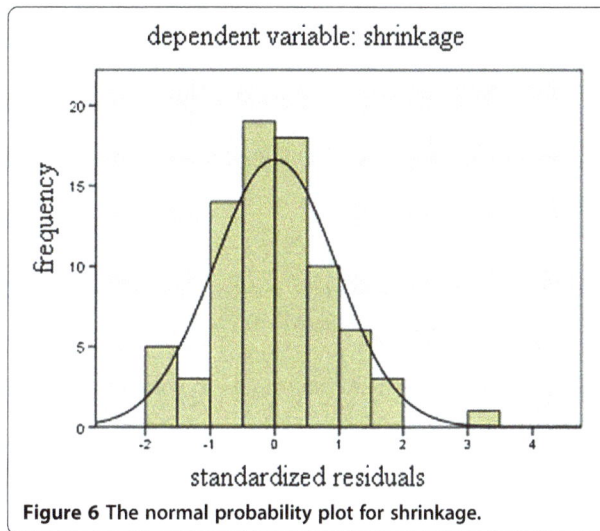
Figure 6 The normal probability plot for shrinkage.

subsection. This subsection will also present how to create the complete dataset for finding the efficient combinations. Finally, 'Determining the efficient frontier of process parameters by DEA' subsection will discuss how to find the efficient frontier by DEA.

Experimental design

The Taguchi experimental design with orthogonal array is an efficient experimental design for fraction factorial design (Rose 1989; Montgomery 2005). Because there are nine process parameters considered in this research, complete experimental design is just too expensive to execute. Therefore, this research adopts the Taguchi experimental design with orthogonal array, L27, to perform the experiment on Moldflow. The experimental results of L27 is shown in Appendix A. Three levels of each process parameters are assigned to lower bound, mid-point, and upper bound of the range of operation listed in Table 2; for example, levels 1, 2 and 3 of injection time x_1 are 0.5, 1 and 1.5 s, respectively.

Finding significant process parameters by ANOVA

In order to simplify the regression equations (and thus simplify the following procedure), ANOVA is firstly executed to find significant process parameters to affect the parts' three quality indices which will only be considered in the regression equations. The results are shown in Table 3. Each node represents the warpage at this node, each edge means the shrinkage at this edge, and the volume is the volumetric shrinkage at ejection in Table 3. The symbol '◎' means the corresponding process parameter significantly affects the quality index under the significant level 0.05 in the figure. The last column of Table 3, U, is remarked by '•' if the corresponding process parameter significantly affects at least one quality index. From Table 3, only four process parameters are significant to affect at least one quality index, and

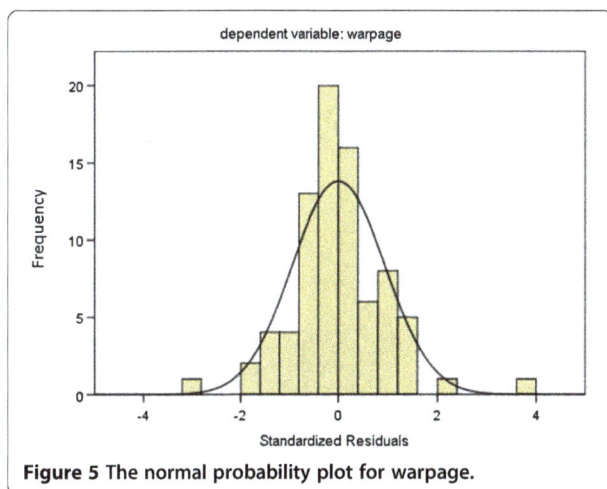
Figure 5 The normal probability plot for warpage.

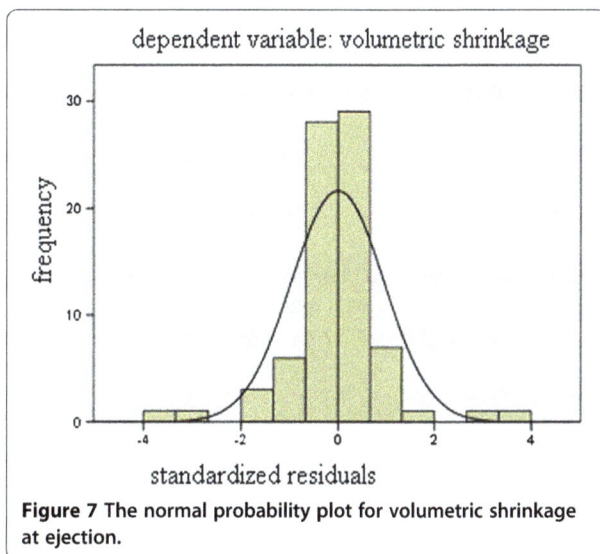
Figure 7 The normal probability plot for volumetric shrinkage at ejection.

Table 4 Levels of process parameters

Process parameter	L1	L2	L3	L4	L5	L6	L7
Injection pressure (MPa)	100	110	120	130	140		
Injection time (s)	0.5	0.67	0.84	1	1.17	1.34	1.5
Packing pressure (MPa)	80	86.67	93.34	100	106.67	113.34	120
Packing time (s)	7.5	8.75	10	11.25	12.5		

Table 5 DMUs on the efficient frontier

DMU	Score
560	100
840	100
871	100
1,151	100
1,186	100
1,187	100
1,188	100
1,189	100
1,190	100

hence only these four parameters are considered in the regression analysis in the next subsection.

Setting up the regression response model to create the complete dataset

To obtain a more complete efficient frontier for the process parameters, more data are required. The regression model, the response surface model, is utilized to create more data. In order to have better forecasting accuracy of the regression model, the complete experiment design with four significant process factors is executed again on Moldflow before the regression equations are established. The results are shown in the Appendix B.

The results of the complete experiment design with four significant process factors are then utilized to set up the second-order response surface model by the regression analysis of statistics software, SPSS. Three response surface equations for three quality indices are found below:

$$\text{Warp} = 1.45 - 0.011\,\text{INP} - 0.238\,\text{INT} - 0.00823\,\text{PP}$$
$$+ 0.000046\,\text{INP2} + 0.108\,\text{INT2} + 0.000016\,\text{PP2}$$
$$- 0.00176\,\text{INP}*\text{INT} + 0.00001\,\text{INP}*\text{PP}$$
$$+ 0.00211\,\text{INT}*\text{PP}, \tag{1}$$

$$\text{Shrink} = 1.32 - 0.0145\,\text{INP} + 0.35\,\text{INT} + 0.000086\,\text{INP2}$$
$$+ 0.0603\,\text{INT2} - 0.000013\,\text{PP2} - 0.00593\,\text{INP}*\text{INT}$$
$$- 0.000026\,\text{INP}*\text{PP} + 0.00122\,\text{INT}*\text{PP}, \tag{2}$$

and

$$\text{Volume} = 27.0 - 0.348\,\text{INP} + 0.00165\,\text{INP2} + 2.74\,\text{INT2}$$
$$- 0.00035\,\text{PP2} - 0.0751\,\text{INP}*\text{INT} - 0.000613\,\text{INP}*\text{PT}$$
$$+ 0.0369\,\text{INT}*\text{PP} + 0.0839\,\text{INT}*\text{PT}. \tag{3}$$

The normal probability plots are provided in Figures 5, 6, 7 to justify the validity of the regression analysis.

To find a more complete efficient frontier of process parameters, more data are required. Regressed response

surface equations, Equations 1, 2, 3, are exploited to create more data points. Because DEA software, Banxia Frontier Analyst 3, has the limitation on the maximal number of data points (also called decision making units (DMUs)), the design of data points to be created is explained below. Based on the results of ANOVA, because injection time and packing pressure are more significant than the other two process parameters, there are seven levels selected for these two process parameters and five levels for the other two parameters. Therefore, there are $5 \times 7 \times 7 \times 5 = 1{,}225$ data points to be created by Equations 1, 2, 3. The levels of each process parameter are listed in Table 4. Note that in Table 4, all levels of each parameter all fall its range of operation in Table 2.

Determining the efficient frontier of process parameters by DEA

DEA is a technique to evaluate the relative efficiency of many DMUs by analyzing multiple inputs and multiple outputs of each DMU. Its goal is to find the efficient DMUs, also called efficient frontier in the literature of DEA. This research uses the standard DEA Charnes, Cooper, and Rhodes (CCR) (Charnes et al. 1978) model to find the efficient frontier of DMUs which is created in the previous subsection. The mathematical model of DEA CCR is briefly outlined below. Suppose that there are K DMUs, each of which consumes N inputs and

Table 6 The reference counts of the efficient DMUs

DMU	Reference count
560	773
840	624
871	402
1,151	182
1,186	47
1,187	0
1,188	0
1,189	0
1,190	0

Table 7 Levels of process parameters for efficient DMUs

		DMU				
		560	840	871	1,151	1,186
Process parameter	INP	120	130	130	140	140
	INT	0.67	0.84	1	1.17	1.34
	PP	120	120	120	120	120
	PT	12.5	12.5	7.5	7.5	7.5
	COTI	19	19	19	19	19
	COTE	25	25	25	25	25
	MOO	5	5	5	5	5
	MET	275	275	275	275	275
	MOTE	70	70	70	70	70
Quality index (forecasting)	Warp.	0.0964	0.0930	0.0906	0.0997	0.1065
	Shrink.	0.1397	0.1076	0.0814	0.0436	0.0126
	Vol.	1.9019	1.9418	2.0416	2.3027	2.5442
Quality index (real)	Warp.	0.0796	0.0720	0.0780	0.0820	0.1114
	Shrink.	0.1232	0.9250	0.0835	0.0597	−0.0948
	Vol.	2.0650	1.9610	2.0680	2.8260	3.3670

produces M outputs. Then, the DEA CCR model can be established as follows:

$$\text{Max } H_{K'} = \frac{\sum\limits_{m=1}^{M} U_m Y_{k'm}}{\sum\limits_{n=1}^{N} V_n X_{k'n}}$$

such that

$$\frac{\sum\limits_{m=1}^{M} U_m Y_{km}}{\sum\limits_{n=1}^{N} V_n X_{kn}} \leq 1$$

$$U_m, V_n \geq \varepsilon > 0$$
$$m = 1, 2, ..., M$$
$$n = 1, 2, ..., N$$
$$k = 1, 2, ..., K,$$

where Y_{km} is the value of the mth output generated by the kth DMU, X_{kn} is the value of the nth input consumed by the kth DMU, V_n and U_m are X_{kn}'s and Y_{km}'s weights, respectively, whose values are determined by solving the model, H_k is the relative efficiency of the kth DMU, and ε is a small positive number. After solving the CCR DEA model, a DMU is on the efficient frontier if its relative efficiency, H_k, is equal to 1.

Three quality indices, warpage, shrinkage, and volumetric shrinkage at ejection, are considered in this paper. Among these three indices, warpage is treated as the output, while shrinkage and volumetric shrinkage at ejection are two inputs in the DEA model. Because the output in DEA model needs to be maximized and warpage is apparently the minimized quality index, the transformation, 1-warpage, is adopted.

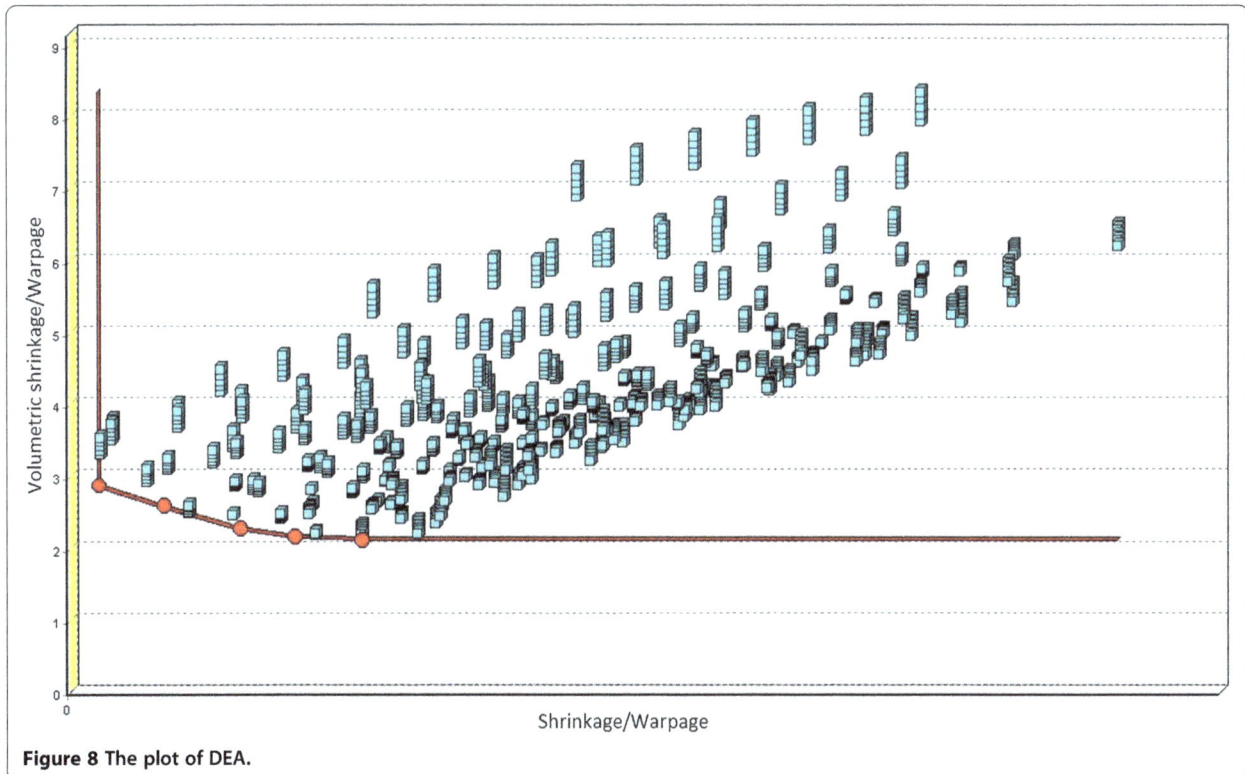

Figure 8 The plot of DEA.

Table 8 Relatively efficient DMUs

DMU	Score
44	100
45	100
840	*100*
871	*100*
1,151	*100*
54	97.82
52	97.80
53	97.13
43	96.88
560	*94.19*

DMUs in italics represent the efficient ones suggested in this study.

The DEA software, Banxia Frontier Analyst 3, is used to find the efficient frontier of process parameters. The dataset used is the dataset of 1,225 data points created in the previous subsection. Each data point is treated as a DMU. After running Banxia Frontier Analyst 3, data points on the efficient frontier are found in Table 5. There are nine DMUs on the efficient frontier, among which five DMUs have at least one reference count as shown in Table 6. Therefore, these five DMUs with positive reference counts are treated as the efficient frontier of process parameters in this paper. The levels of each process parameter for these five DMUs are shown in Table 7, where the forecasting value of a quality index is its value derived from the corresponding regression equations and the real value of a quality index means its value by re-running Moldflow on this combination of process parameters. The plot of DEA results is shown in Figure 8.

To verify the efficiency of five DMUs found in this paper, we re-run Moldflow on each DMU and then the results are compared with those of $3^4 = 81$ data points which are utilized to set up the regression equations in 'Setting up the regression response model to create the complete dataset' subsection. The comparison is accomplished by executing DEA on 5 efficient DMUs found in this paper and 81 data points. The results are shown in Tables 8 and 9. From Table 9, it can be observed that

Table 9 Efficient DMUs with positive counts

DMU	Reference count
840	62
1,151	12
871	9
45	9
44	1

among five efficient DMUs found in this paper, three DMUs are still on the efficient frontier and one DMU is relatively highly efficient with 94.18% DEA score. Only DMU 1,186 is not quite efficient with 71.28% DEA score, and this may be due to the error of the regression equation at this DMU. It is fair to suggest that the error induced by the regression equation at most of the points is fairly small. Therefore, the efficient frontier of process parameters found by this paper with only 108 (=27 + 81) repeats of experiments can really provide good combinations of process parameters for decision making.

Summary and conclusions

Part quality for plastic injection molding is often evaluated by multiple interrelated quality indices, and each quality index is highly related with process parameters. This paper proposes a method of finding the complete efficient frontier of process parameters with only a few times of experiments when multiple quality indices are considered for plastic injection molding. The thin front cover of a digital camera is provided as the example of executing the method. Based on the literature, nine process parameters are considered in this research. The experimental design with the Taguchi orthogonal L27 is used to run the experiment on Moldflow. ANOVA is then executed to find significant parameters to affect the part's quality indices, and the results show that four out of nine parameters are significant with the significant level 0.05. In order to set up the complete efficient frontier of DEA analysis, more data are required, and the regression equations are used to create them. To have good accuracy of the multiple regressed equations, the complete experimental design with 3^4 times (only four significant process parameters are considered) of experiments is again executed on Moldflow. The multiple regression equations are then set up and are used to produce the dataset for DEA analysis. The results of DEA analysis shows that the five combinations are on the efficient frontier.

To show the efficiency of these combinations suggested in this paper, DEA analysis is again conducted on them as well as the results of the experiments of 3^4 times used for establishing multiple regression equations. The results show that only one combination is not as efficient mainly because of the error of the regressed equations at this combination. Hence, the method proposed here is believed indeed can find the efficient frontier of process parameters with only a few times of experiments.

The classic DEA method, CCR, is used in this paper; in the future, some other DEA methods, such as BCC, can be used, and the performance of each method can be compared. Another possible future research topic is to evaluate the performance of Moldflow analysis.

Appendix A

The results of L27 are presented in Table 10.

Table 10 Experimental results of L27 array

	N1991	N1962	N1955	N2261	E1	E2	E3	E4	Volume
1	0.2222	0.2118	0.2151	0.203	0.4014	0.1292	0.3728	0.171	4.384
2	0.2137	0.2027	0.2077	0.1953	0.3856	0.1226	0.3598	0.1619	4.346
3	0.2185	0.2082	0.2126	0.2001	0.395	0.1273	0.3681	0.1666	4.379
4	0.1436	0.1322	0.1392	0.125	0.2532	0.0529	0.2346	0.0913	3.281
5	0.1513	0.1416	0.1466	0.133	0.2675	0.0643	0.2475	0.1006	3.468
6	0.1416	0.1301	0.137	0.1223	0.2491	0.0516	0.2308	0.0896	3.154
7	0.1032	0.0902	0.098	0.0821	0.173	0.0035	0.1579	0.0379	3.936
8	0.0927	0.0798	0.0877	0.0693	0.1497	-0.0153	0.1321	0.0187	4.095
9	0.0986	0.0867	0.0947	0.0783	0.165	-0.0014	0.1503	0.0341	3.973
10	0.1602	0.1564	0.1612	0.1521	0.2877	0.0609	0.2717	0.1046	4.337
11	0.1455	0.1416	0.1412	0.1356	0.262	0.0564	0.2432	0.0959	3.938
12	0.1414	0.1365	0.1354	0.1301	0.2526	0.0504	0.2309	0.0896	3.989
13	0.0697	0.0593	0.0776	0.0559	0.083	-0.0475	0.0739	-0.013	2.063
14	0.0677	0.0593	0.0718	0.0542	0.091	-0.0379	0.0797	-0.0041	2.018
15	0.0735	0.0618	0.0824	0.0575	0.0873	-0.0491	0.0779	-0.0126	2.086
16	0.1367	0.1358	0.1243	0.1288	0.2529	0.0685	0.2254	0.1026	3.593
17	0.1352	0.134	0.1249	0.1289	0.2506	0.0647	0.2261	0.0994	3.369
18	0.131	0.1303	0.1192	0.1246	0.243	0.059	0.2161	0.0937	3.526
19	0.1446	0.1537	0.1601	0.1062	0.2521	0.0288	0.2252	0.0535	7.185
20	0.1339	0.1483	0.1488	0.1027	0.247	0.0297	0.2139	0.0475	6.509
21	0.1216	0.1423	0.144	0.1021	0.2177	0.0251	0.1904	0.0434	6.474
22	0.1563	0.1809	0.1834	0.1507	0.3036	0.0629	0.2963	0.0983	7.855
23	0.15	0.1893	0.1672	0.1452	0.3118	0.0546	0.2718	0.0897	7.12
24	0.1786	0.2015	0.1828	0.1496	0.35	0.0686	0.3018	0.106	7.41
25	0.0673	0.0611	0.0939	0.0325	0.0558	-0.0288	0.0621	-0.0166	3.294
26	0.0951	0.0826	0.1094	0.051	0.1015	-0.0014	0.1003	-0.0039	3.055
27	0.0794	0.0698	0.0991	0.0393	0.0651	-0.0164	0.0728	-0.0138	3.21

Appendix B

The results of 81 are presented in Table 11.

Table 11 Moldflow execution results

	Warpage	Shrinkage	Volume shrinkage at ejection
1	0.2185	0.395	4.379
2	0.2182	0.3944	4.378
3	0.2169	0.3922	4.364
4	0.1611	0.2867	3.611
5	0.161	0.2863	3.61
6	0.1601	0.2847	3.596
7	0.1154	0.1973	2.724
8	0.1156	0.1969	2.723
9	0.1145	0.1951	2.706
10	0.2006	0.3635	4.08
11	0.2003	0.3629	4.079
12	0.1998	0.3622	4.078
13	0.1451	0.2556	3.288
14	0.1452	0.2551	3.287
15	0.1448	0.2545	3.286
16	0.0982	0.1632	2.25
17	0.098	0.1629	2.251
18	0.0978	0.1626	2.249
19	0.2271	0.4148	4.56
20	0.2264	0.4134	4.516
21	0.2253	0.4117	4.503
22	0.1612	0.2931	3.683
23	0.1607	0.2919	3.642
24	0.1596	0.2905	3.627
25	0.1111	0.1965	2.683
26	0.1103	0.195	2.642
27	0.1092	0.1931	2.264
28	0.1952	0.3589	4.869
29	0.195	0.3589	4.879
30	0.1981	0.3646	4.958
31	0.1472	0.265	4.016
32	0.1471	0.2651	4.023
33	0.1487	0.2679	4.073
34	0.1068	0.1776	3.158
35	0.107	0.1776	3.163
36	0.1084	0.1805	3.2
37	0.136	0.2522	3.551
38	0.1335	0.2447	3.509
39	0.1332	0.2471	3.506
40	0.0982	0.1743	2.926
41	0.096	0.1696	2.882
42	0.0957	0.1691	2.883
43	0.078	0.0835	2.068
44	0.0775	0.0779	2.048

Table 11 Moldflow execution results *(Continued)*

45	0.0778	0.0772	2.054
46	0.1528	0.2846	3.924
47	0.153	0.2847	3.939
48	0.1517	0.2827	3.921
49	0.1068	0.197	3.146
50	0.1074	0.1973	3.155
51	0.1063	0.1952	3.138
52	0.0674	0.1021	2.015
53	0.0678	0.1027	2.028
54	0.0677	0.1007	2.014
55	0.2068	0.3438	7.623
56	0.1449	0.2523	3.853
57	0.2079	0.3452	6.844
58	0.1528	0.2605	4.782
59	0.1465	0.2469	4.573
60	0.1712	0.2751	6.899
61	0.1617	0.2321	7.921
62	0.1614	0.2345	6.631
63	0.1618	0.2355	8.349
64	0.1954	0.3269	7.239
65	0.1214	0.2301	3.745
66	0.1751	0.3106	6.779
67	0.1362	0.1936	4.24
68	0.1366	0.194	4.355
69	0.1352	0.1922	4.284
70	0.108	0.0802	3.177
71	0.1049	0.0744	3.16
72	0.1049	0.0746	3.165
73	0.1021	0.1438	3.199
74	0.1003	0.1455	3.186
75	0.1004	0.1425	3.197
76	0.1011	0.078	3.2
77	0.0982	0.0732	3.206
78	0.0981	0.0721	3.21
79	0.1165	−0.063	3.203
80	0.1137	−0.0656	3.22
81	0.1132	−0.0662	3.224

Acknowledgements

The authors would like to thank financial support from the research project 98-2221-E-029-019, National Science Council of Taiwan. The authors are grateful to the expert anonymous reviewers and the editor-in-chief whose comments and suggestions considerably improved this article.

Author details

[1]Department of Computer Science and Information Management, Providence University, Taichung, Republic of China (Taiwan). [2]Department of Industrial Engineering and Enterprise Information, Tunghai University, Taichung, Republic of China (Taiwan).

References

Box GEP, Wilson KB (1951) On the experimental attainment of optimum condition. J Royal Statistic Soc 13:1–45

Castro CE, Ríos MC, Castro JM, Lilly B (2007) Multiple criteria optimization with variability considerations in injection molding. Polym Eng Sci 47(4):400–409

Charnes A, Cooper WW, Rhodes E (1978) Measuring the efficiency of decision making units. Eur J Oper Res 2:429–444

Chen ZB, Turng LS (2005) A review of current developments in process and quality control for injection molding. Adv Polym Technol 24(3):165–182

Chen RS, Lee HH, Yu CY (1997) Application of Taguchi's method on the optimal process design of an injection molded PC/PBT automobile bumper. Compos Struct 39:209–214

Chen WL, Huang CY, Hung CW (2010) Optimization of plastic injection molding process by dual response surface method with non-linear programming. Computations Engineering 27(8):951–966

Chiang KT, Chang FP (2007) Analysis of shrinkage and warpage in an injection-molded part with a thin shell feature using the response surface methodology. Int J Adv Manuf Technol 35:468–479

Erzurumlu T, Ozcelik B (2006) Minimization of warpage and sink index in injection-molded thermoplastic parts using Taguchi optimization method. Mater Des 27:853–861

Fan SK (2000) Quality improvement of chemical–mechanical wafer planarization process in semiconductor manufacturing using a combined generalized linear modeling - non-linear programming approach. Int J Prod Res 13:3011–3029

Huang MC, Tai CC (2001) The effective factors in the warpage problem of an injection-molded part with a thin shell feature. J Mater Process Technol 110:1–9

Kim BH, Lee BH (1997) Variation of part wall thicknesses to reduce warpage of injection-molded part: robust design against process variability. Polymer Plast Tech Eng 36(5):791–807

Kurtaran H, Erzurumlu T (2006) Efficient warpage optimization of thin shell plastic parts using response surface methodology and genetic algorithm. Int J Adv Manuf Technol 27:468–472

Kurtaran H, Ozcelik B, Erzurumlu T (2005) Warpage optimization of a bus ceiling lamp base using neural network model and genetic algorithm. J Mater Process Technol 169(3):314–319

Liao SJ, Hsieh WH, Wang JT, Su YC (2004a) Shrinkage and warpage prediction of injection-molded thin-wall parts using artificial neural networks. Polym Eng Sci 44:2029–2040

Liao SJ, Chang DY, Chen HJ, Ho JR, Yau HT, Hsieh WH (2004b) Optimal process conditions of shrinkage and warpage of thin-wall parts. Polym Eng Sci 44:917–928

Liu SJ, Chen CF (2002) Significance of processing parameters on the warpage of rotationally molded parts. J Reinf Plast Compos 21(8):723–733

Mackerle J (2005) Finite element modelling of ceramics and glass, an addendum – a bibliography (1998–2004). Eng Comput 22(3):297–373

Montgomery DC (2005) Introduction to Statistical Quality Control. Wiley, New York

Oktem H, Erzurumlu T, Uzman I (2007) Application of Tauguchi optimization technique in determining plastic injection molding process parameters for a thin-shell part. Mater Des 27:1271–1278

Ozcelik B, Erzurumlu T (2005) Determination of effecting dimensional parameters on warpage of thin shell plastic parts using integrated response surface method and genetic algorithm. Int Comm Heat Mass Tran 32(8):1085–1094

Ozcelik B, Erzurumlu T (2006) Comparison of the warpage optimization in the plastic injection molding using ANOVA, neural network model and genetic algorithm. J Mater Process Technol 171:437–445

Rose PJ (1989) Taguchi Techniques for Quality Engineering. McGraw-Hill, New York

Shoemaker J (2006) Moldflow Design Guide: A Resource for Plastics Engineers. Hanser Gardner Publications, Cincinnati

Walsh SF (1993) Shrinkage and warpage prediction for injection molded components. J Reinf Plast Compos 12:769–777

8

Retailer's inventory system in a two-level trade credit financing with selling price discount and partial order cancellations

A Thangam

Abstract

In today's fast marketing over the Internet or online, many retailers want to trade at the same time and change their marketing strategy to attract more customers. Some of the customers may decide to cancel their orders partially with a retailer due to various reasons such as increase in customer's waiting time, loss of customer's goodwill on retailer's business, and attractive promotional schemes offered by other retailers. Even though there is a lag in trading and order cancellation, this paper attempts to develop the retailer's inventory model with the effect of order cancellations during advance sales period. The retailer announces a price discount program during advance sales period to promote his sales and also offers trade credit financing during the sales periods. The retailer availing trade credit period from his supplier offers a permissible delay period to his customers. The customer who gets an item is allowed to pay on or before the permissible delay period which is accounted from the buying time rather than from the start period of inventory sales. This accounts for significant changes in the calculations of interest payable and interest earned by the retailer. The retailer's total cost is minimized so as to find out the optimal replenishment cycle time and price discount policies through a solution procedure. The results derived in mathematical theorems are implemented in numerical examples, and sensitivity analyses on several inventory parameters are obtained.

Keywords: Inventory; Advance sales; Price discount; Two-echelon trade credit

Introduction

In today's business era, retailers have the dominant power of controlling or affecting another member's decision in a supply chain. A retailer has the ability to offer an effective promotional effect such as price discount and credit period. The retailers often try to stimulate the demand by offering price discounts. Price discounts could improve economic benefits to consumers and influence consumers' beliefs about the brand which will increase consumers' purchase intentions. In real life, there are situations in which the retailer announces price discount offers to the customers who can commit their orders before the selling period. Due to the booming in IT, customers can easily commit their orders prior to selling period and the estimation error in demand can also be reduced. This situation is adopted more in the selling of musical disks, apparel, video games, or books. In dairy-product manufacturing scheme prior to selling season, the retailer offers price discount to the customers who register their orders via email or phone call.

Certainly, the credit period facility would promote the purchases, and it attracts new customers who consider trade credit policy as a type of price reduction. To handle the risks of trade credit situations, retailer collects a higher interest from his customers when they did not settle the payment within the credit period time. In this paper, supplier offers the retailer a trade credit period t_1. The retailer offers his customers a credit period t_2, and he receives the revenue from t_2 to $T + t_2$, where T is the cycle time at the retailer. Under this situation, three cases such as $t_1 \leq T$, $T \leq t_1 \leq T + t_2$, and $T + t_2 \leq t_1$ are to be considered. Customers under advance sales booking system may cancel their reservations due to various reasons such as increase in customer's waiting time, loss of customer's goodwill on retailer's business, and attractive

Correspondence: thangamgri@yahoo.com
Department of Mathematics, Pondicherry University – Community College, Lawspet -08, 605 008, Pondicherry, India

promotional schemes offered by other retailers. Here, we consider the partial order cancellations during advance sales period.

This paper investigates retailer inventory system in which customers are partially canceling their orders during advance sales period. The customers commit their orders before the selling period. Among the committed orders, a fraction of the orders are cancelled. The customer who receives an item at time 't' will remit at time $t + t_2$ due to his availability of trade credit period t_2. So, the retailer gets revenue from earning the interest on customer's payment during the period from t_2 to t_1, and he pays interest during the period t_2 to $T + t_2$. So, the cases in Tsao (2009) are to be reconsidered as $t_1 \leq T$, $T \leq t_1 \leq T + t_2$, and $T + t_2 \leq t_1$. The retailer also earns interest from standby orders during the advance sales period. During the normal sales period, all customers receive their orders at the time of their purchase. With the help of derived mathematical theorems in the model, a simple solution procedure is provided to find the optimal solution. Numerical examples are given to illustrate the solution process, and sensitivity analyses are performed for various inventory key parameters.

Literature review
During the past few years, many researchers have studied inventory models for permissible delay in payments. Goyal (1985) was the first proponent for developing an economic order quantity (EOQ) model under the conditions of permissible delay in payments. Shah (1993) considered a stochastic inventory model when items in the inventory deteriorate and delays in payments are permissible. Aggarwal and Jaggi (1995) extended Goyal's model (1985) to allow the inventory to have deteriorating items. Jamal et al. (1997) further generalized Aggarwal and Jaggi's model (1995) to allow for shortages. Hwang and Shinn (1997) developed a model considering exponentially deteriorating items and found decision policy for selling price and lot size. Teng (2002) amended Goyal's model (1985) by considering the difference between unit price and unit cost and established an easy analytical closed-form solution to the problem. Chang et al. (2003) constructed a mathematical model for an EOQ inventory with deteriorating items and supplier credits are linked to ordering quantity. Chung and Huang (2003) generalized Goyal's EOQ model (1985) to an economic production quantity (EPQ) model in which the selling price is the same as the purchase cost. Huang (2003) extended Goyal's model (1985) to the case in which the supplier offers the retailer the permissible delay period M (i.e., the upstream trade credit), and the retailer in turn provides the trade credit period N (with $N < M$) to his customers (i.e., the downstream trade credit). Ouyang et al. (2006) developed an EOQ model for

deteriorating items under trade credits. Teng and Goyal (2007) amended Huang's model (2003) by complementing his shortcomings. Liao (2007) established an EPQ model for deteriorating items under permissible delay in payments. Chang et al. (2008) reviewed the contributions on the literature in modeling of inventory lot sizing under trade credits. Ho et al. (2008) developed an integrated supplier–buyer inventory model with the assumption that demand is sensitive to retail price and the supplier adopts a two-part trade credit policy. Huang and Hsu (2008) have developed an inventory model under two-level trade credit policy by incorporating partial trade credit option at the customers of the retailer. Liao (2008) developed an EOQ model with non-instantaneous receipt and exponentially deteriorating items under two-level trade credit financing. Teng and Chang (2009) extended Huang's model (2007) by relaxing the assumption $N < M$. Jaggi et al. (2008) developed a simple EOQ model in which the retailer's demand is linked to credit period. Thangam and Uthayakumar (2009) developed an EPQ model for perishable items under two-level trade credit policy when demand depends on selling price and credit period. Teng (2009) developed an EOQ model for a retailer who receives a full trade credit from its supplier and offers a partial trade credit to its bad credit customers or a full trade credit to its good credit customers. Teng et al. (2009) developed a mathematical model for an EOQ inventory with two warehouses and solved the problem by an arithmetic–geometric inequality method. Tsao (2009) developed a model by considering advance sales discount and trade credits. In the paper of Tsao (2009), he considered a strategy, namely advance sales discount (ASD) program, that the customers can commit their orders at a discount price prior to the selling season. He considers the cases such as $T \geq t_1$, $t_2 \leq T \leq t_1$, and $T \leq t_2$. Chen and Kang (2010a) considered trade credit and imperfect quality in an integrated vendor-buyer supply chain model. Concurrently, Chen and Kang (2010b) developed an integrated vendor-buyer inventory model with two-level trade credits and price negotiation scheme. Chang et al. (2010) have extended Liao's model (2008) by considering the case $M < N$ also. Hu and Liu (2010) established an EPQ model with permissible delay in payments and allowable shortages. Cárdenas-Barrón et al. (2010) developed a model which considers the advantage of a one-time discount offer with allowed backorders. Balkhi (2011) has developed a finite horizon inventory model with deteriorating items under inflation and time value of money when shortages are not allowed. Thangam and Uthayakumar (2011) have built a mathematical model for a retailer under two-level trade credit and two-payment methods. Tsao (2011) developed an EOQ model by considering trade credit and logistics risk. Teng et al. (2011) extended an EOQ model for stock-dependent demand to supplier's trade credit with a progressive payment scheme. Skouri et al. (2011) studied supply chain models for deteriorating items with ramp-type demand rate under permissible delay in payments. Jaggi et al. (2012) developed an EOQ model under two-levels of trade credit policy when demand is influenced by credit period. Liao et al. (2012) have developed a two-warehouse lot-sizing model with order-dependent trade credit period. Tsao and Sheen (2012) have developed a multi-item supply chain model with trade credit periods and weight freight cost. Thangam (2012) developed a two-level trade credit financing model for a supply chain with deteriorating items and advance payment scheme. Teng et al. (2012a) developed vendor-buyer inventory models with trade credit financing under a non-cooperative and integrated environments. Concurrently, Teng et al. (2012b) proposed an EOQ model with trade credit financing for increasing demand. Min et al. (2012) established an EPQ model with inventory level-dependent demand and permissible delay in payments. Tsao (2012) considered manufacturer's production and warranty decisions for an imperfect production system under system maintenance and trade credit. Teng et al. (2013) have developed a two-level trade credit financing model with timely increasing demand at the retailer. Feng et al. (2013) have developed an EPQ inventory model with supplier's cash discount and two-level trade credit financing. Ouyang et al. (2013) have developed a mathematical model with two-level trade credit financing in which trade credit offer depends on the amount of ordering quantity. Taleizadeh et al. (2013) have developed an EOQ model with perishable items, special sale offers, and shortages. Chung and Cárdenas-Barrón (2013) presented a simplified solution procedure to an EOQ model for deteriorating items by Min et al. (2010) with stock-dependent demand and two-level trade credit. Chern et al. (2013) established Stackelberg solution in a vendor-buyer supply chain model with permissible delay in payments. Ouyang and Chang (2013) proposed an optimal production lot with imperfect production process under permissible delay in payments and complete backlogging. Chen et al. (2013a) established the retailer's optimal EOQ when the supplier offers conditionally permissible delay in payments linked to order quantity. Concurrently, Chen et al. (2013b) attempted to overcome some shortcomings of mathematical model and expressions in Liao et al. (2012). Jaggi et al. (2013) established an EOQ inventory model with defective items under allowable shortages and trade credit.

Mathematical model formulation

We follow the same notations as in Tsao (2009) and introduce a new notation, namely δ (fixed) referring the rate at which the orders are cancelled during advance sales period:

p unit retailer price,
c unit purchase cost,
A ordering cost per order,
H unit inventory holding cost,
t_1 retailer's credit period provided by supplier,
t_2 customer's credit period provided by retailer,
r price discount,
δ the rate at which the orders are cancelled during advance sales period,
I_{p} the interest paid per dollar per unit time,
I_{e} the interest earned per dollar per unit time,
T replenishment cycle time,
D_1 annual demand rate for the retailer, say retailer 1, to whom optimum decision policy is considered,
D_2 annual demand rate for other retailers,
Y_1 fraction of retailer 1's customers who use advance sales discount program,
Y_2 fraction of other retailers' customers who use advance sales discount program,
$(1 - Y_1)D_1$ annual demand of customers who are not using advance sales discount program.

We follow the assumptions as in Tsao (2009) and include other assumptions that

1. The orders during advance sales period are partially cancelled.
2. The customers who gets an item at time 't' pays at time $t + t_2$ and so the retailer earns interest from the revenue obtained during the time t_2 to t_1 instead of time $t = 0$ to $t = t_1$. Retailer starts paying interest for the items in stock at the rate I_{p}.

Assumptions as in Tsao (2009) are as follow:

1. The problem considers an inventory system with single item.
2. The retailer offers price discount r to his customers if they can commit their orders prior to the sales period.
3. Y_1 percentage of the retailer 1's customers use advance sales discount program and Y_2 percentage of other retailer's customers use advance sales discount program.

The objective is to minimize the annual total cost incurred at the retailer, TC(T) = Annual ordering cost + Annual stock holding cost + Annual interest payable – Annual interest earned:

1. Annual ordering cost = A/T,
2. Annual holding cost = $\frac{(1-Y_1)D_1TH}{2}$,
3. Annual interest earned by the retailer.

Case 1: when $t_1 \leq T$

For this case, please see Figure 1.

Since $(1 - \delta)(Y_1 D_1 + Y_2 D_2)$ number of orders are stand on, the interest earned during the period from t_2 to t_1 due to advance sales discount (ASD) program is $(1 - \delta)(Y_1 D_1 + Y_2 D_2)\, p(1 - r)\, I_e\, (t_1 - t_2)T$. The interest earned during the normal sales period is $p I_e (1 - Y_1) D_1 \left[\frac{(t_1 - t_2)^2}{2}\right]$.

The annual interest earned is $(1 - \delta)(Y_1 D_1 + Y_2 D_2) p(1-r) I_e (t_1 - t_2) + \frac{p I_e (1 - Y_1) D_1}{2T} (t_1 - t_2)^2$.

Case 2: when $T \leq t_1 \leq T + t_2$

For this case, see Figure 2.

Interest earned due to ASD program is $(1 - \delta)(Y_1 D_1 + Y_2 D_2)\, p(1 - r)\, I_e\, (t_1 - t_2)T$. The interest earned during the normal sales period is $p I_e (1 - Y_1) D_1 \left[\frac{(t_1 - t_2)^2}{2}\right]$. The annual interest earned is $(1 - \delta)(Y_1 D_1 + Y_2 D_2) p(1-r) I_e (t_1 - t_2) + \frac{p I_e (1 - Y_1) D_1}{2T} (t_1 - t_2)^2$.

Case 3: when $T + t_2 \leq t_1$ (see Figure 3)

The interest earned due to ASD program is $(1 - \delta)(Y_1 D_1 + Y_2 D_2) p(1-r) I_e (t_1 - t_2)T$. The interest earned during the normal sales period is $p I_e (1 - Y_1) D_1 \left[\frac{T^2}{2} + T(t_1 - T - t_2)\right]$. The annual interest earned by the retailer is $(1 - \delta)(Y_1 D_1 + Y_2 D_2) p(1-r) I_e (t_1 - t_2) + p I_e (1 - Y_1) D_1 \left[t_1 - t_2 - \frac{T}{2}\right]$.

Annual interest payable by the retailer

Case 1: when $t_1 \leq T$

See Figure 1 for this case. The interest payable for the items in stock is $c I_p (1 - Y_1) D_1 \left[\frac{(T - t_1)^2}{2}\right]$. The interest payable for the items, which are sold but not paid yet, is $p I_p (1 - Y_1) D_1 t_2 \left[T - t_1 + \frac{t_2}{2}\right]$. Therefore, the annual interest payable is $\frac{c I_p (1 - Y_1) D_1}{2T} (T - t_1)^2 + \frac{p I_p (1 - Y_1) D_1 t_2}{T} \left[T - t_1 + \frac{t_2}{2}\right]$.

Case 2: when $T \leq t_1 \leq T + t_2$

Please see Figure 2 for this case. Since there is no stock on hand, the retailer does not need to pay interest for the items in stock. However, he pays interest for the items which are sold, but not paid yet. Therefore, the annual interest payable is $\frac{p I_p (1 - Y_1) D_1}{2T} \left[T + t_2 - t_1\right]^2$.

Case 3: when $T + t_2 \leq t_1$

Since the retailer pays off his items at time t_1, which is later than the time $T + t_2$ at which he receives all payment from his customers, there is no interest payable by the retailer. Therefore, the total cost TC(T) incurred at the retailer is

$$TC(T) = \begin{cases} TC_1(T) & \text{if} & t_1 \leq T \\ TC_2(T) & \text{if} & T \leq t_1 \leq T + t_2\,, \\ TC_3(T) & \text{if} & T + t_2 \leq t_1 \end{cases}$$

where

$$TC_1(T) = \frac{1}{2T}\left[2A + (1 - Y_1)D_1\left[c I_p t_1^2 - p I_e (t_1 - t_2)^2 - p I_p t_2 (2t_1 - t_2)\right]\right]$$
$$+ \frac{T}{2}\left[(1 - Y_1)D_1(c I_p + H)\right] + (1 - Y_1)D_1\left[p I_p t_2 - c I_p t_1\right]$$
$$- (1 - \delta)(Y_1 D_1 + Y_2 D_2)P(1-r)I_e(t_1 - t_2),$$

$$(1)$$

$$TC_2(T) = \frac{1}{2T}\left[2A - (1 - Y_1)D_1 p (I_e - I_p)(t_1 - t_2)^2\right]$$
$$+ \frac{T}{2}\left[(1 - Y_1)D_1(H + p I_p)\right] - p I_p (1 - Y_1)D_1[t_1 - t_2]$$
$$- (1 - \delta)(Y_1 D_1 + Y_2 D_2)P(1-r)I_e(t_1 - t_2),$$

$$(2)$$

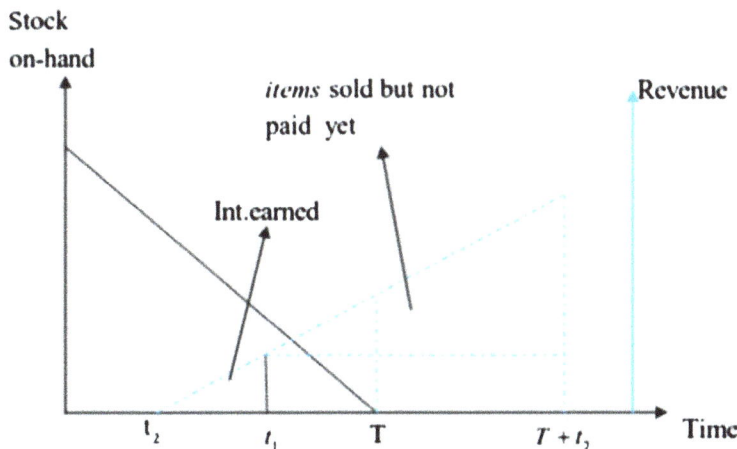

Figure 1 Interest earned and interest payable for the case $t_1 \leq T$.

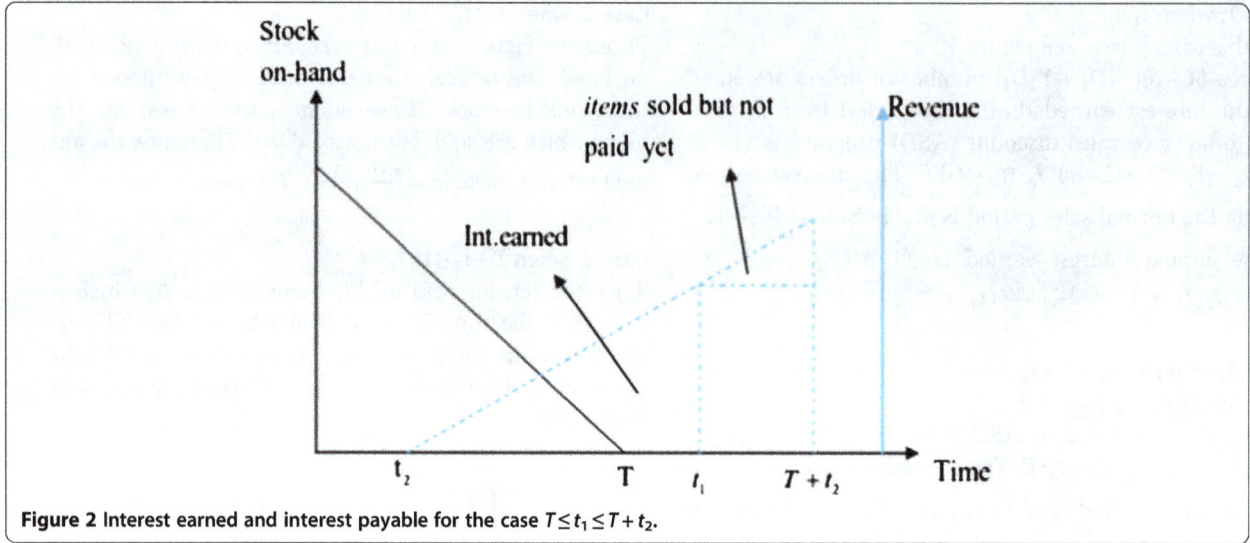

Figure 2 Interest earned and interest payable for the case $T \leq t_1 \leq T + t_2$.

$$TC_3(T) = \frac{A}{T} + \frac{T}{2}\left[(1-Y_1)D_1(H + pI_e)\right]$$
$$-pI_e(1-Y_1)D_1[t_1-t_2]$$
$$-(1-\delta)(Y_1D_1 + Y_2D_2)P(1-r)I_e(t_1-t_2). \tag{3}$$

Optimal solutions

When retail price discount rate r is fixed

The first-order and second-order derivatives of $TC_i(T)$, $i = 1,2,3$, are as follows:

$$\frac{dTC_1(T)}{dT} = \frac{-1}{2T^2}[2A + (1-Y_1)D_1(cI_p{t_1}^2 - pI_e(t_1-t_2)^2$$
$$-pI_pt_2(2t_1-t_2))]$$
$$+ \frac{1}{2}\left[(1-Y_1)D_1(H + cI_p)\right]$$

$$\frac{d^2TC_1(T)}{dT^2} = \frac{1}{T^3}[2A + (1-Y_1)D_1(cI_p{t_1}^2 - pI_e(t_1-t_2)^2$$
$$-pI_pt_2(2t_1-t_2))]$$

$$\frac{dTC_2(T)}{dT} = \frac{-1}{2T^2}\left[2A - p(1-Y_1)D_1(t_1-t_2)^2(I_e - I_p)\right]$$
$$+ \frac{1}{2}\left[(1-Y_1)D_1(H + pI_p)\right]$$

$$\frac{d^2TC_2(T)}{dT^2} = \frac{1}{T^3}\left[2A - p(I_e - I_p)(1-Y_1)D_1(t_1-t_2)^2\right]$$

$$\frac{dTC_3(T)}{dT} = \frac{-A}{T^2} + \frac{1}{2}\left[(1-Y_1)D_1(H + pI_e)\right]$$

$$\frac{d^2TC_3(T)}{dT^2} = \frac{2A}{T^3} > 0.$$

From the above, we observe that $TC_1(T)$ is a convex function on T if $[2A + (1-Y_1)D_1(cI_p{t_1}^2 - pI_e(t_1-t_2)^2 - pI_pt_2(2t_1-t_2))] > 0$. If $[2A + (1-Y_1)D_1(cI_p{t_1}^2 - pI_e(t_1-t_2)^2 - pI_pt_2(2t_1-t_2))] < 0$, then $TC_1(T)$ is a concave function on T

and $\frac{dTC_1(T)}{dT}$ is an increasing function on $[t_1, \infty)$. Therefore, minimum $TC_1(T)$ is attained at $T_1^* = t_1$ when $TC_1(T)$ is a concave function of T.

$TC_2(T)$ is a convex function on T if $[2A - p(I_e - I_p)(1 - Y_1)D_1(t_1 - t_2)^2] > 0$. If $[2A - p(I_e - I_p)(1 - Y_1)D_1(t_1 - t_2)^2] < 0$, then $TC_2(T)$ is a concave function on T and $\frac{dTC_2(T)}{dT}$ is an increasing function on $[t_1 - t_2, t_1]$. Therefore, minimum $TC_2(T)$ is attained at $T_2^* = t_1 - t_2$ when $TC_1(T)$ is a concave function of T.

The optimal cycle times T_i^* ($i = 1,2,3$) are obtained by solving $\frac{dTC_i(T)}{dT} = 0$ ($i = 1,2,3$) respectively.

$$T_1^* = \left[\frac{2A + (1-Y_1)D_1(cI_p{t_1}^2 - pI_e(t_1-t_2)^2 - pI_pt_2(2t_1-t_2))}{(1-Y_1)D_1(H + cI_p)}\right]^{\frac{1}{2}} \tag{4}$$

$$T_2^* = \left[\frac{2A - p(1-Y_1)D_1(t_1-t_2)^2(I_e - I_p)}{(1-Y_1)D_1(H + pI_p)}\right]^{\frac{1}{2}} \tag{5}$$

$$T_3^* = \left[\frac{2A}{(1-Y_1)D_1(H + pI_e)}\right]^{\frac{1}{2}}. \tag{6}$$

Ensuring the condition that $T_1^* \geq t_1$, we have $2A \geq (1 - Y_1)D_1(H{t_1}^2 + pI_e(t_1 - t_2)^2 + pI_pt_2(2t_1 - t_2))$ if and only if $T^* = T_1^*$. Ensuring the condition that $T_2^* \leq t_1 \leq T_2^* + t_2$, we have $T^* = T_2^*$ if and only if $2A \geq (1 - Y_1)D_1(t_1 - t_2)^2(H + pI_e)$ and $2A \leq (1 - Y_1)D_1(H{t_1}^2 + pI_e(t_1 - t_2)^2 + pI_pt_2(2t_1 - t_2))$. Ensuring the condition that $T_3^* + t_2 \leq t_1$, we have $T^* = T_3^*$ if and only if $2A \leq (1 - Y_1)D_1(t_1 - t_2)^2(H + pI_e)$.

Let

$$\Delta_1 = (1-Y_1)D_1(H{t_1}^2 + pI_e(t_1-t_2)^2 + pI_pt_2(2t_1-t_2)),$$
$$\Delta_2 = (1-Y_1)D_1(t_1-t_2)^2(H + pI_e).$$

It is to observe that $\Delta_1 - \Delta_2 = (1-Y_1)D_1[H({t_1}^2 - (t_1 - t_2)^2) + pI_pt_2(2t_1 - t_2)] \geq 0$ and so $\Delta_1 \geq \Delta_2$.

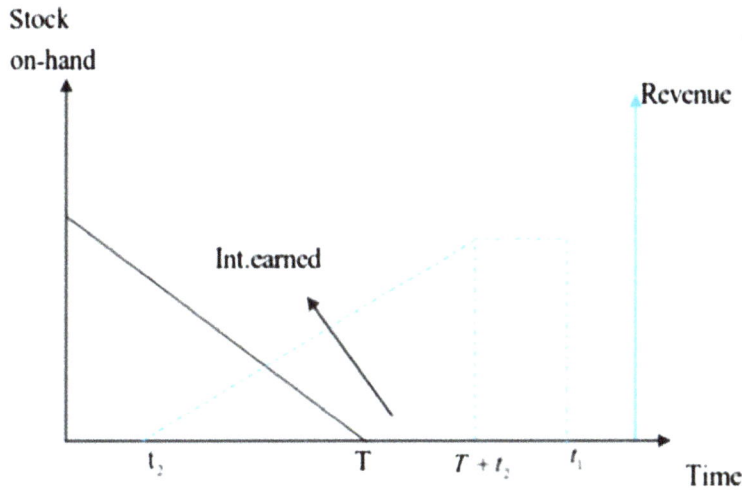

Figure 3 Interest earned and interest payable for the case $T + t_2 \leq t_1$.

Theorem 1

1. If $2A \geq \Delta_1$, then $T^* = T_1^*$.
2. If $2A \leq \Delta_1$ and $2A \geq \Delta_2$, then $T^* = T_2^*$.
3. If $2A \leq \Delta_2$, then $T^* = T_3^*$.

Proof

(1) $2A \geq \Delta_1$ is implied from $t_1 \leq T_1^*$. Since $TC_1'(T)$ is an increasing function on $[t_1, \infty)$, $TC_1'(t_1) \leq TC_1'(T_1^*)$. So, we have $TC_1'(t_1) \leq 0$, $TC_1'(T_1^*) = 0$, and $TC_1'(T) > 0$ for $T \in [T_1^*, \infty)$. Thus, $TC_1(T)$ is decreasing on $[t_1, T_1^*]$ and increasing on $[T_1^*, \infty)$.

Since $2A \geq \Delta_1$ can also be implied from $T_2^* \geq t_1$. Since $TC_2'(T)$ is an increasing function, $TC_2'(T_2^*) \geq TC_2'(t_1)$ which implies that $TC_2'(t_1) \leq 0$. $t_1 - t_2 \leq t_1$ implies that $TC_2'(t_1 - t_2) \leq TC_2'(t_1)$. Hence, $TC_2(T)$ is decreasing on $[t_1 - t_2, t_1]$.

$2A \geq \Delta_1$ implies $2A \geq \Delta_2$. $2A \geq \Delta_2$ is implied from $T_3^* \geq t_1 - t_2$. Since $TC_3'(T)$ is an increasing function, $TC_3'(t_1 - t_2) \leq TC_3'(T_3^*)$ which implies that $TC_3'(t_1 - t_2) \leq 0$. Since $TC_3'(0) \leq 0$, $TC_3(T)$ is decreasing on $[0, t_1 - t_2]$. From the discussions, we have

(1) $TC_1(T)$ is decreasing on $[t_1, T_1^*]$ and increasing on $[T_1^*, \infty)$.
(2) $TC_2(T)$ is decreasing on $[t_1 - t_2, t_1]$.
(3) $TC_3(T)$ is decreasing on $[0, t_1 - t_2]$.

Therefore, $TC(T)$ attains minimum at $T^* = T_1^*$ and $TC^*(T) = TC_1(T_1^*)$.

(2) Let $2A \geq \Delta_1$ and $2A \geq \Delta_2$. $2A \geq \Delta_1$ is implied from $t_1 \geq T_1^*$. So, $TC_1'(t_1) \geq 0$. Since $TC_1'(T)$ is an increasing function on $[t_1, \infty]$, $TC_1'(T) \geq 0$ for $T \in [t_1, \infty)$. Therefore $TC_1(T)$ is increasing on $[t_1, \infty]$. $2A \leq \Delta_1$ can also be implied from $t_1 \geq T_2^*$. $2A \geq \Delta_2$ is implied from $t_1 - t_2 \leq T_2^*$. Since $TC_2'(T)$ is an increasing function, TC_2'

$(t_1) \geq TC_2'(T_2^*)$ and $TC_2'(t_1 - t_2) \leq TC_2'(T_2^*)$. So, $TC_2'(t_1) \geq 0$ and $TC_2'(t_1 - t_2) \leq 0$. Since $TC_2'(T_2^*) = 0$, $TC_2(T)$ is decreasing on $[t_1 - t_2, T_2^*]$ and increasing on $[T_2^*, t_1]$. $2A \geq \Delta_2$ can be implied from $T_3^* \geq t_1 - t_2$. Since $TC_3'(T)$ is an increasing function, $TC_3'(T_3^*) \geq TC_3'(t_1 - t_2)$ which implies that $TC_3'(t_1 - t_2) \leq 0$. Since $TC_3'(T)$ is increasing on $[0, t_1 - t_2]$, $TC_3'(T) \leq 0$ for $T \in [0, t_1 - t_2]$. Therefore, $TC_3(T)$ is decreasing on $T \in [0, t_1 - t_2]$. Hence, we have

(1) $TC_1(T)$ is decreasing on $[t_1, \infty)$.
(2) $TC_2(T)$ is decreasing on $[t_1 - t_2, T_2^*]$ and increasing on $[T_2^*, t_1]$.
(3) $TC_3(T)$ is decreasing on $[0, t_1 - t_2]$.

Therefore, $TC(T)$ attains minimum at $T^* = T_2^*$ and $TC^*(T) = TC_2(T_2^*)$.

(3) $2A \leq \Delta_2$ implies that $2A \leq \Delta_1$. $2A \leq \Delta_1$ implies that $TC_1(T)$ is increasing on $[t_1, \infty)$. $2A \leq \Delta_2$ can be implied from $t_1 - t_2 \geq T_2^*$. Since $TC_2'(T)$ is increasing, $TC_2'(t_1 - t_2) \geq TC_2'(T_2^*)$. Therefore, $TC_2'(t_1 - t_2) \geq 0$. $2A \leq \Delta_1$ is implied from $t_1 \geq T_2^*$. Thus, $TC_2'(t_1) \geq TC_2'(T_2^*)$. Therefore, $TC_2'(t_1) \geq 0$. Therefore, $TC_2(T)$ is increasing on $[t_1 - t_2, t_1]$. $2A \leq \Delta_2$ implies that $TC_3(T)$ is a convex function on $[0, t_1 - t_2]$. Therefore, $TC_3(T)$ is decreasing on $[0, T_3^*]$ and increasing on $[T_3^*, t_1 - t_2]$. From the above discussion, we have

(1) $TC_1(T)$ is increasing on $[t_1, \infty)$.
(2) $TC_2(T)$ is increasing on $[t_1, t_2, t_1]$.
(3) $TC_3(T)$ is decreasing on $[0, T_3^*]$ and increasing on $[T_3^*, t_1 - t_2]$.

Hence, $TC(T)$ attains minimum at $T^* = T_3^*$ and $TC^*(T) = TC_3(T_3^*)$.

When retail price discount rate is endogenous

Here, the retailer determines the optimal replenishment cycle time T^* and the optimal price discount r^* to minimize $TC(T)$. With the consideration that the retailer 1's demand due to advance sales discount program and the fraction of other retailers' customers who switch to retailer 1 under advance sales discount program are linearly increasing with retail price discount r, let $Y_1(r) = \alpha.r$ and $Y_1(r) = \beta.r$ as in Tsao (2009). The problem here is to minimize

$$TC(T,r) = \begin{cases} TC_1(T,r) & \text{if} & t_1 \leq T \\ TC_2(T,r) & \text{if} & T \leq t_1 \leq T + t_2 \\ TC_3(T,r) & \text{if} & T + t_2 \leq t_1 \end{cases} \quad (7)$$

To solve this problem, the closed form solution for each $r_i(T)$ is found by solving $\frac{\partial TC_i(T,r)}{\partial r} = 0$, $i = 1,2,3$. Substituting these $r_i(T)$ to the corresponding $TC_i(T,r)$, $TC_i(T,r)$ reduces to single variable function as $TC_i(T)$. The optimal value of T_i^* is determined by solving the $\frac{dTC_i(T,r_i(T))}{\partial T} = 0$, $i = 1,2,3$. The optimal values of T^* and r^* are such that $TC(T^*,r^*) = \min\{TC_1(T_1^*,r_1^*), TC_2(T_2^*,r_2^*), TC_3(T_3^*,r_3^*)\}$.

The second derivative of $TC_i(T,r)$ with respect to r is $\frac{\partial^2 TC_i(T,r)}{\partial r^2} = 2(1-\delta)(\alpha D_1 + \beta D_2)pI_e(t_1-t_2) > 0$. Thus, $TC_i(T,r)$, $i = 1,2,3$, is a convex function of r for a fixed value of T. Solving the partial differential equation $\frac{\partial TC_i(T,r)}{\partial r} = 0$, $i = 1,2,3$, we get

$$r_1(T) = \frac{1}{4(1-\delta)(\alpha D_1 + \beta D_2)pI_e(t_1-t_2)}$$
$$\times \begin{bmatrix} 2(1-\delta)(\alpha D_1 + \beta D_2)pI_e(t_1-t_2) + \frac{\alpha D_1}{T}\left(\begin{matrix} cI_pt_1^2 - pI_e(t_1-t_2)^2 \\ -pI_pt_2(2t_1-t_2) \end{matrix} \right) \\ + \alpha D_1 T(H + cI_p) + 2\alpha D_1(pI_pt_2 - cI_pt_1) \end{bmatrix}$$

$$(8)$$

$$r_2(T) = \frac{1}{4(1-\delta)(\alpha D_1 + \beta D_2)pI_e(t_1-t_2)}$$
$$\times \begin{bmatrix} 2(1-\delta)(\alpha D_1 + \beta D_2)pI_e(t_1-t_2) + \frac{\alpha D_1}{T}\left((I_e + I_p)p(t_1-t_2)^2 \right) \\ + \alpha D_1 T(H + cI_p) - 2\alpha D_1 pI_p(t_1-t_2) \end{bmatrix}$$

$$(9)$$

$$r_3(T) = \frac{1}{4(1-\delta)(\alpha D_1 + \beta D_2)pI_e(t_1-t_2)}$$
$$\times [\alpha D_1 TH - 2\alpha D_1 pI_e(t_1-t_2-T/2)$$
$$+ 2(1-\delta)(\alpha D_1 + \beta D_2)pI_e(t_1-t_2)]$$

$$(10)$$

Theorem 2

(a) If

$$-\frac{\alpha D_1}{2T}\left(cI_pt_1^2 - PI_e(t_1-t_2)^2 - pI_pt_2(2t_1-t_2) \right) - \frac{\alpha D_1}{2}T(H + cI_p)$$
$$- \alpha D_1\left(PI_pt_2 - cI_pt_1 \right) + (1-\delta)(\alpha D_1 + \beta D_2)PI_e(t_1-t_2) > 0,$$

then the unique optimal solution $r_1^*(T)$ lies in the interval $(0,1)$.

(b) If

$$-\frac{\alpha D_1}{2T}\left((I_e + I_p)p(t_1-t_2)^2 \right) - \frac{\alpha D_1}{2}T(H + pI_p) + \alpha D_1 pI_p(t_1-t_2)$$
$$+ (1-\delta)(\alpha D_1 + \beta D_2)pI_e(t_1-t_2) > 0,$$

then the unique optimal solution $r_2^*(T)$ lies in the interval $(0,1)$.

(c) If

$$-\frac{\alpha D_1}{2T}\left((I_e + I_p)p(t_1-t_2)^2 \right) - \frac{\alpha D_1}{2}T(H + pI_p) + \alpha D_1 pI_p(t_1-t_2)$$
$$+ (1-\delta)(\alpha D_1 + \beta D_2)pI_e(t_1-t_2) > 0,$$

then the unique optimal solution $r_3^*(T)$ lies in the interval $(0,1)$.

Proof

(a) Let

$$G_1(r) = \frac{\partial TC_1(T,r)}{\partial r} = -\frac{\alpha D_1}{2T}$$
$$\times \left(cI_pt_1^2 - pI_e(t_1-t_2)^2 - pI_pt_2(2t_1-t_2) \right)$$
$$-\frac{\alpha D_1}{2}T(H + cI_p) - \alpha D_1\left(pI_pt_2 - cI_pt_1 \right)$$
$$- (1-\delta)(\alpha D_1 + \beta D_2)p(1-2r)I_e(t_1-t_2).$$

Since $TC_1(T,r)$ is a convex function of r, $G_1(r)$ is an increasing function of r.

$$G_1(0) = -\frac{\alpha D_1}{2T}\left(cI_pt_1^2 - pI_e(t_1-t_2)^2 - pI_pt_2(2t_1-t_2) \right)$$
$$-\frac{\alpha D_1}{2}T(H + cI_p) - \alpha D_1\left(pI_pt_2 - cI_pt_1 \right)$$
$$- (1-\delta)(\alpha D_1 + \beta D_2)pI_e(t_1-t_2) < 0$$

$$G_1(1) = -\frac{\alpha D_1}{2T}\left(cI_pt_1^2 - PI_e(t_1-t_2)^2 - pI_pt_2(2t_1-t_2) \right)$$
$$-\frac{\alpha D_1}{2}T(H + cI_p) - \alpha D_1\left(PI_pt_2 - cI_pt_1 \right)$$
$$+ (1-\delta)(\alpha D_1 + \beta D_2)PI_e(t_1-t_2)$$

If $G_1(1) > 0$, then $r_1^*(T)$ lies in the interval $(0,1)$.

(b) Let

$$G_2(r) = \frac{\partial TC_2(T,r)}{\partial r} = -\frac{\alpha D_1}{2T}\left((I_e + I_p)p(t_1-t_2)^2 \right)$$
$$-\frac{\alpha D_1}{2}T(H + pI_p) + \alpha D_1 pI_p(t_1-t_2)$$
$$-(1-\delta)(\alpha D_1 + \beta D_2)p(1-2r)I_e(t_1-t_2).$$

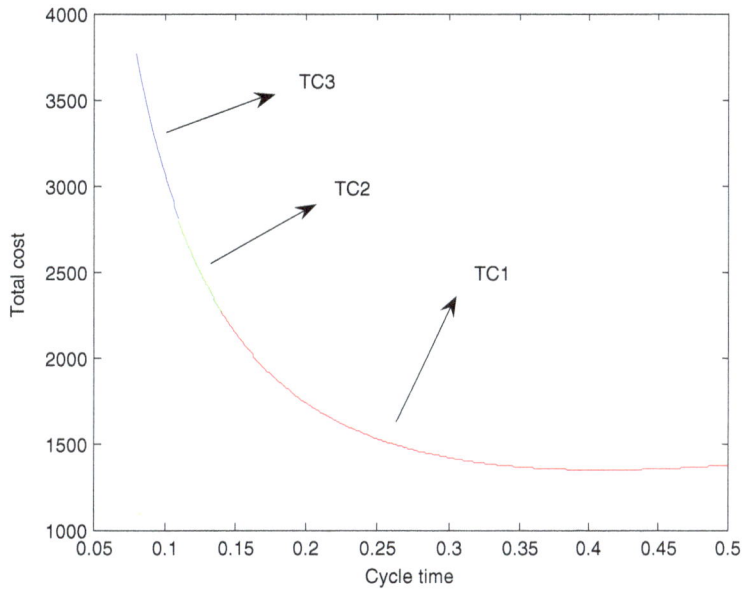

Figure 4 Graphic representation of TC_i for Example 1.

Since $TC_2(T,r)$ is a convex function of r, $G_2(r)$ is an increasing function of r.

$$G_2(0) = -\frac{\alpha D_1}{2T}\left(\left(I_e + I_p\right)p(t_1-t_2)^2\right) - \frac{\alpha D_1}{2}TH$$
$$-\frac{\alpha D_1 pI_p}{2}\left(T-2(t_1-t_2)\right) + \alpha D_1 pI_p(t_1-t_2)$$
$$-(1-\delta)(\alpha D_1 + \beta D_2)pI_e(t_1-t_2) < 0$$

$$G_2(1) = -\frac{\alpha D_1}{2T}\left(\left(I_e + I_p\right)p(t_1-t_2)^2\right) - \frac{\alpha D_1}{2}T\left(H + pI_p\right)$$
$$+\alpha D_1 pI_p(t_1-t_2) + (1-\delta)(\alpha D_1 + \beta D_2)pI_e(t_1-t_2).$$

If $G_2(1) > 0$, then $r_2^*(T)$ lies in the interval $(0,1)$.

(c) Similar to (b)

First, we find $r_1(T)$, $r_2(T)$, and $r_3(T)$ using Equations (8), (9), and (10), respectively. Substituting $r_1(T)$, $r_2(T)$, and $r_3(T)$ into $TC_1(r,T)$, $TC_2(r,T)$, and $TC_3(r,T)$, respectively, each

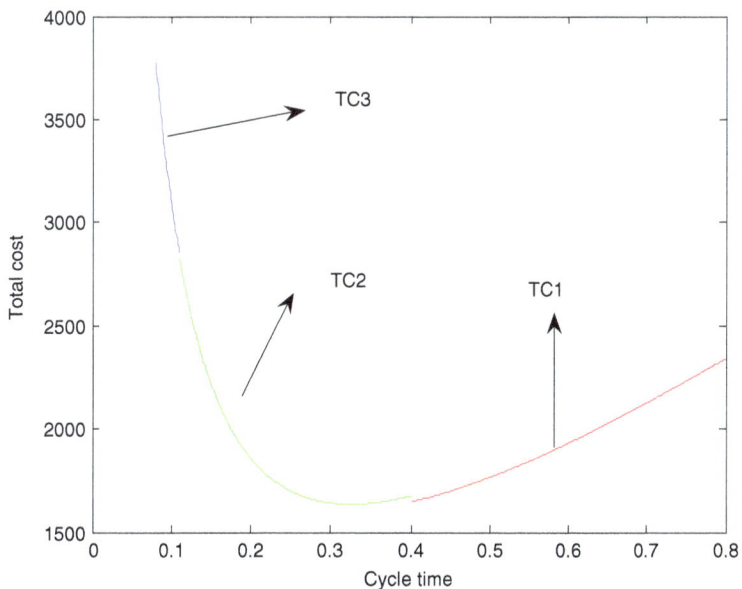

Figure 5 Graphic representation of TC_i for Example 2.

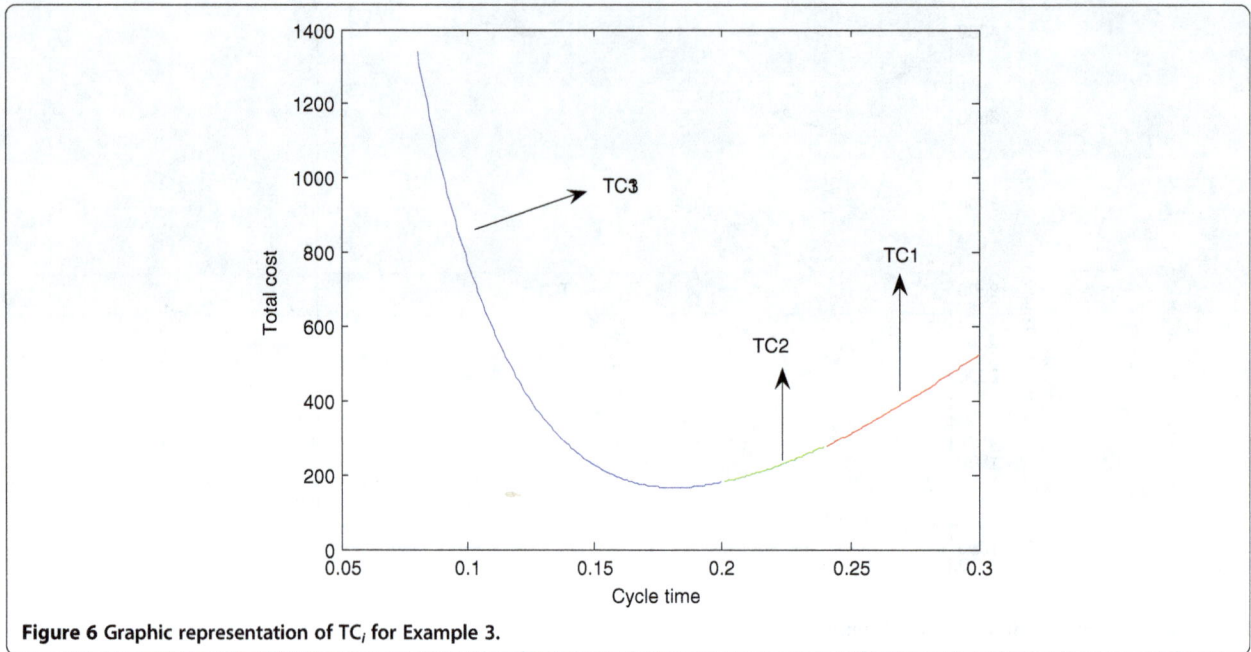

Figure 6 Graphic representation of TC_i for Example 3.

$TC_i(r,T)$ becomes a function of T alone rather than a function of r and T, since r is a function of T. Solving the equations $\frac{dTC_i}{dT} = 0, (i = 1, 2, 3)$, we get the optimal solutions T_1^*, T_2^*, and T_3^*.

Find the minimum of $\{TC_1(r_1(T_1^*),T_1^*),\ TC_2(r_2(T_2^*),T_2^*),$ $TC_3(r_3(T_3^*),T_3^*)\}$ and the resultant is the optimal cost and corresponding T_i^* and $r_i(T_i^*)$ are the optimal solutions. While we are solving the differential equation $\frac{dTC_i}{dT} = 0$, if we get multiple solutions, then we have to check the $\frac{d^2TC_i}{dT^2} > 0$ for optimality.

Numerical analysis

Here, we find the optimal solutions for various cases to illustrate the solution procedures and obtain sensitivity analysis on order cancellation rate (δ), ordering cost (A), and holding cost (H), trade credit periods t_1 and t_2.

Example 1

Let $A = 300$, $D_1 = 2,000$, $D_2 = 2,000$, $\delta = 0.1$, $H = 1$, $I_p = 0.15$, $I_e = 0.2$, $t_1 = 0.14$, $t_2 = 0.10$, $p = 11$, $c = 10$. First, let r be fixed, say $r = 0.56$. We get $\Delta_1 = 453.80$ and $\Delta_2 = 7.25$. Clearly, $2A > \Delta_1$; by Theorem 1, we obtain that $T_1^* = 0.4096$, and the total cost $TC_1 = 1,347.60$. If r is a decision variable, then we utilize the solution procedure in 'When retail price discount rate is endogenous' section. The optimal solutions are $T_1^* = 0.5560$, $r_1^* = 0.8963$, and $TC_1^* = 1,214.80$. A graphic representation of TC_i is shown in Figure 4.

Example 2

Let $A = 300$, $D_1 = 3,000$, $D_2 = 3,000$, $\delta = 0.1$, $H = 1$, $I_p = 0.15$, $I_e = 0.2$, $t_1 = 0.14$, $t_2 = 0.10$, $p = 11$, $c = 10$. First, let r be fixed, say $r = 0.56$. We get $\Delta_1 = 680.7$ and $\Delta_2 = 10.9$. Clearly, $\Delta_1 > 2A$ and $2A < \Delta_2$; by Theorem 1, we obtain

Table 1 Sensitivity analysis with respect to the model parameters

Parameter	Value	T^*	r^*	TC^*
δ	0.1	0.4899	0.8613	1,486.70
	0.2	0.4710	0.8893	1,457.00
	0.25	0.4609	0.9055	1,441.70
	0.3	0.4504	0.9237	1,426.00
	0.4	0.4278	0.9671	1,393.60
H	1	0.4899	0.8613	1,486.70
	1.2	0.4667	0.8712	1,539.30
	1.25	0.4613	0.8736	1,552.10
	1.3	0.4561	0.8760	1,564.80
	1.4	0.4461	0.8808	1,589.70
A	300	0.4899	0.8613	1,486.70
	310	0.4953	0.8656	1,508.40
	315	0.4979	0.8677	1,519.10
	320	0.5005	0.8698	1,529.70
	330	0.5057	0.8740	1,550.70
t_1	0.11	0.3359	0.8465	1,735.60
	0.12	0.3358	0.8465	1,684.90
	0.13	0.3357	0.8465	1,633.90
	0.14	0.3354	0.8465	1,582.50
	0.15	0.3351	0.8465	1,530.80
t_2	0.05	0.3333	0.6468	1,320.30
	0.06	0.3339	0.6468	1,373.50
	0.07	0.3343	0.6468	1,426.30
	0.08	0.3348	0.6468	1,478.70
	0.09	0.3351	0.6468	1,530.80

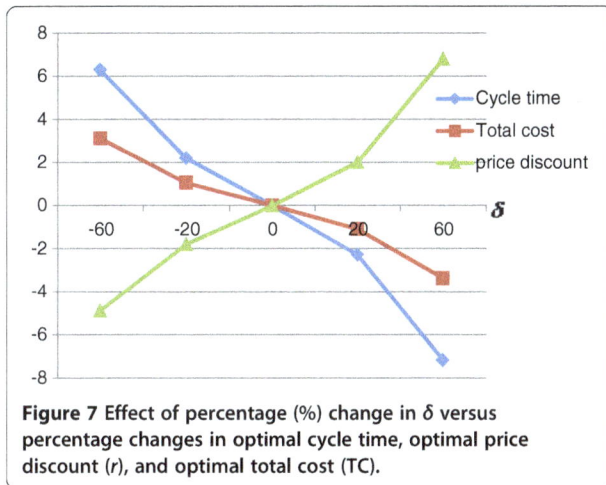

Figure 7 Effect of percentage (%) change in δ versus percentage changes in optimal cycle time, optimal price discount (r), and optimal total cost (TC).

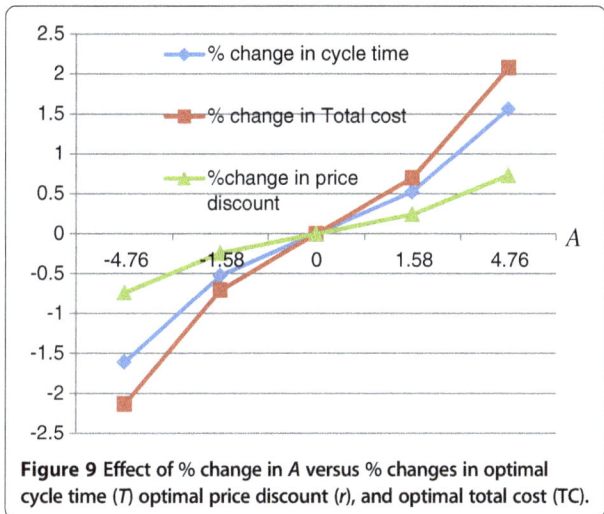

Figure 9 Effect of % change in A versus % changes in optimal cycle time (T) optimal price discount (r), and optimal total cost (TC).

$T_2^* = 0.3354$ and the total cost $TC_2 = 1,582.50$. If r is a decision variable, then we utilize the solution procedure in 'When retail price discount rate is endogenous' section. The optimal solutions are $r_2^* = 0.8613$, $T_2^* = 0.4899$, and $TC_1^* = 1,486.70$. A graphic representation of TC_i is shown in Figure 5.

Example 3

Let $A = 300$, $D_1 = 8,000$, $D_2 = 8,000$, $\delta = 0.1$, $H = 1$, $I_p = 0.15$, $I_e = 0.2$, $t_1 = 0.28$, $t_2 = 0.09$, $p = 11$, $c = 10$. First, let r be fixed, say $r = 0.56$. We get $\Delta_1 = 5,292$ and $\Delta_2 = 655$. Clearly, $\Delta_2 > 2A$; by Theorem 1, we obtain $T_3^* = 0.1818$, and the total cost $TC_2 = 165.53$. If r is a decision variable, then we utilize the solution procedure in 'When retail price discount rate is endogenous' section. The optimal solutions are $r_2^* = 0.05611$, $T_2^* = 0.5573$, and $TC_1^* = 245.80$. A graphic representation of TC_i is shown in Figure 6.

Sensitivity analysis

Here, we consider the data as in numerical Example 2. Sensitivity analysis on various parameters is presented in Table 1. It is also illustrated in Figures 7, 8, 9, 10, and 11.

Based on the results in Table 1 and Figures 7, 8, 9, 10 and 11, the following are observed:

- When partial order cancellation rate δ increases, the cycle time and total cost are decreased where as the optimal price discount rate increases marginally in order to increase the sales marginally.
- When holding cost H increases, the retailer will increase the price discount and shorten the cycle time. Total cost marginally increases due to the marginal increase in price discount rate offered by the retailer.
- When the ordering cost A increases, the retailer will increase his price discount rate and the inventory cycle

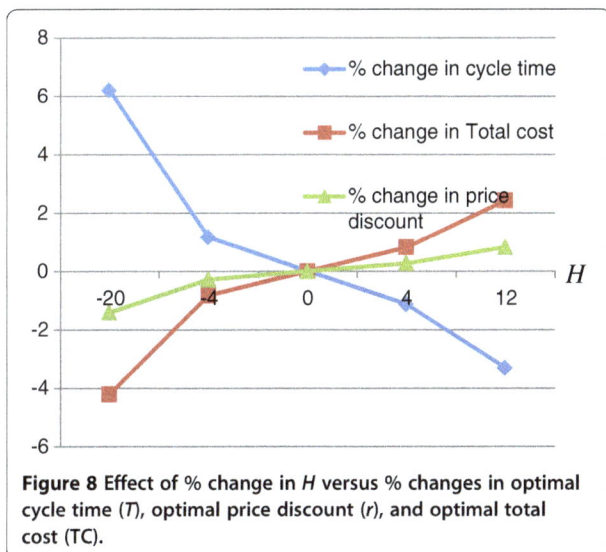

Figure 8 Effect of % change in H versus % changes in optimal cycle time (T), optimal price discount (r), and optimal total cost (TC).

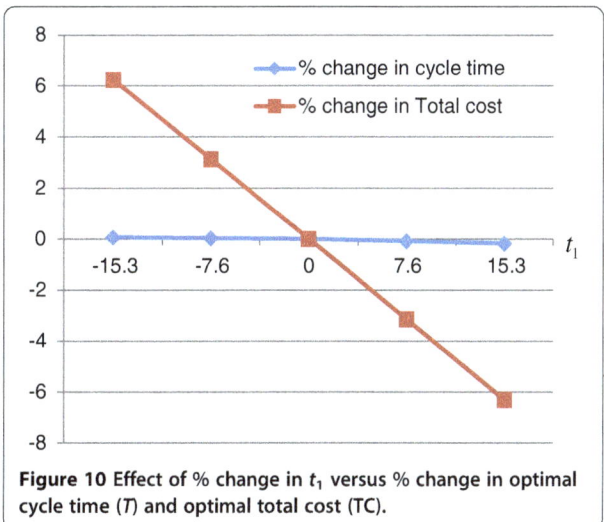

Figure 10 Effect of % change in t_1 versus % change in optimal cycle time (T) and optimal total cost (TC).

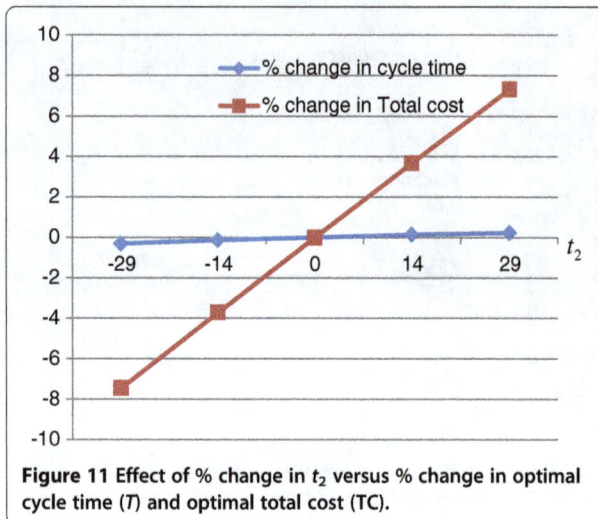

Figure 11 Effect of % change in t_2 versus % change in optimal cycle time (T) and optimal total cost (TC).

time is also increased. To reduce the frequency of replenishment, the retailer lengthens the cycle time.

- When supplier provides a longer credit period t_1, the retailer replenishes the goods more often. In other words, the retailer will minimize the inventory cycle time to take advantage of longer credit period.
- When the retailer provides longer credit period t_2, the retailer's cycle time will be increased. Thus, the retailer will replenish the goods not often to decrease the loss.

Conclusions and future research

Unlike the existing research in two-level trade credit EOQ models, this paper considers partial order cancellation during advance sales period in the retailer's inventory system. The objective behind this consideration is that the market customers' decisions upon their orders play a vital role in trade. Since the payment time of the customers has an impact in the interest earned and interest payable by the retailer, the total costs are estimated under the investigation of exact payment time. The solution procedures are obtained for two cases: (a) when price discount rate is fixed and (b) when price discount rate is a decision variable. Using the derived mathematical theorems, the optimal solutions for price discount and replenishment cycle time are found out for both cases when price discount rate is fixed and endogenous. The sensitivity analyses are made for various inventory parameters.

In future research, one can implement the effect of inflation, perishability to this paper. Considering the trade credit periods t_1 and t_2 as decision variables could be a good extension of this paper. Further, one can improve the paper by considering integrated supplier-retailer inventory system.

Competing interests
The author declare that he has no competing interests.

Acknowledgements
The author thanks two anonymous reviewers for their constructive comments to improve the paper.

References
Aggarwal SP, Jaggi CK (1995) Ordering policies of deteriorating items under permissible delay in payment. J Oper Res Soc 46:658–662

Balkhi ZT (2011) Optimal economic ordering policy with deteriorating items under different supplier trade credits for finite horizon case. Int J Prod Econ 133:216–223

Cárdenas-Barrón LE, Smith NR, Goyal SK (2010) Optimal order size to take advantage of a one-time discount offer with allowed backorders. Appl Math Model 34(6):1642–1652

Chang C-T, Ouyang L-Y, Teng J-T (2003) An EOQ model for deteriorating items under supplier credits linked to ordering quantity. Appl Math Model 27:983–996

Chang CT, Teng JT, Goyal SK (2008) Inventory lot-size models under trade credits: a review. Asia Pac J Oper Res 25:89–112

Chang CT, Teng JT, Chern MS (2010) Optimal manufacturer's optimal policies for deteriorating items in a supply chain with upstream and downstream trade credits. Int J Prod Econ 127:197–202

Chen LH, Kang FS (2010a) Integrated inventory models considering the two-level trade credit policy and price negotiation scheme. Eur J Oper Res 205:47–58

Chen LH, Kang FS (2010b) Coordination between vendor and buyer considering trade credit and items of imperfect quality. Int J Prod Econ 123:52–61

Chen S-H, Cárdenas-Barrón LE, Teng JT (2013a) Retailer's economic order quantity when the supplier offers conditionally permissible delay in payments link to order quantity. Int J Prod Econ, doi:10.1016/j.ijpe.2013.05.032

Chen S-C, Chang C-T, Teng J-T (2013b) A comprehensive note on "Lot-sizing decisions for deteriorating items with two warehouses under an order-size-dependent trade credit". Int Trans Oper Res: , doi:10.1111/itor.12045

Chern M-S, Chern M-S, Pan Q, Teng J-T, Chan Y-L, Chen S-C (2013) Stackelberg solution in a vendor-buyer supply chain model with permissible delay in payments. Int J Prod Econ 1:397–404

Chung KJ, Cárdenas-Barrón LE (2013) The simplified solution procedure for deteriorating items under stock-dependent demand and two-level trade credit in the supply chain management. Appl Math Model 37(7):4653–4660

Chung KJ, Huang YF (2003) The optimal cycle time for EPQ inventory model under permissible delay in payments. Int J Prod Econ 84:307–318

Feng H, Li J, Zhao D (2013) Retailer's optimal replenishment and payment policies in the EPQ model under cash discount and two-level trade credit policy. Appl Math Model 37:3322–3339

Goyal SK (1985) Economic order quantity under conditions of permissible delay in payments. J Oper Res Soc 36:335–338

Ho CH, Ouyang LY, Su CH (2008) Optimal pricing, shipment and payment policy for an integrated supplier–buyer inventory model with two-part trade credit. Eur J Oper Res 187:496–510

Hu F, Liu D (2010) Optimal replenishment policy for the EPQ model with permissible delay in payments and allowable shortages. Appl Math Model 34(10):3108–3117

Huang YF (2003) Optimal retailer's ordering policies in the EOQ model under trade credit financing. J Oper Res Soc 54:1011–1015

Huang YF (2007) Economic order quantity under conditionally permissible delay in payments. Eur J Oper Res 176:911–924

Huang YF, Hsu KH (2008) An EOQ model under retailer partial trade credit policy in supply chain. Int J Prod Econ 112:655–664

Hwang H, Shinn SW (1997) Retailer's pricing and lot sizing policy for exponentially deteriorating products under the condition of permissible delay in payments. Comp Oper Res 24:539–547

Jaggi CK, Goyal SK, Goel SK (2008) Retailer's optimal replenishment decisions with credit-linked demand under permissible delay in payments. Eur J Oper Res 190:130–135

Jaggi CK, Kapur PK, Goyal SK, Goel SK (2012) Optimal replenishment and credit policy in EOQ model under two-levels of trade credit policy when demand is influenced by credit period. Int J Assurance and Eng Manage 3(4):352–359, doi:10.1007/s13198-012-0106-9

Jaggi CK, Goel SK, Mittal M (2013) Credit financing in economic ordering policies for defective items with allowable shortages. Appl Math Comput 219:5268–5282.

Jamal AMM, Sarker BR, Wang S (1997) An ordering policy for deteriorating items with allowable shortage and permissible delay in payment. J Oper Res Soc 48:826–833

Liao J-J (2007) On an EPQ model for deteriorating items under permissible delay in payments. Appl Math Model 31(3):393–403

Liao JJ (2008) An EOQ model with noninstantaneous receipt and exponentially deteriorating items under two-level trade credit. Int J Prod Econ 113:852–861

Liao JJ, Huang KN, Chung KJ (2012) Lot-sizing decisions for deteriorating items with two warehouses under an order-size-dependent trade credit. Int J Prod Econ 137:102–115

Min J, Zhou YW, Zhao J (2010) An inventory model for deteriorating items under stock-dependent demand and two-level trade credit. Appl Math Model 34:3273–3285

Min J, Zhou YW, Liu G-Q, Wang S-D (2012) An EPQ model for deteriorating items with inventory-level-dependent demand and permissible delay in payments. Int J Syst Sci 43(6):1039–1053

Ouyang L-Y, Chang C-T (2013) Optimal production lot with imperfect production process under permissible delay in payments and complete backlogging. Int J Prod Econ 144(2):610–617

Ouyang L-Y, Teng J-T, Chen L-H (2006) Optimal ordering policy for deteriorating items with partial backlogging under permissible delay in payments. J Glob Optim 34:245–271

Ouyang L-Y, Yang C-T, Chan YL, Cárdenas-Barrón LE (2013) A comprehensive extension of the optimal replenishment decisions under two levels of trade credit policy depending on the order quantity. Appl Math Comput 224(1):268–277

Shah NH (1993) Probabilistic time-scheduling model for an exponentially decaying inventory when delay in payment is permissible. Int J Prod Econ 32:77–82

Skouri K, Konstantaras I, Papachristos S, Teng J-T (2011) Supply chain models for deteriorating products with ramp type demand rate under permissible delay in payments. Expert Syst Appl 38:14861–14869

Taleizadeh AA, Mohammadi B, Cárdenas-Barrón LE, Samimi H (2013) An EOQ model for perishable product with special sale and shortage. Int J Prod Econ 145(1):318–338

Teng JT (2002) On the economic order quantity under conditions of permissible delay in payments. J Oper Res Soc 53:915–918

Teng JT (2009) Optimal ordering policies for a retailer who offers distinct trade credits to its good and bad credit customers. Int J Prod Econ 119:415–423

Teng JT, Chang CT (2009) Optimal manufacturer's replenishment policies in the EPQ model under two-levels of trade credit policy. Eur J Oper Res 195:358–363

Teng JT, Goyal SK (2007) Optimal ordering policies for a retailer in a supply chain with up-stream and down-stream trade credits. J Oper Res Soc 58:1252–1255

Teng J-T, Chen J, Goyal SK (2009) A comprehensive note on an inventory model under two levels of trade credit and limited storage space derived without derivatives. Appl Math Model 33:4388–4396

Teng J-T, Krommyda IP, Skouri K, Lou K-R (2011) A comprehensive extension of optimal ordering policy for stock-dependent demand under progressive payment scheme. Eur J Oper Res 215:97–104

Teng J-T, Chang C-T, Chern M-S (2012a) Supply chain vendor-buyer inventory models with trade credit financing under an on-cooperative and an integrated environments. Int J Syst Sci 43(11):2050–2061

Teng J-T, Min J, Pan Q (2012b) Economic order quantity model with trade credit financing and non-decreasing demand. Omega 40:328–335

Teng JT, Yang HL, Chern MS (2013) An inventory model for increasing demand under two levels of trade credit linked to order quantity. Appl Math Model 37:7624–7632

Thangam A (2012) Optimal price discounting and lot-sizing policies for perishable items in a supply chain under advance payment scheme and two-echelon trade credits. Int J Prod Econ 139:459–472

Thangam A, Uthayakumar R (2009) Two-echelon trade credit financing for perishable items in a supply chain when demand depends on both credit period and selling price. Comput Ind Eng 57:773–786

Thangam A, Uthayakumar R (2011) Two-echelon trade credit financing in a supply chain with perishable items and two different payment methods. Int J Oper Res 11(4):365–382

Tsao YC (2009) Retailers optimal ordering and discounting policies advance sales discount and trade credits. Comput Ind Eng 56:208–215

Tsao YC (2011) Replenishment policies considering trade credit and logistics risk. Sci Iran 18:753–758

Tsao Y-C (2012) Determination of production runtime and warranty length under system maintenance and trade credits. Int J Syst Sci 43(12):2351–2360

Tsao YC, Sheen GJ (2012) A multi-item supply chain with credit periods and weight freight cost discounts. Int J Prod Econ 135:106–115

9

Threshold *F*-policy and *N*-policy for multi-component machining system with warm standbys

Kamlesh Kumar[*] and Madhu Jain

Abstract

This study is concerned with threshold *F*-policy and *N*-policy for controlling the arrivals and service in the queueing scenario of a machining system, having active and redundant components. For both *F*-policy and *N*-policy models, the queue size distributions are determined by the recursive method. Various performance measures, namely the average number of failed units in the system, probability that the server is busy or idle in the system, etc., are established using the queue size distribution.

Keywords: Machining systems; *F*-policy; *N*-policy; Start-up time; Warm standbys; Recursive method; Queue size

Background

In many real-life day-to-day queueing situations as well as industrial systems such as production, communication, transportation, and manufacturing systems, the controlling *F*-policy and *N*-policy are being used as cost-effective approaches. Multi-component machines are playing a vital role for solving our daily life problems by reducing the time component. Whenever a machine fails, it causes not only delay in the expected production but also reduction in expected profit. In a multi-component machining system, the *F*-policy states that failed units are not allowed to enter the system when they reach the capacity of the system. As soon as the queue length of failed units is decreased up to a threshold parameter value *F*, then the server takes some start-up time and allows the failed units to enter the system for repair. However, the *N*-policy states that the server will start the service to the failed units only if there are *N* or more failed units accumulated in the system. The facility of standbys in the machining system is provided for utilizing the proper capacity and desired level of the reliability/availability of the machining system. The smoothness of any machining system can be enhanced by standby support so that the machines can work properly in spite of the failure of some components which are unavailable due to physical/technical constraints.

The provision of a 'serviceman' along with standby units to replace the operating units is suggested for minimizing the interference of the machining systems.

Machine interference problems with spares are widely studied by many authors in different frameworks using queueing theoretic approaches. The maximum profit with the utilization of any machining system can be obtained by providing proper combination of maintainability and standby support to the system which may improve the system's reliability under unavoidable techno-economic constraints. It is worthwhile to cite some important contribution in this direction. The analytic solutions of a single-server queueing system with a warm type of standbys were given by Gopalan (1975). The concept of standby support in machine repair problems was incorporated by many researchers, namely Sivazlian and Wang (1989), Gupta and Rao Srinivasa (1996), Wang and Kuo (2000), and many more. Jain (1997) developed a (*m*, *M*) machine repair problem model with state-dependent rates and standby support. Jain et al. (2004b) have proposed a bilevel control policy model for machine repair systems. Ke and Wang (2007) obtained the steady-state probabilities of the number of failed machines in the system and other performance measures for the machine repair problem with vacations and two types of spares. A survey report on the machine interference problem has been presented by Haque and Armstrong (2007). Jain et al. (2008) investigated a multi-component repairable system with mixed

---
* Correspondence: kamleshkumarkundan@gmail.com
Department of Mathematics, Indian Institute of Technology Roorkee, Roorkee, Haridwar, Uttarakhand 247667, India

standbys (warm and cold). Hajeeh and Jabsheh (2009) analyzed a multi-component machining system having two modes of failure. Jain et al. (2010) also published a survey article on various aspects of machine repair problems by emphasizing the practical importance of Markov queueing models. Recently, Yuan and Meng (2011) analyzed the reliability of a warm standby repairable system with priority constraint. The reliability and availability analysis of four series configurations with warm and cold standbys was studied by Hajeeh (2011). More recently, Ke and Wu (2012) have done an investigation on machine repair problem with standbys support.

In queueing modeling, the N-policy concept is mainly incorporated to maintain the techno-economic constraints more effectively. The N-policy is applied by many researchers in queueing problems of a variety of scenarios for providing better cost-effective service to the arrivals. The N-policy utilizes the server's utility properly with no wastage of available resources (or servers). Firstly, Yadin and Naor (1963) introduced the N-policy concept in queueing modeling. Jain et al. (2004a) considered the N-policy model of a machine repairable system and derived the explicit expressions of the reliability function using Laplace transforms. Zhang and Tian (2004) obtained stationary distributions of queue length and waiting time for the threshold N-policy model. The threshold N-policy model of a degraded multi-component machining system with multiple standbys was studied by Jain and Upadhyaya (2009). Jain and Agrawal (2009) proposed the N-policy model for an unreliable server $M^X/M/1$ queueing system with server breakdowns. The N-policy model for a machine repair problem was described by Jain and Bhargava (2009). Sharma (2012) developed a cost model for the machine repair system with N-policy and solved the governing equations by the recursive method. Jain et al. (2012b) investigated the performance of a multi-component machining system by developing an N-policy model. They explored the sensitivity and cost analysis for a machining system with different characteristic parameters and provided the numerical results.

Sometimes the server may take setup time before starting the service in the system; this time is defined as the start-up time of the server. Many researchers have used this concept in the field of queueing modeling of machining systems. In the modeling of queueing systems, the threshold F-policy is used for controlling the arrivals in the system. The arrivals are not allowed in the system whenever the number of arrivals reaches the capacity of the system. In such systems, the service is started only when the buffer, i.e., the capacity of the system, is full, and the arrivals are allowed when the queue length decreases to the threshold value F. Any system which contains comparatively small number of customers in the system allows more pleasurable environment which reduces the waiting

time, discomforts in the service, and load on the server. Gupta (1995) first introduced the concept of F-policy and gave an interrelationship between N-policy and F-policy models. Wang et al. (2008) considered a G/M/1/K queueing system with F-policy and start-up time by employing the recursive method. Wang and Yang (2009) presented a matrix analytic solution for developing the steady-state solution of a control F-policy M/G/1/K model with exponential start-up time. Yang et al. (2010) considered the F-policy to study the optimization and sensitivity analysis of a queueing system with single vacation. Kuo et al. (2011) demonstrated that the solution algorithm for an F-policy G/M/1/K queue with start-up time can be derived using the N-policy M/G/1/K queue with start-up time. More recently, Jain et al. (2012a) studied the effect of different parameters on various performance measures in the M/M/2/K queuing system with (N, F) policy with multi-optional phase repair and start-up.

In this paper, we analyze the performance measures of F-policy and N-policy models of machine repair problem with warm standby support. In our study, we employ the recursive method to determine the steady-state probabilities of the systems. The model description including assumptions and notations is given in the 'Model description' section. The steady-state difference equations governing the models and the recursive method to solve these equations for obtaining the queue size distributions are given in the 'F-policy model' section. The performance measures using queue size distribution are derived in the 'Performance measures' section. In order to discuss the further extension and to highlight the notable features of the investigation done, concluding remarks are given in the 'Conclusion' section.

Model description

In order to study the threshold F-policy and N-policy of a multi-component machining system with warm standbys with a single server, we develop the Markovian model by the birth-death process. To formulate the mathematical model, we construct the governing equations in terms of probabilities using the appropriate rates of inflow and outflow. We develop a (m, M) model for a multi-component system under the assumption that the system fails when there are $L = M + S - m + 1$ $(m = 1, 2,..., M)$ or more failed units in the system. The following assumptions and notations are used to formulate our model:

- In the F-policy model, the server starts the service when the number of failed units in the system reaches its capacity L. At this time, no failed unit is allowed to queue up in the system until the number of failed units attains the threshold value F.
- In the N-policy model, the server starts the service when there are N or more failed units accumulated in the system. The server leaves the

system when it becomes empty, i.e., no failed unit is available in the system.

- It is assumed that the interfailure time and repair time of the failed units and the start-up time of the server are exponentially distributed.
- The server takes some start-up time before providing the service to the failed units. The discipline of rendering the repair is considered according to the first-come first-served discipline.
- When any operating unit fails, it is replaced by an available standby unit. When all the standbys are used, the system may also work till m-operating units function properly.
- When all the standby units are exhausted, the failure rate of the remaining operating units increases due to stress and the system works in degrading mode due to increased load on the system.
- If failure of any unit occurs in case when the system has total L failed units (i.e., the available failed units are equal to the capacity of the system) in the system, it is not permitted to enter the system.

Some notations used for the model formulation are as follows:

- M Total number of operating units in the system
- S Total number of standby units in the system
- α Failure rate of the standby unit
- λ Failure rate of the operating unit
- λ_d Degraded failure rate of remaining operating units ($\lambda_d \geq \lambda$) when there are less than M but more than m operating units in the system
- μ Repair rate of the server
- β Set-up rate to start allowing failed units for repair in the system
- γ Start-up rate to start the repairing of the failed units in the system

The steady-state probabilities for the system states are defined as follows:

- $P_{n,j}$ Probability that there are n failed units in the system and the failed units are either allowed ($j = 1$) or not allowed ($j = 0$) for repair in the case of the F-policy model
- $Q_{n,j}$ Probability that there are n failed units in the system and the server is either busy ($j = 1$) or idle ($j = 0$) in the case of the N-policy model

The state-dependent failure rate λ_n is given by

$$\lambda_n = \begin{cases} M\lambda + (S-n)\alpha & ; \quad 0 \leq n < S \\ (M+S-n)\lambda_d & ; \quad S \leq n < L \\ 0 & ; \quad \text{otherwise} \end{cases}$$

F-policy model
The governing equations
In this section, we construct the steady-state difference equations for the F-policy Markovian model of the machine repair problem. The governing equations are constructed by taking appropriate transition rates (see Figure 1) as follows:

1. *For $j = 0$: when failed units (i.e., arrivals) are not allowed in the system.*
   In this case, to construct the governing equation for state $(0, 0)$, we equate the outflow from state $(0, 0)$ to the inflow from $(1, 0)$. Thus, we obtain

$$\mu P_{1,0} = \beta P_{0,0}. \tag{1}$$

In a similar manner, by equating the inflow rate from state $(n + 1, 0)$ to state $(n, 0)$ and the outflow rate from state $(n, 0)$ to state $1 \leq n \leq L$, we get

$$\mu P_{n+1,0} = (\mu + \beta)P_{n,0}; 1 \leq n \leq F \tag{2}$$

$$P_{n+1,0} = P_{n,0}; F + 1 \leq n \leq L-1 \tag{3}$$

$$\lambda_{L-1}P_{L-1,1} = \mu P_{L,0}. \tag{4}$$

2. *For $j = 1$: when failed units (i.e., arrivals) are allowed for repair in the system.*
   We construct the equations for state $(0, 1)$ using the inflows from states $(1, 1)$ and $(0, 0)$ to $(0, 1)$ = outflow from state $(0, 1)$. Thus,

$$\mu P_{1,1} + \beta P_{0,0} = \lambda_0 P_{0,1}. \tag{5}$$

Similarly, we consider the inflow from states $(n - 1, 1)$, $(n + 1, 1)$, and $(n, 0)$ to $(n, 1)$ = outflow from state $(n, 1)$, $1 \leq n \leq F$, and obtain

$$\lambda_{n-1}P_{n-1,1} + \mu P_{n+1,1} + \beta P_{n,0} \\ = (\lambda_n + \mu)P_{n,1}; 1 \leq n \leq F. \tag{6}$$

Again, using the inflow from states $(n - 1, 1)$ and $(n + 1, 1)$ to $(n, 1)$ = outflow from state $(n, 1)$, where $F + 1 \leq n \leq L - 1$, we get

$$\lambda_{n-1}P_{n-1,1} + \mu P_{n+1,1} = (\lambda_n + \mu)P_{n,1}; \\ F + 1 \leq n \leq L-2. \tag{7}$$

Further balancing the inflow = outflow for state $(L - 1, 0)$, we get

$$\lambda_{L-2}P_{L-2,1} = (\lambda_{L-1} + \mu)P_{L-1,1}; F \neq L-1. \tag{8}$$

The normalization condition is given by

$$\sum_{j=0}^{1} \sum_{n=0}^{L-1} P_{n,j} + P_{L,0} = 1. \tag{9}$$

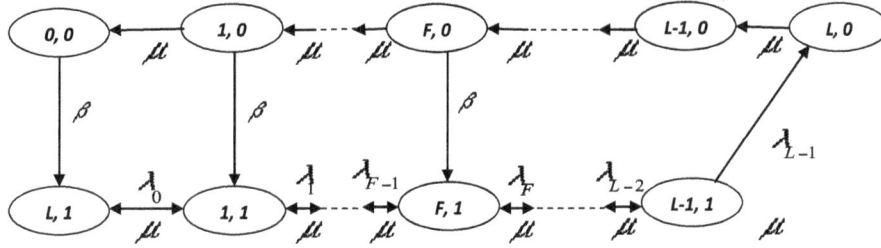

Figure 1 State transition rate diagram for the *F*-policy model.

Queue size distribution for F-policy

The main task of getting the solution of governing equations is to develop the steady-state probabilities of all the states. The probabilities at steady states can be evaluated by the well-known recursive method for the set of governing difference equations (Equations 1 to 9) of the *F*-policy model. Now, first we solve Equations 1 to 3 recursively and obtain the steady-state probabilities as

$$P_{n,0} = \delta(1+\delta)^{n-1}P_{0,0}; 1 \le n \le F \tag{10}$$

$$P_{n,0} = \delta(1+\delta)^{F}P_{0,0}; F+1 \le n \le L \tag{11}$$

where $\delta = \frac{\beta}{\mu}$.

We find $P_{L-1,1}$ from Equation 4 using Equation 11:

$$P_{L-1,1} = \frac{\mu\delta(1+\delta)^{F}}{\lambda_{L-1}}P_{0,0}. \tag{12}$$

Now, from Equations 8 and 12, we get

$$P_{L-2,1} = \frac{(\lambda_{L-1}+\mu)\mu\delta(1+\delta)^{F}}{\lambda_{L-2}\lambda_{L-1}}P_{0,0}. \tag{13}$$

Putting $n = L - 2, L - 3,..., F + 1$ in Equation 7, we get

$$P_{n,1} = \frac{\delta_F}{\prod_{i=n}^{L-1}\lambda_i}\sum_{m=0}^{L-n-1}\left\{\mu^m\prod_{i=n+m+1}^{L-1}(\lambda_i)\right\}; F \le n \le L-1 \tag{14}$$

where $\delta_F = \mu(1+\delta)^F\delta P_{0,0}$, and for $p > q$, we take

$$\prod_{i=p}^{q}\lambda_i = 1; p > q. \tag{15}$$

In Equation 6, we put $n = F, F - 1, F - 2,..., 1$ and get

$$P_{F-1,1} = \frac{\delta_F}{\prod_{i=F-1}^{L-1}\lambda_i}\left[\sum_{m=0}^{L-F}\left\{\mu^m\prod_{i=m+F}^{L-1}(\lambda_i)\right\}\right] - \frac{\delta\delta_F}{\lambda_{F-1}(1+\delta)} \tag{15a}$$

$$P_{F-2,1} = \frac{\delta_F}{\prod_{i=F-2}^{L-1}\lambda_i}\left[\sum_{m=0}^{L-(F-1)}\left\{\mu^m\prod_{i=m+F-1}^{L-1}(\lambda_i)\right\}\right] - \frac{(\lambda_{F-1}+\mu)\delta\delta_F}{\lambda_{F-2}\lambda_{F-1}(1+\delta)} - \frac{\delta\delta_F}{\lambda_{F-2}(1+\delta)^2} \tag{15b}$$

$$P_{F-3,1} = \frac{\delta_F}{\prod_{i=F-3}^{L-1}\lambda_i}\left[\sum_{m=0}^{L-(F-2)}\left\{\mu^m\prod_{i=m+F-2}^{L-1}(\lambda_i)\right\}\right] - \frac{(\lambda_{F-2}\lambda_{F-1}+\mu\lambda_{F-1}+\mu^2)\delta\delta_F}{\lambda_{F-3}\lambda_{F-2}\lambda_{F-1}(1+\delta)} - \frac{(\lambda_{F-2}+\mu)\delta\delta_F}{\lambda_{F-3}\lambda_{F-2}(1+\delta)^2} - \frac{\delta\delta_F}{\lambda_{F-3}(1+\delta)^3} \tag{15c}$$

....
....

$$P_{0,1} = \frac{\delta_F}{\prod_{i=1}^{L-1}\lambda_i}\left[\sum_{m=0}^{L-2}\left\{\mu^m\prod_{i=m+2}^{L-1}(\lambda_i)\right\}\right] - \frac{\delta\delta_F}{(1+\delta)\prod_{i=1}^{F-1}\lambda_i}\sum_{m=0}^{F-2}\left\{\mu^m\prod_{i=m+2}^{L-1}(\lambda_i)\right\} - ... - \frac{\delta\delta_F}{(1+\delta)^{F-2}\prod_{i=1}^{2}\lambda_i}\sum_{m=0}^{1}\left\{\mu^m\prod_{i=m+2}^{2}(\lambda_i)\right\} - \frac{\delta\delta_F}{\lambda_1(1+\delta)^{F-1}}. \tag{15d}$$

Now, in general, we obtain

$$P_{n,1} = \frac{\delta_F}{\prod_{i=n}^{L-1}\lambda_i}\left[\sum_{m=0}^{L-n-1}\left\{\mu^m\prod_{i=n+m+1}^{L-1}(\lambda_i)\right\}\right] - \delta\delta_F\sum_{k=1}^{F-n}\left[\frac{\sum_{m=0}^{F-n-k}\left\{\mu^m\prod_{i=n+m+1}^{F-k}(\lambda_i)\right\}}{(1+\delta)^k\prod_{i=n}^{F-k}(\lambda_i)}\right]; 0 \le n \le F-1. \tag{16}$$

Now, we substitute the values of $P_{n,j}$, $1 \le n \le L$ and $j = 0$, 1, from Equations 10, 11, 14, and 16 in the normalizing Equation 9 and obtain the value for $P_{0,0}$ as

$$P_{0,0}^{-1} = (1+\delta)^F \{1 + \delta(L-F)\}$$
$$+\mu\delta(1+\delta)^F \sum_{n=0}^{L-1}\left[\frac{1}{\prod_{i=n}^{L-1}\lambda_i}\sum_{m=0}^{L-n-1}\left\{\mu^m\prod_{i=n+m+1}^{L-1}(\lambda_i)\right\}\right]$$
$$-\mu\delta^2(1+\delta)^F\sum_{n=0}^{F-1}\left[\sum_{k=1}^{F-n}\left\{\frac{\sum_{m=0}^{F-n-k}\left(\mu^m\prod_{i=n+m+1}^{F-k}\lambda_i\right)}{(1+\delta)^k\prod_{i=n}^{F-k}\lambda_i}\right\}\right]. \tag{17}$$

N-policy model
The governing equations
Now, we construct the steady-state difference equations for the N-policy model using the appropriate birth-death rates (see Figure 2). For different system states, we equate the outflows to the inflows and get the balance equations in a similar manner as obtained for the F-policy model.

1. *For $j = 0$: when the server is idle in the system.*
   The steady-state equation for state $(0, 0)$, i.e., when no failed unit is present in the system, is obtained using the outflow from state $(0, 0)$ that equals the inflow from state $(1, 1)$ to $(0, 0)$ as

   $$\mu Q_{1,1} = \lambda_0 Q_{0,0}. \tag{18}$$

   Further, for $1 \le n \le N - 1$, we obtain

   $$\lambda_n Q_{n,0} = \lambda_{n-1}Q_{n-1,0}; 1 \le n \le N-1. \tag{19}$$

   Similarly, for other states, we get

   $$(\lambda_n + \gamma)Q_{n,0} = \lambda_{n-1}Q_{n-1,0}; N \le n \le L-1 \tag{20}$$

   $$\gamma Q_{L,0} = \lambda_{L-1}Q_{L-1,0}. \tag{21}$$

2. *For $j = 1$: when the server is busy in the system.*
   Applying the outflow from state $(n, 1)$ that equals the inflow from different states to state $(n, 1)$, we get

   $$(\lambda_1 + \mu)Q_{1,1} = \mu Q_{2,1}; N \ne 1 \tag{22}$$

$$(\lambda_n + \mu)Q_{n,1} = \lambda_{n-1}Q_{n-1,1} + \mu Q_{n+1,1}; 2 \le n \le N-1 \tag{23}$$

$$(\lambda_n + \mu)Q_{n,1} = \lambda_{n-1}Q_{n-1,1} + \mu Q_{n+1,1} + \gamma Q_{n,0}; N \le n \le L-1 \tag{24}$$

$$\mu Q_{L,1} = \lambda_{L-1}Q_{L-1,1} + \gamma Q_{L,0}. \tag{25}$$

The normalization condition is given by

$$Q_{0,0} + \sum_{j=0}^{1}\sum_{n=1}^{L}Q_{n,j} = 1. \tag{26}$$

Queue size distribution for the N-policy model
We use the recursive method to evaluate the steady-state probabilities for the N-policy model from Equations 18 to 26. By solving the steady-state Equation 19 recursively, we obtain

$$Q_{n,0} = \frac{\lambda_0}{\lambda_n}Q_{0,0}; 1 \le n \le N-1. \tag{27}$$

Now, in Equation 20, we put $n = N, N + 1, N + 2,...,$ $L - 1$ and get

$$Q_{n,0} = \frac{\lambda_0 Q_{0,0}}{\lambda_n}\prod_{i=N}^{n}\theta_i; N \le n \le L-1 \tag{28}$$

where $\theta_i = \dfrac{\lambda_i}{\lambda_i + \gamma}$.

Equation 21 yields

$$Q_{L,0} = \frac{\lambda_{L-1}}{\gamma}Q_{L-1,0} = \frac{\theta_0 P_{0,0}}{1-\theta_0}\prod_{i=N}^{L-1}\theta_i. \tag{29}$$

We find $Q_{1,1}$ from Equation 18 as

$$Q_{1,1} = \frac{\lambda_0}{\mu}Q_{0,0}. \tag{30}$$

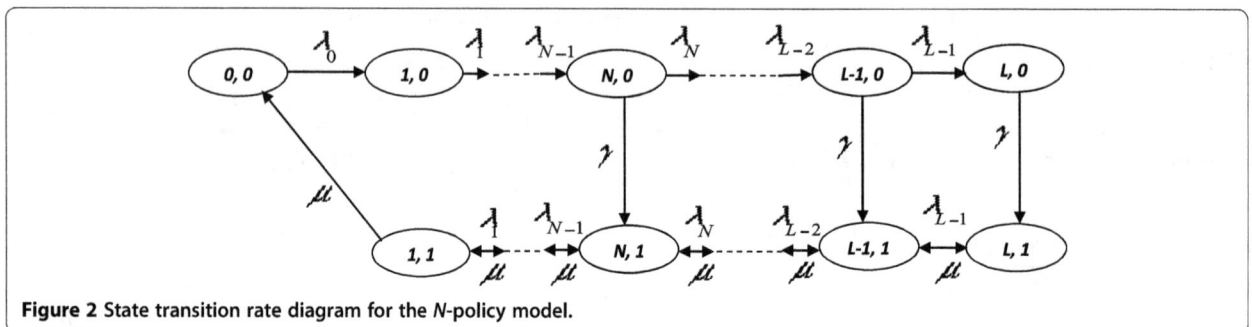

Figure 2 State transition rate diagram for the *N*-policy model.

From Equation 22, we get

$$Q_{2,1} = \frac{(\lambda_1 + \mu)}{\mu^2} \lambda_0 Q_{0,0}. \tag{31}$$

In Equation 23, substituting $n = 2, 3,..., N - 1$ and using Equations 30 and 31, we obtain the steady-state probabilities as

$$Q_{n,1} = \frac{\lambda_0 Q_{0,0}}{\mu^n} \left[\sum_{m=0}^{n-1} \left\{ \mu^m \prod_{i=m+1}^{n-1} (\lambda_i) \right\} \right]; 1 \leq n \leq N. \tag{32}$$

Further, we find the values of $Q_{N+1,1}, Q_{N+2,1}, Q_{N+3,1},...,$ $Q_{L,1}$ by putting $n = N, N + 1, N + 2, N + 3,..., L - 1$ in Equation 24 as

$$Q_{N+1,1} = \frac{\lambda_0 Q_{0,0}}{\mu^{N+1}} \left[\sum_{m=0}^{N} \left\{ \mu^m \prod_{i=m+1}^{N} (\lambda_i) \right\} \right] - \frac{\lambda_0 Q_{0,0}}{\mu} (1 - \theta_N) \tag{33a}$$

$$Q_{N+2,1} = \frac{\lambda_0 Q_{0,0}}{\mu^{N+2}} \left[\sum_{m=0}^{N+1} \left\{ \mu^m \prod_{i=m+1}^{N+1} (\lambda_i) \right\} \right] - \frac{\lambda_0 Q_{0,0}}{\mu^2} \left\{ \lambda_{N+1}(1-\theta_N) + \mu(1-\theta_N\theta_{N+1}) \right\} \tag{33b}$$

$$Q_{N+3,1} = \frac{\lambda_0 Q_{0,0}}{\mu^{N+3}} \left[\sum_{m=0}^{N+2} \left\{ \mu^m \prod_{i=m+1}^{N+2} (\lambda_i) \right\} \right] - \frac{\lambda_0 Q_{0,0}}{\mu^3} \left\{ \begin{array}{l} \lambda_{N+2}\lambda_{N+1}(1-\theta_N) + \mu\lambda_{N+2}(1-\theta_N\theta_{N+1}) \\ + \mu^2(1-\theta_N\theta_{N+1}\theta_{N+2}) \end{array} \right\} \tag{33c}$$

$$Q_{N+4,1} = \frac{\lambda_0 Q_{0,0}}{\mu^{N+4}} \left[\sum_{m=0}^{N+3} \left\{ \mu^m \prod_{i=m+1}^{N+3} (\lambda_i) \right\} \right] - \frac{\lambda_0 Q_{0,0}}{\mu^4} \left\{ \begin{array}{l} \lambda_{N+3}\lambda_{N+2}\lambda_{N+1}(1-\theta_N) \\ + \mu\lambda_{N+3}\lambda_{N+2}(1-\theta_N\theta_{N+1}) \\ + \mu^2\lambda_{N+3}(1-\theta_N\theta_{N+1}\theta_{N+2}) \\ + \mu^3(1-\theta_N\theta_{N+1}\theta_{N+2}\theta_{N+3}) \end{array} \right\} \tag{33d}$$

....
....

$$Q_{L,1} = \frac{\lambda_0 Q_{0,0}}{\mu^L} \left[\sum_{m=0}^{L-1} \left\{ \mu^m \prod_{i=m+1}^{L-1} (\lambda_i) \right\} \right] - \frac{\lambda_0 Q_{0,0}}{\mu^{L-N}} \left\{ \begin{array}{l} \lambda_{L-1}...\lambda_{N+2}\lambda_{N+1}(1-\theta_N) \\ + \mu\lambda_{L-1}...\lambda_{N+2}(1-\theta_N\theta_{N+1}) \\ +...+ \mu^{L-N-1}(1-\theta_N\theta_{N+1}\theta_{N+2}...\theta_{L-1}) \end{array} \right\}. \tag{33e}$$

In general, we get

$$Q_{n,1} = \frac{\lambda_0 Q_{0,0}}{\mu^n} \left[\left[\sum_{m=0}^{n-1} \left\{ \mu^m \prod_{i=m+1}^{n-1} (\lambda_i) \right\} \right] - \mu^N \left[\sum_{m=N}^{n-1} \left\{ \mu^{m-N} \left(1 - \prod_{i=N}^{m} (\theta_i) \right) \prod_{i=m+1}^{n-1} (\lambda_i) \right\} \right] \right]; N + 1 \leq n \leq L. \tag{34}$$

Now, we substitute the values of $Q_{n,j}$, $1 \leq n \leq L$ and $j = 0, 1$, use $Q_{0,1} = 0$ in Equation 26, and find the value of $Q_{0,0}$ as follows:

$$Q_{0,0}^{-1} = \sum_{n=0}^{N-1} \left(1 + \frac{\lambda_0}{\lambda_n} \right) + \sum_{n=N}^{L-1} \left(\frac{\lambda_0}{\lambda_n} \prod_{i=N}^{n} \theta_i \right) + \frac{\theta_0}{1-\theta_0} \prod_{i=N}^{L-1} \theta_i$$
$$+ \sum_{n=1}^{L} \left[\frac{\lambda_0}{\mu^n} \sum_{m=0}^{n-1} \left\{ \mu^m \prod_{i=m+1}^{n-1} (\lambda_i) \right\} \right]$$
$$- \sum_{n=N+1}^{L} \left[\frac{\lambda_0}{\mu^{n-N}} \left[\sum_{m=N}^{n-1} \left\{ \mu^{m-N} \left(1 - \prod_{i=N}^{m} (\theta_i) \right) \prod_{i=m+1}^{n-1} (\lambda_i) \right\} \right] \right]. \tag{35}$$

The steady-state probabilities evaluated by the recursive method can be used to derive various performance measures for both F-policy and N-policy systems.

Performance measures

We establish some important performance measures for the F-policy and N-policy models by using the steady-state probabilities obtained for different system states. In order to examine the system's behavior, the quantitative assessment of the performance measures is the main objective and key component of the performance modeling of any queueing system including the machine repair system. Here we are interested to derive system characteristics such as the expected number of failed units, probability that the server is busy or idle in the system, probability of blocking of the system, etc. Using the steady-state probabilities, we derive various performance measures for both models as given below.

F-policy model

In this subsection, we find the expressions for some performance measures in terms of steady-state probabilities such that results can be useful to predict the behavior of the machine repair system operating under the F-policy:

1. The expected number of failed units in the system is given by

$$E(N_F) = \sum_{n=1}^{L} n P_{n,0} + \sum_{n=1}^{L-1} n P_{n,1}$$
$$= \left[\frac{\left\{ 1-(1+\delta)^F(1-F\delta) \right\}}{\delta} + \frac{(L-F)(L+F+1)}{2} \delta(1+\delta)^F \right.$$
$$+ \mu\delta(1+\delta)^F \left[\sum_{n=1}^{L-1} \left[\frac{n}{\prod_{i=n}^{L-1} \lambda_i} \sum_{m=0}^{L-n-1} \left\{ \mu^m \prod_{i=n+m+1}^{L-1} (\lambda_i) \right\} \right] \right]$$
$$\left. - \delta \sum_{n=1}^{F-1} \left[n \sum_{k=1}^{F-n} \left\{ \frac{\sum_{m=0}^{F-n-k} \left(\mu^m \prod_{i=n+m+1}^{F-k} \lambda_i \right)}{(1+\delta)^k \prod_{i=n}^{F-k} \lambda_i} \right\} \right] \right] P_{0,0}. \tag{36}$$

2. The probability that the server is idle is obtained as

$$P(I_F) = \sum_{j=0}^{1} P_{0,j} = \left[\left[\frac{1}{\prod_{i=0}^{L-1} \lambda_i} \sum_{m=0}^{L-1} \left\{ \mu^m \prod_{i=m+1}^{L-1} (\lambda_i) \right\} - \delta \sum_{k=1}^{F} \left[\frac{\sum_{m=0}^{F-k} \left\{ \mu^m \prod_{i=m+1}^{F-k} (\lambda_i) \right\}}{(1+\delta)^k \prod_{i=0}^{F-k} (\lambda_i)} \right] \right] \mu\delta(1+\delta)^F + 1 \right] P_{0,0}. \quad (37)$$

3. The probability that server is busy is obtained as

$$P(B_F) = \sum_{n=1}^{L} P_{n,0} + \sum_{n=1}^{L-1} P_{n,1} \quad (38)$$

$$= \left[\mu\delta(1+\delta)^F \left[\sum_{n=1}^{L-1} \left[\frac{1}{\prod_{i=n}^{L-1} \lambda_i} \sum_{m=0}^{L-n-1} \left\{ \mu^m \prod_{i=n+m+1}^{L-1} (\lambda_i) \right\} \right] \right. \right.$$
$$\left. \left. - \delta \sum_{n=1}^{F-1} \left[\sum_{k=1}^{F-n} \left\{ \frac{\sum_{m=0}^{F-n-k} \left(\mu^m \prod_{i=n+m+1}^{F-k} \lambda_i \right)}{(1+\delta)^k \prod_{i=n}^{F-k} \lambda_i} \right\} \right] \right] \right.$$
$$\left. + \left[(1+\delta)^F \{1 + \delta(L-F)\} - 1 \right] \right] P_{0,0}.$$

4. The probability that the server takes start-up time before starting the service to failed units is

$$P(ST_F) = \sum_{n=0}^{F} P_{n,0} = (1+\delta)^F P_{0,0} \quad (39)$$

5. The probability that the system is blocked (i.e., the failed unit is not allowed to join the queue) is

$$P(SB_F) = \sum_{n=0}^{L} P_{n,0} = (1+\delta)^F [1 + \delta(L-F)] P_{0,0}. \quad (40)$$

6. The probability of the buildup state is obtained as

$$P(BS_F) = \sum_{n=F+1}^{L} P_{n,0} = \delta(1+\delta)^F (L-F) P_{0,0}. \quad (41)$$

7. The expected number of operating units in the system is obtained for two cases as follows:
(a) Case 1: when $S < F$

$$E(O_F) = M - \left[\frac{(1+\delta)^S}{\delta} \left[1 - \{1 - \delta(F-S)\}(1+\delta)^{F-S} \right] - \frac{\delta}{2} (1+\delta)^F (L-F-1)(L+F-2S) \right.$$
$$\left. + \mu\delta(1+\delta)^F \left[\sum_{n=S+1}^{L-1} \left[\frac{(n-S)}{\prod_{i=n}^{L-1} \lambda_i} \sum_{m=0}^{L-n-1} \left\{ \mu^m \prod_{i=n+m+1}^{L-1} (\lambda_i) \right\} \right] - \delta \sum_{n=S+1}^{F-1} \left[(n-S) \sum_{k=1}^{F-n} \left\{ \frac{\sum_{m=0}^{F-n-k} \left(\mu^m \prod_{i=n+m+1}^{F-k} \lambda_i \right)}{(1+\delta)^k \prod_{i=n}^{F-k} \lambda_i} \right\} \right] \right] \right] P_{0,0} \quad (42)$$

(b) Case 2: when $S \geq F$

$$E(O_F) = M \left[\frac{\delta}{2} \left\{ (1+\delta)^F (L-S)(L-S+1) \right\} + \mu\delta(1+\delta)^F \sum_{n=S+1}^{L-1} \left[\frac{(n-S)}{\prod_{i=n}^{L-1} \lambda_i} \sum_{m=0}^{L-n-1} \left\{ \mu^m \prod_{i=n+m+1}^{L-1} (\lambda_i) \right\} \right] \right] P_{0,0} \quad (43)$$

8. The expected number of warm standby units in the system is obtained as follows:
(a) Case 1: when $S < F$

$$E(S_F) = \left[(1-S) + \frac{1}{\delta} (1+\delta) \left\{ (1+\delta)^{S-2} - 1 \right\} \right.$$
$$\left. + \mu\delta(1+\delta)^F \left[\sum_{n=1}^{S-1} \left[\frac{(S-n)}{\prod_{i=n}^{L-1} \lambda_i} \sum_{m=0}^{L-n-1} \left\{ \mu^m \prod_{i=n+m+1}^{L-1} (\lambda_i) \right\} \right] \right. \right.$$
$$\left. \left. - \delta \sum_{n=1}^{S-1} \left[(S-n) \sum_{k=1}^{F-n} \left\{ \frac{\sum_{m=0}^{F-n-k} \left(\mu^m \prod_{i=n+m+1}^{F-k} \lambda_i \right)}{(1+\delta)^k \prod_{i=n}^{F-k} \lambda_i} \right\} \right] \right] \right] P_{0,0} \quad (44)$$

(b) Case 2: when $S \geq F$

$$E(S_F) = \left[(1-S) + \frac{1}{\delta} (1+\delta) \left\{ (1+\delta)^{F-2} - 1 \right\} \right.$$
$$+ \frac{1}{2} (S-F)(1+\delta)^F \{\delta(S-F-1) + 2\}$$
$$+ \mu\delta(1+\delta)^F \left[\sum_{n=1}^{S-1} \left[\frac{(S-n)}{\sum_{i=n}^{L-1} \lambda_i} \sum_{m=0}^{L-n-1} \left\{ \mu^m \prod_{i=n+m+1}^{L-1} (\lambda_i) \right\} \right] \right.$$
$$\left. \left. - \delta \sum_{n=1}^{F-1} \left[(S-n) \sum_{k=1}^{F-n} \left\{ \frac{\sum_{m=0}^{F-n-k} \left(\mu^m \prod_{i=n+m+1}^{F-k} \lambda_i \right)}{(1+\delta)^k \prod_{i=n}^{F-k} \lambda_i} \right\} \right] \right] \right] P_{0,0} \quad (45)$$

The value of $P_{0,0}$ is given by Equation 17.

N-policy model

In the 'F-policy model' section, we have determined the steady-state probabilities for different states of the machine interference system under the N-policy. Now, we evaluate some key performance measures as follows:

1. The expected number of failed units in the system is given by

$$E(N_N) = \sum_{j=0}^{1} \sum_{n=1}^{L} nQ_{n,j} = Q_{0,0} \sum_{j=0}^{1} \sum_{n=1}^{L} nF_{n,j}$$

where $Q_{n,j} = Q_{0,0}F_{n,j}$ and

$$F_{n,j} = \left[\sum_{n=1}^{N-1} n\left(\frac{\lambda_0}{\lambda_n}\right) + \sum_{n=N}^{L-1} n\left(\frac{\lambda_0}{\lambda_n}\prod_{i=N}^{n}\theta_i\right) + \frac{L\theta_0}{1-\theta_0}\prod_{i=N}^{L-1}\theta_i \right]$$
$$+ \sum_{n=1}^{L} \left[\frac{n\lambda_0}{\mu^n}\sum_{m=0}^{n-1}\left\{\mu^m\prod_{i=m+1}^{n-1}(\lambda_i)\right\} \right]$$
$$- \sum_{n=N+1}^{L} \left[\frac{n\lambda_0}{\mu^{n-N}}\left[\sum_{m=N}^{n-1}\left\{\mu^{m-N}\left(1-\prod_{i=N}^{m}(\theta_i)\right)\prod_{i=m+1}^{n-1}(\lambda_i)\right\}\right] \right]. \tag{46}$$

2. The probability that the server is busy is obtained as

$$P(B_N) = \sum_{n=1}^{L} Q_{n,1} = \left[\sum_{n=1}^{L}\left[\frac{\lambda_0}{\mu^n}\sum_{m=0}^{n-1}\left\{\mu^m\prod_{i=m+1}^{n-1}(\lambda_i)\right\}\right] \right.$$
$$\left. - \sum_{n=N+1}^{L}\left[\frac{\lambda_0}{\mu^{n-N}}\left[\sum_{m=N}^{n-1}\left\{\mu^{m-N}\left(1-\prod_{i=N}^{m}(\theta_i)\right)\prod_{i=m+1}^{n-1}(\lambda_i)\right\}\right]\right] \right] Q_{0,0}. \tag{47}$$

3. The probability that the server is idle is obtained as

$$P(I_N) = \sum_{n=0}^{L} Q_{n,0}$$
$$= \left[\sum_{n=0}^{N-1}\left(1+\frac{\lambda_0}{\lambda_n}\right) + \sum_{n=N}^{L-1}\left(\frac{\lambda_0}{\lambda_n}\prod_{i=N}^{n}\theta_i\right) + \frac{\theta_0}{1-\theta_0}\prod_{i=N}^{L-1}\theta_i \right] Q_{0,0}. \tag{48}$$

4. The probability that the server takes start-up time before starting the repair of the failed units is

$$P(ST_N) = \sum_{n=N}^{L} Q_{n,0} = \left[\sum_{n=N}^{L-1}\left(\frac{\lambda_0}{\lambda_n}\prod_{i=N}^{n}\theta_i\right) + \frac{\theta_0}{1-\theta_0}\prod_{i=N}^{L-1}\theta_i \right] Q_{0,0} \tag{49}$$

5. The probability of buildup state is

$$P(BS_N) = \sum_{n=1}^{N-1} Q_{n,0} = \lambda_0 Q_{0,0} \sum_{n=1}^{N-1}\left(\frac{1}{\lambda_n}\right) \tag{50}$$

6. The expected number of operating units in the system is obtained for two cases:
(a) Case 1: when $S < N$

(b) Case 2: when $S \geq N$

$$P(O_N) = \left[\sum_{n=S+1}^{L-1}(n-S)\left(\frac{\lambda_0}{\lambda_n}\prod_{i=N}^{n}\theta_i\right) + \frac{(L-S)\theta_0}{1-\theta_0}\prod_{i=N}^{L-1}\theta_i \right.$$
$$+ \sum_{n=S+1}^{L}\left[\frac{(n-S)\lambda_0}{\mu^n}\sum_{m=0}^{n-1}\left\{\mu^m\prod_{i=m+1}^{n-1}(\lambda_i)\right\}\right]$$
$$\left. - \sum_{n=S+1}^{L}\left[\frac{(n-S)\lambda_0}{\mu^{n-N}}\left[\sum_{m=N}^{n-1}\left\{\mu^{m-N}\left(1-\prod_{i=N}^{m}(\theta_i)\right)\prod_{i=m+1}^{n-1}(\lambda_i)\right\}\right]\right] \right] Q_{0,0} \tag{52}$$

7. The expected number of warm standby units in the system is determined for two cases as follows:
(a) Case 1: for $S < N$

$$P(S_N) = \left[\sum_{n=1}^{S-1}(S-n)\left(\frac{\lambda_0}{\lambda_n}\right) \right.$$
$$\left. + \sum_{n=1}^{S-1}\left[\frac{(S-n)\lambda_0}{\mu^n}\sum_{m=0}^{n-1}\left\{\mu^m\prod_{i=m+1}^{n-1}(\lambda_i)\right\}\right] \right] Q_{0,0} \tag{53}$$

(b) Case 2: for $S \geq N$

$$P(S_N) = \left[\sum_{n=1}^{N-1}(S-n)\left(\frac{\lambda_0}{\lambda_n}\right) + \sum_{n=N}^{S-1}(S-n)\left(\frac{\lambda_0}{\lambda_n}\prod_{i=N}^{n}\theta_i\right) \right.$$
$$+ \sum_{n=1}^{S-1}\left[\frac{(S-n)\lambda_0}{\mu^n}\sum_{m=0}^{n-1}\left\{\mu^m\prod_{i=m+1}^{n-1}(\lambda_i)\right\}\right]$$
$$\left. - \sum_{n=N+1}^{S-1}\left[\frac{(S-n)\lambda_0}{\mu^{n-N}}\left[\sum_{m=N}^{n-1}\left\{\mu^{m-N}\left(1-\prod_{i=N}^{m}(\theta_i)\right)\prod_{i=m+1}^{n-1}(\lambda_i)\right\}\right]\right] \right] Q_{0,0} \tag{54}$$

The value of $Q_{0,0}$ is given by Equation 35.

Conclusion

In this paper, we have explored the concepts of the F-policy and N-policy for multi-component machining systems with warm standbys. The steady-state probability distributions established for both the F-policy and N-policy are further used to establish some performance measures such as the expected number of failed machines in the system, probability that the server is busy or idle in the system, throughput, etc. The explicit expressions of various performance measures are provided which may be further used for the improvement and performance evaluation of many real-time machining systems. The provision of warm types of standbys is a general assumption as in a

$$P(O_N) = \left[\sum_{n=S+1}^{N-1}(n-S)\left(\frac{\lambda_0}{\lambda_n}\right) + \sum_{n=N}^{L-1}(n-S)\left(\frac{\lambda_0}{\lambda_n}\prod_{i=N}^{n}\theta_i\right) + \frac{(L-S)\theta_0}{1-\theta_0}\prod_{i=N}^{L-1}\theta_i + \sum_{n=1}^{L}\left[\frac{(n-S)\lambda_0}{\mu^n}\sum_{m=0}^{n-1}\left\{\mu^m\prod_{i=m+1}^{n-1}(\lambda_i)\right\}\right] \right.$$
$$\left. - \sum_{n=N+1}^{L}\left[\frac{(n-S)\lambda_0}{\mu^{n-N}}\left[\sum_{m=N}^{n-1}\left\{\mu^{m-N}\left(1-\prod_{i=N}^{m}(\theta_i)\right)\prod_{i=m+1}^{n-1}(\lambda_i)\right\}\right]\right] \right] Q_{0,0} \tag{51}$$

special case when the failure rate is zero or the same as that of operating units, and it facilitates results for the cold standby case. The study of control policy-based models in the present investigation will be helpful in the quantitative assessment of the system's reliability and other mean characteristics of many embedded systems such as computer networks, manufacturing systems, transportation systems, etc. In the future, we can further extend our study by considering the mixed type of standbys facility and bulk failure to make it more versatile.

Competing interests
The authors declare that they have no competing interests.

Authors' contributions
KK and MJ conceived the idea of extension of earlier existing results for machine repair problems with standbys provisioning due to a lot of applications in real time systems. The formulation of governing equations and derivation of performance measures are carried out by KK. MJ made the final language corrections of the manuscript. All authors read and approved the final manuscript.

Acknowledgements
The authors are highly thankful to the learned referees and editor for their valuable suggestions for the improvement in this paper. One of the authors (Kamlesh Kumar) would like to acknowledge the financial assistantship in the form of JRF/SRF from Council of Scientific and Industrial Research (CSIR) Delhi, India.

References
Gopalan MN (1975) Probabilistic analysis of a single-server *n*-unit system with (*n*-1) warm standbys. Operations Research 23(3):591–598

Gupta SM (1995) Interrelationship between controlling arrival and service in queueing systems. Computer & Operations Research 22(10):1005–1014

Gupta UC, Rao Srinivasa TSS (1996) Theory and methodology on the M/G/l machine interference model with spares. European Journal of Operational Research 89(1):164–171

Hajeeh MA (2011) Reliability and availability of a standby system with common cause failure. International Journal of Operational Research 11(3):343–363

Hajeeh M, Jabsheh F (2009) Multiple component series systems with imperfect repair. International Journal of Operational Research 4(2):125–141

Haque L, Armstrong MJ (2007) A survey of the machine interference problem. European Journal of Operational Research 179(2):469–482

Jain M (1997) An (m, M) machine repair problem with spares and state dependent rates: a diffusion process approach. Microelectronics Reliability 37(6):929–933

Jain M, Agrawal PK (2009) Optimal policy for bulk queue with multiple types of server breakdown. International Journal of Operational Research 4(1):35–54

Jain M, Bhargava C (2009) N-policy machine repair system with mixed standbys and unreliable server. Quality Technology & Quantitative Management 6(2):171–184

Jain M, Upadhyaya S (2009) Threshold *N*-policy for degraded machining system with multiple type spares and multiple vacations. Quality Technology & Quantitative Management 6(2):185–203

Jain M, Rakhee, Maheshwari S (2004a) N-policy for a machine repair system with spares and reneging. Applied Mathematical Modelling 28(6):513–531

Jain M, Rakhee, Singh M (2004b) Bilevel control of degraded machining system with warm standbys, setup and vacation. Applied Mathematical Modelling 28(12):1015–1026

Jain M, Sharma GC, Sharma R (2008) Performance modeling of state dependent system with mixed standbys and two modes of failure. Applied Mathematical Modelling 32(5):712–724

Jain M, Sharma GC, Pundhir RS (2010) Some perspectives of machine repair problems. International Journal of Engineering 23(3 & 4):253–268

Jain M, Sharma GC, Sharma R (2012a) Optimal control of (N, F) policy for unreliable server queue with multi-optional phase repair and start-up. International Journal of Mathematics in Operational Research 4(2):152–174

Jain M, Shekhar C, Shukla S (2012b) Queueing analysis of a multi-component machining system having unreliable heterogeneous servers and impatient customers. American Journal of Operational Research 2(3):16–26

Ke JC, Wang KH (2007) Vacation policies for machine repair problem with two type spares. Applied Mathematical Modelling 31(5):880–896

Ke JC, Wu CH (2012) Multi-server machine repair model with standbys and synchronous multiple vacation. Computers and Industrial Engineering 62(1):296–305

Kuo CC, Wang KH, Pearn WL (2011) The interrelationship between N-policy M/G/1/K and F-policy G/M/1/K queues with startup time. Quality Technology & Quantitative Management 8(3):237–251

Sharma DC (2012) Machine repair problem with spares and N-policy vacation. Research Journal of Recent Sciences 1(4):72–78

Sivazlian BD, Wang KH (1989) System characteristics and economic analysis of the G/G/R machine repair problem with warm standbys using diffusion approximation. Microelectronics Reliability 29(5):829–848

Wang KH, Kuo CC (2000) Cost and probabilistic analysis of series system with mixed standby components. Applied Mathematical Modelling 24(12):957–967

Wang KH, Yang DY (2009) Controlling arrivals for a queueing system with an unreliable server: Newton-Quasi method. Applied Mathematics and Computation 213(1):92–101

Wang KH, Kuo CC, Pearn WL (2008) A recursive method for the F-policy G/M/1/K queueing system with an exponential startup time. Applied Mathematical Modelling 32(6):958–970

Yadin M, Naor P (1963) Queueing systems with a removable service station. Operational Research Quarterly 14(4):393–405

Yang DY, Wang KH, Wu CH (2010) Optimization and sensitivity analysis of controlling arrivals in the queueing system with single working vacation. Journal of Computational and Applied Mathematics 234(2):545–556

Yuan L, Meng XY (2011) Reliability analysis of a warm standby repairable system with priority in use. Applied Mathematical Modelling 35(9):4295–4303

Zhang ZG, Tian N (2004) The N threshold policy for the GI/M/1 queue. Operations Research Letters 32(1):77–84

10

A novel risk-based analysis for the production system under epistemic uncertainty

Mehran Khalaj[1*], Fereshteh Khalaj[2] and Amineh Khalaj[3]

Abstract

Risk analysis of production system, while the actual and appropriate data is not available, will cause wrong system parameters prediction and wrong decision making. In uncertainty condition, there are no appropriate measures for decision making. In epistemic uncertainty, we are confronted by the lack of data. Therefore, in calculating the system risk, we encounter vagueness that we have to use more methods that are efficient in decision making. In this research, using Dempster-Shafer method and risk assessment diagram, the researchers have achieved a better method of calculating tools failure risk. Traditional statistical methods for recognizing and evaluating systems are not always appropriate, especially when enough data is not available. The goal of this research was to present a more modern and applied method in real world organizations. The findings of this research were used in a case study, and an appropriate framework and constraint for tools risk were provided. The research has presented a hopeful concept for the calculation of production systems' risk, and its results show that in uncertainty condition or in case of the lack of knowledge, the selection of an appropriate method will facilitate the decision-making process.

Keywords: Dempster-Shafer theory; Epistemic uncertainty; Risk analysis; Risk assessment diagram

Introduction

Over the last few decades, different methods of decision making in uncertainty condition have been considered. Among the suggested approaches, evidence theory which is also called Dempster-Shafer theory (DST) introduces a stronger framework for our incomplete knowledge presentation and expression. The different type of available information calls for the development of a different method to represent and propagate the associated uncertainty. Resorting to probability theory to address this issue is questionable (Baraldi et al. 2013). The use of evidence theory started with Dempster's work by description of the accounting principles of the upper and lower probabilities, and mathematical theory of evidence was defined precisely by Shafer (1976). However, in the last decades, Bayesian inferences (Bayes 1763) based on previous applications are valid, but the Dempster-Shafer studies as techniques of modeling in uncertainty condition have had a lot of applications; various perspectives for uncertainty management have been proposed. Buchanan and Shortcliffe (1975)

proposed a model that manages uncertainty and has almost certain factors. Hence, when our knowledge is incomplete, uncertain approaches toward application are more appropriate. Fedrizzi and Kacprzyk (1988) encouraged studies in the setting of fuzzy preference using interval value for expressing experts' views and judgments presented through cumulative distribution function. Each approach toward uncertainty management has its own advantages and disadvantages (Lee et al. 1987). For example, Walley (1987) and Caselton and Luo (1992) discuss the problems dealt with Bayes common analysis which are due to the unreliability of information, and Klir (1989) has criticized the probable presentation of uncertainty for knowledge inference. The DST of Dempster-Shafer (Dempster 1967) upon multidimensional sources in which information is obtained from some various sources had lots of application, and some justifications for the appropriateness of this method for the inference of knowledge have been indicated. Dempster-Shafer theory has been applied successfully in various domains such as face recognition (Ip and Ng 1994), and so far, it has had also broad applications in the discussions of diagnosis, statistical classification (Denoeux 1995), data fusion (Telmoudi and Chakhar 2004), environmental impact assessment (Wang et al. 2006), knowledge reduction

---
* Correspondence: mkhalaj@rkiau.ac.ir
[1]Department of Industrial Engineering, Robat Karim Branch, Islamic Azad University, Tehran, Iran
Full list of author information is available at the end of the article

(Wu et al. 2005), organizational self-assessment (Siow et al. 2001), regression analysis (Monney 2003), multi-criterion decision-making analyses (Bauer 1997; Beynon et al. 2001), pattern classification (Binaghi and Madella 1999; Binaghi et al. 2000), reasoning and logic (Benferaht et al. 2000), medical diagnosis (Yen 1989), safety analysis (Liu et al. 2004; Wang and Yang 2001), expert systems (Beynon et al. 2001; Biswas et al. 1988), target identification (Buede and Girardi 1997), and uncertainty (Klir and Wierman 1998). In this study, researchers have applied the theory of Dempster-Shafer as well in accounting for the tools failure risk in a production organization. When failure time is unknown, loss of production system will be occurred. Several methods can be used to avoid the risk of failure in factories risk, for instance, all approach such as reliability centered maintenance (Knezevic 1997 and Moubray 1991), risk-based inspection (Chang et al. 2005), risk-based maintenance (Khan and Haddara 2003), risk-based decision making (Carazas and Souza 2010), is related to risk management tools. In the theoretical concept, we have applied tools, such as Dempster-Shafer and fuzzy theory, that provide significant patterns about the risk and reliability, and they can be extracted from the data which originated in a factory. If we do not have enough available data, we can use fuzzy and precise numbers together to calculate the risk. These results in discovering new knowledge about the failure risk to the factory.

In real situations, what happens to production systems is unpredictable. Hence, in decision making, we always encounter a risk especially with incomplete information. Although various studies have been done concern the use of the Dempster-Shafer's theory in systems recognition, accounting, and decision making, yet we encounter some problems in the application of this theory in the systems risk assessment and making administrative decisions in real production systems. This was the main reason of doing this research. The purpose of the research, therefore, was to present a composite approach for more recognition and applied tools risk assessment and its operational evidence is provided by accounting for the tools risk of a production organization.

Uncertainty and information incompleteness

There are various forms of uncertainty in the amount of uncertainties affecting the operation of the engineers, scientists, and decision makers. Each group focuses on different types of uncertainty; it is necessary before any decision being made that their type is identified and classified depending upon the amount and the correctness of available information. In the previous decades, the probability theory has been used as a device for accounting for modeling uncertainty. With regard to the limitations of the probability theory, the approaches and methods of describing and determining the uncertainty condition now are developed, and other than probability theory, some other different theories are proposed and used (Parsons 2001). Due to the limitations of the probability theory, its use in risk assessment is not always appropriate, especially when lack of information causes uncertainty. There are many approaches for accounting for uncertainty and deviation, for example, mathematics models and simulation tools. When we encounter lack or incompleteness of information, some other theories such as fuzzy set theory (Zimmermann 1991), possibility theory (Dubois and Prade 1988), Dempster-Shafer theory, and upper and lower probabilities (Dempster 1967) are more powerful.

In this research, the researchers' focus has been on the use of Dempster-Shafer theory in recognizing and accounting for uncertainty and deciding and determination of the system risk. In the real world, the static mathematical models are resulted from a system full of non-deterministic nature; their parameters become non-unique uncertainties and a chain of uncertainties has been made in this model. Among the available approaches and methods, the Dempster-Shafer theory provides an appropriate recognition tool in uncertainty condition resulted from the lack of information. This is why it uses all available data, and it also determines and quantifies its distinctive goal precisely.

The main sources of uncertainty should be identified and analyzed. Uncertainty is classified in two main groups - epistemic uncertainty and aleatory uncertainty. There are many differences between aleatory uncertainties and epistemic uncertainty; aleatory uncertainty is usually named as stochastic uncertainty, inherent uncertainty, or irreducible uncertainty. Epistemic uncertainty is usually named as subjective uncertainty or reducible uncertainty. When there is inherent variation of the physical system, we encounter aleatory uncertainty. Generally, inherent variation is due to the random nature of the input data, and it is possible to represent it mathematically by a probability distribution if sufficient experimental data is accessible. In non-deterministic systems, some factors such as ignorance, lack of knowledge, or incomplete information lead to the occurrence of epistemic uncertainty. The Dempster-Shafer theory which is also called evidence theory provides an appropriate picture of epistemic uncertainty. The advantage of using evidence theory is in its success of quantifying the degree of uncertainty when the amount of available information is small. Like most of the current theories of uncertainty, evidence theory provides two types of uncertain measure which are known as belief and plausibility.

Identification of the sources of uncertainty is the first step of a methodology in quantification of epistemic uncertainty or aleatory uncertainty. Uncertainty can occur in every phase of the modeling and simulation. Detailed up-to-date descriptions on the various forms of uncertainty are given in the literature (Oberkampf et al. 2001). The act of quantifying uncertainty and variability is significant, and researchers have proposed different mathematical models to properly illustrate them. Once enough data is on hand, every type of variability can be represented mathematically via probability distribution functions. There is approximately a general concurrence on this concern.

In most cases, when maintenance managers try to determine the best policy for their systems, they only consider the cost criterion as the most important and the only criterion to be taken into account. This is a very dangerous point of view (Faghihinia and Mollaverdi 2012). Generally, the most important focus of the current study is to quantify epistemic uncertainty in maintenance engineering models and repair, particularly to be used in multidisciplinary systems and to apply it for design optimization.

Operational risk's management

Operational risk's management is the central core of any management strategy that is related to production properties. The control of operational risk depends on understanding it, measuring it, and knowing how to reduce it. Accordingly, the process of managing operational risk consists of some stages as follows: risk assessment and risk prioritizing; identification of possible failure scenarios (each failure scenario is estimated from the probability of failure and the degree of consequences); prioritizing risks according to their magnitude; estimating and dealing with the total risk. Assessing and dealing with the degree of risk mean managing and selecting appropriate risk response strategy. For example, tolerable risk is accepted, and if the estimated risk is not tolerable, it is reduced through appropriate risk reduction approaches.

In order to estimate the risk, we should assess the risk, and for assessment, we need the failure probability of expected risk and the degree of consequences. Then, for risk estimation, we use the integration and interaction of the failure probability and the amount of loss. According to a classical definition (Henley and Kumamoto 1981; Vose 2003), the risk of failure R_i is defined by

$$R_i = P_f \times C, \tag{1}$$

where P_f is the probability of failure and C is the cost given failure for tools operation. According to the experimental definition of probability suppose that N shows pieces of equipment, N_f shows pieces of equipment in which is not succeed prior to time a, since only breakdown prior to time a is related with losses. If the amount of experiments N is adequately large, then the entire loss caused by failures throughout N trials is $N_f \times C$. The anticipated loss is

$$P_f = \lim_{N \to \infty} (N_f/N) \tag{2}$$
$$R_i = (N_f/N) \times C.$$

In Equation 2, P_f is approximates the failure probability of the machine prior to time a, also R_i is risk of systems. Looking over the traditional analysis of risk, what is important in assessment is having primary data for analyzing and changing the data to the information which determine the risk. If this data will not be available or the least appropriate and precise data be available, the provided analysis would not be realistic, and probably, the outcome results could not be reliable. Therefore, quantitative assessment of risk will be an appropriate device. The risk management is a decision-making process which lack of the data leads to weak decision making. Therefore, whenever we could not use these tails, it is appropriate to use semiquantitative or qualitative approaches.

Although qualitative approach of the risk estimation needs less effort to gather information, because of its subjectivity, it is not reliable. In the complete uncertainty condition, using the qualitative approach of risk assessment is suitable and applicable. What is investigated in this research is the condition we have information about the past but it is not sufficient for statistical application and decision making including the risk. Our purpose is to use all the data in order to increase the precision of decisions. If the data is available, it should not be ignored, and whenever the small data are analyzed and converted to information appropriately, they provide appropriate measure for decision making.

An overview of Dempster-Shafer theory

There are various methods of decision making under uncertain conditions. Among these methods, the Dempster-Shafer theory is a powerful method for showing and representing uncertainty of our incomplete knowledge. Theory of evidence allows one to combine evidence from different sources and arrive at a degree of belief that takes into account all the available evidence. The theory was first developed by the Dempster's work in explaining the principles of calculating the upper and lower probabilities (Dempster 1967) and then its mathematical theory developed by Shafer (1976). This theory had had a wide range of application as a model under uncertain conditions. Briefly, the Dempster-Shafer theory may be summarized as follows. The primary step consists in developing a frame of discernment Θ which is a finite set of mutually exclusive hypotheses. Then, we can define the m function which assigns an evidential weight to each subset A of Θ. This function is also called basic probability assignment.

The basic probability assignment (bpa or m) is different from the classical definition of probability. It is defined by mapping over the interval [0–1], in which the basic assignment of the null set $m(\emptyset)$ is zero, and the summation of basic assignments in a given set A is '1'. The basic probability assignment is called a focal point for each element for which $m(A) \neq 0$ is true. This can be represented by

$$
\begin{aligned}
&m(A) \rightarrow [0, 1] \\
&m(\phi) = 0 \\
&\sum\nolimits_{A \subseteq \Theta} m(A) = 1.
\end{aligned}
\tag{3}
$$

The lower and upper bounds of an interval can be determined through a basic probability assignment, which includes the probability of the set bounded by two non-additive measures, namely belief and plausibility. The lower limit of belief for a given set A is defined as the summation of all basic probability assignments of the proper subsets B, in which B is a subset of A. The general relation between bpa and belief can be represented by

$$
\begin{aligned}
&\mathrm{bel}(A) = \sum_{B \subseteq A} m(B) \\
&\mathrm{bel}(\phi) = 0 \\
&\mathrm{bel}(1) = 1.
\end{aligned}
\tag{4}
$$

The upper bound is plausibility, which is the summation of basic probability assignments of subsets of B, for which A (i.e., $B \cap A \neq \emptyset$) is true, the function pl is called plausibility and can be written as

$$
\mathrm{pl}(A) = \sum_{B \cap A \neq \phi} m(B).
\tag{5}
$$

Moreover, the following relationship is true for the belief function and the plausibility function under all circumstances.

$$
\begin{aligned}
&\mathrm{pl}(A) \geq \mathrm{bl}(A) \\
&\mathrm{pl}(\phi) = 0 \\
&\mathrm{pl}(\theta) = 1 \\
&\mathrm{pl}(\neg A) = 1 - \mathrm{bel}(A)
\end{aligned}
\tag{6}
$$

The belief interval represents a range where the probability may lie. It is determined by reducing the interval between plausibility and belief. The narrow uncertainty band represents more precise probabilities. The probability is uniquely determined if $\mathrm{bel}(A) = \mathrm{pl}(A)$, and for the classical probability theory, all probabilities are unique. If $U(A)$ has an interval $[0, 1]$, it means that no information is available, but if the interval is $[1, 1]$, then it means that A has been completely confirmed by $m(A)$.

In reality, a decision maker can often gain access to more than one information source in order to make his or her decisions. The evidence theory constructs a set of hypotheses of known mass values from these information sources and then computes a new set of numbers m that represents combined evidence.

The part of DST that is of direct relevance is Dempster's rule for combination (Zimmermann 1991). When the data comes from different sources, through data fusion and combination, we can summarize the results and simplify

them for decision making. Consider now two pieces of evidence on the same frame H represented by two bpa m1 and m2 Decision maker needs a rule of combination to generate a new bpa. This construction is called Dempster-Shafer rule of combination for group aggregation and can be written as

$$
\left.
\begin{aligned}
&m_{1-2}(A)_{B \cap C = \phi} = \frac{\sum m_1(B) m_2(C)}{(1-k)} \\
&\text{when} \quad A \neq \phi \quad \text{and} \quad m_{1-2}(\phi) = 0 \quad \text{where} \\
&k = \sum_{B \cap C = \phi} m_1(B) m_2(C)
\end{aligned}
\right]
\tag{7}
$$

K represents a basic probability mass associated with conflicts among the sources of evidence. K is often interpreted as a measure of conflict between the sources.

Real case study

In an automobile manufacturing supplier company, different machines are working. An exclusive 5,000-ton hydraulic press machine belongs to this production institute, and there is no replacement for the above machine that in the edited strategy of this company, possessing this machine is considered a competitive advantage and its failures are considered a weak point, because any failures or break in proceedings of this machine will affect the key results of the company's operation. As this machine is essential, so all the parts used in this machine are also considered essential.

Because it is impossible to buy all the spare parts or to make extra or passive systems due to the machine's structural complexity, the reliability of the system is a more concern of the manager of the production institute. Therefore, any break in proceedings will lead to critical losses to the income and significance of this financial institute. One of the other limitations is the store accumulation which should be kept in a proper extent. Buying and providing expensive spare parts will reduce the risk but increase the costs and it is also possible that these parts remain unused for years. The management aims at assessing the likelihood of failures through the whole press machine and since it is not possible to have a replacement for the machine, desires to identify the reliability and the risk in whole press by determining the key parts' risk of break down and decisions are made on it.

To begin the analysis, a part of fault tree has been drawn in Figure 1. In this case study, three parts of the fault tree were investigated. These three parts are the computer which controls all the system; PLC control which is the connector of the hardware and the software of the press; and cushion pumps have the duty of making pressure over the parts pressed. These parts are included as the essential components of the production institute, which may be defected due to exhaustion, and providing their similar

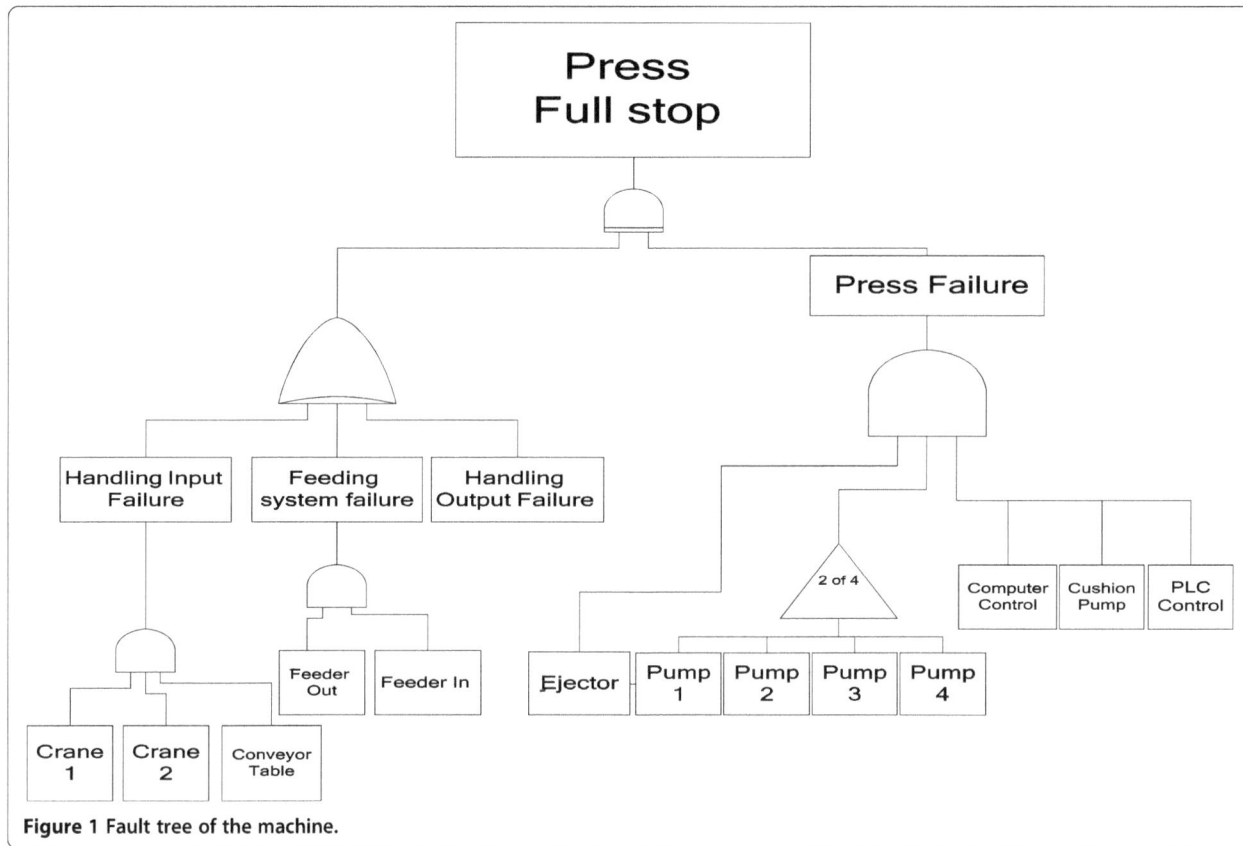

Figure 1 Fault tree of the machine.

ones, imposes some costs to the organization. Over the past 13 years, there are only limited data about the failures of the control computer which is not possible to make direct decisions through statistical methods upon them. The director manager of this institute aims at deciding based on the previous data and the experts' opinions about how to react to the risk resulted from this three-part collection.

Risk-based analysis using Dempster-Shafer theory

Risk has two factors, first index is probability; it shows the probability of occurrence of a risk in a definite period of time. Classification of risks according to their probability of occurrence is also possible. The classification of Table 1 shows a fuzzy categorization of relative probability of an accident or breakdown occurring as a result of uncontrolled risk[a]. This table helps us to understand the importance of an accident according to the probability of its occurrence. It is worth mentioning that in similar categorizations, the

probability of accidents occurring can be defined in a fuzzy way; in this mode, the data expert must gather data in a quite precise way.

With regard to checking the data, we classified the probability of failures into three levels of magnitude: L (Low), in which a breakdown may occur with a low probability in a fixed period of time; M (Medium), in which a breakdown may occur with a medium probability in a fixed period of time; H (High), in which a breakdown may occur with a high probability in a fixed period of time. The set of these occurrences form the set of $\Theta = \{L, M, H\}$. The possible subsets will be eight sets of $\{\phi\},\{L\},\{M\},\{H\},\{L,M\},\{L,H\},\{M,H\}\{L,M,H\}$. Figure 2 shows the basic probability assignment for the breakdowns probability of the production system in fuzzy logic demonstration. On the other hand, risk is the probability of occurrence multiplied by the magnitude of damage (Equation 1). Then, the researchers define fuzzy severity categories according Figure 2, consequences

Table 1 The information failure from sources 1 and 2

Press section		Basic probability assignment													
		Source 1							Source 2						
		L	M	H	L,M	L,H	M,H	L,M,H	L	M	H	L,M	L,H	M,H	L,M,H
Computer	Co	0.5	0.2	0.1	0	0	0.1	0.1	0.6	0.1	0	0	0	0.2	0.1
Cushion	Cu	0.2	0.1	0.2	0	0	0.3	0.4	0.3	0.2	0.1	0.1	0	0.2	0.1
PLC	P	0	0.3	0.1	0	0	0.4	0.2	0.1	0.4	0.2	0	0	0.1	0.2

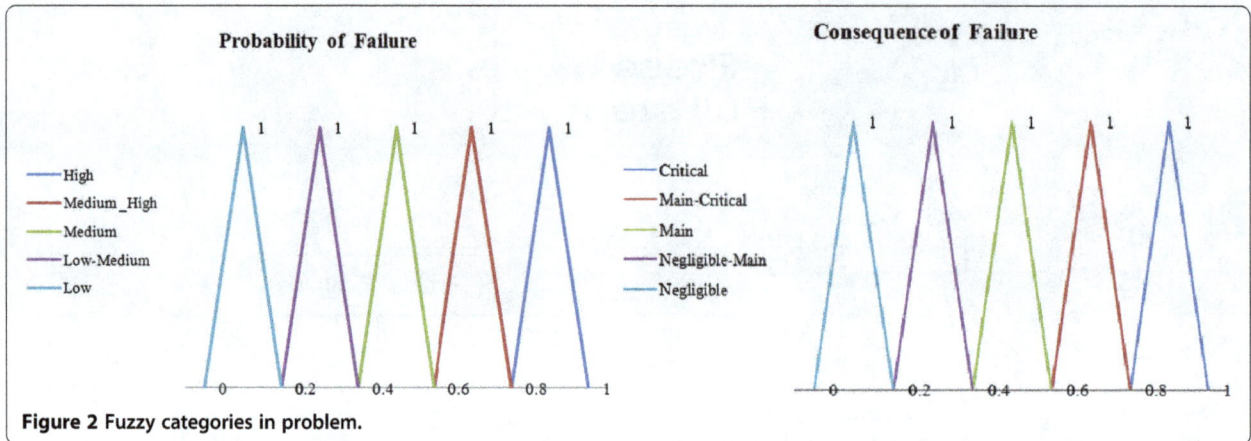

Figure 2 Fuzzy categories in problem.

which are as follows: Negligible (N), a failures which if it happens, it is removable and slight; Main (M), a failures which will cause stopping but it is removable; Critical (C), it is a failures which its consequence is high and if it happens it leads to a crisis. The set of events is put on in the set of $\Theta = \{N, M, C\}$. According to the above supposition, its possible forms are put on the eight subsets of $\{N\}$, $\{M\}$, $\{C\}$, $\{N, M\}$, $\{N, C\}$, $\{M, C\}$, $\{N, M, C\}$.

The lower and upper bounds of an interval can be determined from the basic probability assignment, bpa contains the probability set bounded by two measures belief and plausibility, Table 2 shows the basic probability assignment for the probability breakdowns of the production system.

By applying D-S rule of combination on sources of information for PLC (P), the following data are generated and can be shown as Table 3.

Degree of conflict is $K = 0.27$; therefore, normalization factor is $(1 - K) = 0.73$, in the same way, belief and plausibility functions can be determined by using corresponding equation described earlier from the above analysis. According to Equation 3, it can be written as follows:

$$m_{1-2}(p)_L = 0.63 \qquad m_{1-2}(p)_{L,H} = 0.05$$
$$m_{1-2}(p)_M = 0.16 \qquad m_{1-2}(p)_{M,H} = 0.05$$
$$m_{1-2}(p)_H = 0.04 \qquad m_{1-2}(p)_\Theta = 0.05$$
$$m_{1-2}(p)_{L,M} = 0.00.$$

This method will be similar for other parts namely compute 'Co' and cushion 'Cu' which can be written as follows:

$$m_{1-2}(co)_L = 0.68 \qquad m_{1-2}(cu)_L = 0.34$$
$$m_{1-2}(co)_M = 0.17 \qquad m_{1-2}(cu)_M = 0.35$$
$$m_{1-2}(co)_H = 0.05 \qquad m_{1-2}(cu)_H = 0.23$$
$$m_{1-2}(co)_{L,M} = 0.00 \qquad m_{1-2}(cu)_{L,M} = 0.06$$
$$m_{1-2}(co)_{L,H} = 0.00 \qquad m_{1-2}(cu)_{L,H} = 0.00$$
$$m_{1-2}(co)_{M,H} = 0.08 \qquad m_{1-2}(cu)_{M,H} = 0.26$$
$$m_{1-2}(co)_\Theta = 0.02 \qquad m_{1-2}(cu)_\Theta = 0.06$$
$$k = 0.40 \qquad k = 0.35.$$

The belief and plausibility can be calculated according Equations 4 and 6 after finding the basic probability assignment of each device.

Uncertainty interval for failures probability

The belief interval represents a range in which true probability may lie. It can be determined by subtracting belief from plausibility. The narrow uncertainty band signifies additional strict probabilities. In conditions that $U(A)$ has an interval [0, 1], it denotes that no information is on hand, but once the interval is [1, 1], at that time, it means that A has been entirely confirmed by $m(A)$. The uncertainty interval for failures probability of the PLC is presented in Table 4.

Table 2 The D-S rule of combination information from sources 1 and 2

		$m_1(P)_L$		$m_1(P)_M$		$m_1(P)_H$		$m_1(P)_{L,M}$		$m_1(P)_{L,H}$		$m_1(P)_{M,H}$		$m_1(P)_\Theta$	
		0.4		0.3		0		0		0.1		0.1		0.1	
$m_2(P)_L$	0.5	{L}	0.05	{φ}	0.02	{φ}	0.01	{L}	0	{L}	0	{φ}	0.01	{L}	0.01
$m_2(P)_L$	0	{φ}	0.2	{M}	0.08	{φ}	0.04	{M}	0	{φ}	0	{M}	0.04	{M}	0.04
$m_2(P)_L$	0.1	{φ}	0.1	{φ}	0.04	{H}	0.02	{φ}	0	{H}	0	{H}	0.02	{H}	0.02
$m_2(P)_L$	0	{L}	0	{M}	0	{φ}	0	{L,M}	0	{L}	0	{M}	0	{L,M}	0
$m_2(P)_L$	0	{L}	0	{φ}	0	{H}	0	{L}	0	{L,H}	0	{H}	0	{L,H}	0
$m_2(P)_L$	0	{φ}	0.05	{M}	0.02	{H}	0.01	{M}	0	{H}	0	{M,H}	0.01	{M,H}	0.01
$m_2(P)_L$	0.4	{L}	0.1	{M}	0.04	{H}	0.02	{L,M}	0	{L,H}	0	{M,H}	0.02	Θ	0.02

Table 3 The results of belief function and the plausibility failures of PLC 'P'

Belief function	Amount	Plausibility function	Amount	Uncertainty interval	Amount
$Bel_{1-2}(p)_L$	0.63	$Pl_{1-2}(p)_L$	0.74	$U_{1-2}(p)_L$	[0.63 to 0.74]
$Bel_{1-2}(p)_M$	0.16	$Pl_{1-2}(p)_M$	0.27	$U_{1-2}(p)_M$	[0.16 to 0.27]
$Bel_{1-2}(p)_H$	0.04	$Pl_{1-2}(p)_H$	0.21	$U_{1-2}(p)_H$	[0.04 to 0.21]
$Bel_{1-2}(p)_{L,M}$	0.79	$Pl_{1-2}(p)_{L,M}$	0.96	$U_{1-2}(p)_{L,M}$	[0.79 to 0.96]
$Bel_{1-2}(p)_{L,H}$	0.73	$Pl_{1-2}(p)_{L,H}$	0.84	$U_{1-2}(p)_{L,H}$	[0.73 to 0.84]
$Bel_{1-2}(p)_{M,H}$	0.26	$Pl_{1-2}(p)_{M,H}$	0.37	$U_{1-2}(p)_{M,H}$	[0.26 to 0.37]
$Bel_{1-2}(p)_\Theta$	1.00	$Pl_{1-2}(p)_\Theta$	1.00	$U_{1-2}(p)_\Theta$	[1 to 1]

From the above combinatory analysis, it was found that the failures probability of the PLC is in the Low and Medium levels because the range of [0.79 to 0.96] is acceptable; similarly, we could achieve the risk function of the computer and cushion pumps through their related data.

Uncertainty interval for consequences

Similarity, the lower and upper bounds of consequence impact can be determined from the basic probability assignment; summarizing and simplifying calculation is shown in Table 5.

According to the data of basic probability allocations, Equations 4 and 7 can be used to find the belief and plausibility functions. This method is in the same way that mentioned in the previous section, and it is shown in Table 6. The intervals shown in this table represent the values by which they are approved by the existing data. For example, regarding the cushion, the results of this table show that the breakdown status of the machine is at the level of main and critical. This interval is approved by the probability interval of [0.81 to 0.91].

To determine the failure consequence of each machine from the table of belief and plausibility functions, regarding the computer, the results of this table show that the breakdown status of the consequence impact is at the level of negligible and main. This interval is approved by the probability interval of [0.96 to 0.96]. The interval that has a great belief interval will be chosen since the belief function determines a low probability that is gained through the minimum available data. For example, we can show the calculating consequence of failure as follows (see Table 7).

The risk diagram assessment

First, we have tried to draw the risk assessment diagram which is shown in Figure 3. These diagrams are drawn for the two selected parts of the computer and cushion pumps.

Table 4 The results of uncertainty interval

Press section	Uncertainty interval	Amount
Computer	$U_{1-2}(A)_{L,M}$	[0.85 to 0.97]
Cushion	$U_{1-2}(A)_{M,H}$	[0.85 to 0.0.97]
PLC	$U_{1-2}(A)_{L,M}$	[0.79 to 0.96]

In the risk assessment diagram, the x-axis is divided into five sections of results starting negligible loss until critical condition result. Also, y-axis shows the probability of failure starting with Low probability until High probability in the five sections. If enough data are available, we can use quantitative and precise numbers for probability. But, about the concerned case study, we are in the uncertainty conditions and our data are insufficient. Therefore, the non-deterministic area has been specified in the diagram. The lower bound of this domain is determined by the belief function and the higher bound by the plausibility function. In the domain, the narrower the interval band, it shows stronger probability. An example in the drawing the risk related to the failures probability occurrence in the computer, we get to Figure 3. According to the resulted evidences, probability of failure in the computer has two positions of negligible and main, and will be in the [0.2, 0.4] domain and consequence of failures in the computer has two positions of low and medium; finally, the risk will be in insignificant or minor area. Here is shown the area of the risk under uncertainty condition; for other cases, the risk assessment diagram can be drawn.

Figure 3 shows that the risk of cushion devices is located at the moderate and major level, while the risk of computers is at the level of insignificant and major. There are two solutions if we want to use the risk-based analysis for controlling and reducing the risk of this equipment, (1) decreasing the probability of failure and (2) decreasing the consequence of failure, which will lead to the reduction of failure risk of the equipment. In this section, decreasing the probability of failure about the cushion will not cause decreasing failure risk; therefore, we have to reduce the consequence impact of failure. Figure 3 also shows in the uncertainty condition although it cannot be determine precise number for risk of system but risk area can be obtained.

The analysis of risk diagram assessment

In order to make decisions, the measures should be determined. However, under many conditions, the decision information about alternatives is usually uncertain or fuzzy due to the increasing complexity of the socio-economic

Table 5 The information consequence from sources 1 and 2

Press section		Basic probability assignment												
		Source 1							Source 2						
		L	M	H	L,M	L,H	M,H	L,M,H	L	M	H	L,M	L,H	M,H	L,M,H
Computer	Co	0.2	0.6	0.1	0.1	0	0	0	0.3	0.5	0.1	0	0	0	0.1
Cushion	Cu	0.1	0.1	0.3	0	0	0.3	0.2	0.1	0.1	0.4	0	0	0	0.4
PLC	P	0.3	0.3	0	0.2	0.1	0	0.1	0.3	0.1	0	0.3	0	0	0.3

environment and the vagueness of inherent subjective nature of human thinking (Gui-Wu et al. 2013).

The uncertainty in the nature of fuzzy problems makes the decision makers (DMs) find a solution so that both feasibility and optimality conditions can be satisfied efficiently (Tabrizi and Razmi 2013).

The principles of classifying the risk can be different according to the fuzzy numbers, their nomination, the purposes, aims of each class, etc. The classification presented in the Figure 3 is the risk fuzzy area, by assigning different classes of the system and probable failure. If an appropriate probability distribution cannot be identified for a given situation, it becomes extremely difficult to draw reliable inferences about the given domain of study under investigation (Sundaram and Ramya 2013). It is possible to make a better assessment of the existing conditions and consequently prioritize the controlling actions.

Using the probability and the consequence of risks classification system, it is possible to assess and analyze the risk on the basis of the potential consequences and the probability of their occurrences. From the integration of the above two factors, risk diagram of the danger is resulted which combines the factors in the tables of the consequence and the probability in order to provide a proper device for estimating acceptable level of the risk. With the provision of a two characteristics assessment system for risk occurrence on the basis of the consequence and the probability of the risk, it is possible to classify and assess the risk according to the degree of its acceptability which is called as the risk diagram of the

danger. In Figure 3, this diagram has been determined for two machines in this case study.

From the risk diagram, we find that the failure risk of the cushion for two dimension of square changing between moderate risk to major risk. Other results pertaining to computer is shown in diagram. Risk of computer and PLC is in the insignificant and minor area, therefore does not need improvement. Sometimes, we have to reduce consequence impact of failure, for example, we can decrease consequence impact with added redundancy. On cushion, we can only decrease failure consequence of cushion, therefore decrease risk of failure to moderate and minor risk.

Conclusions

In the discussion of risk management, we can make sound assessments in case we have a reasonable identification about the uncertainty. For identification, it is necessary to know all the factors and activities and understand all the relevant issues. Using the analysis process, the main topic on its factors and the analysis of each will lead to the identification of all related issues. The internal causes of failure include poor management, lack of risk management planning, and failure to adopt a risk limit threshold (Ariful and Des 2012). Most of time in an industry, we have the lack of data to calculate the reliability or make a decision. There are some main questions in any factory how risk-based methods can be used to optimal planning of future, and what the best model is to estimate and forecast future. Especially, a theoretical framework, models, and

Table 6 Calculate the consequence of failure from sources 1 and 2

Consequence impact	Computer			Cushion			PLC		
	Bel	Plu	U_{1-2}	Bel	Plu	U_{1-2}	Bel	Plu	U_{1-2}
Negligible	0.20	0.22	[0.2 to 0.22]	0.09	0.19	[0.09 to 0.19]	0.48	0.72	[0.48 to 0.72]
Main	0.75	0.76	[0.75 to 0.76]	0.12	0.37	[0.12 to 0.37]	0.28	0.48	[0.28 to 0.48]
Critical	0.04	0.04	[0.04 to 0.04]	0.54	0.79	[0.54 to 0.79]	0.00	0.07	[0 to 0.07]
Negligible-Main	0.96	0.96	[0.96 to 0.96]	0.21	0.46	[0.21 to 0.46]	0.93	1.00	[0.93 to 1]
Negligible-Critical	0.24	0.25	[0.24 to 0.25]	0.63	0.88	[0.63 to 0.88]	0.52	0.72	[0.52 to 0.72]
Main-Critical	0.78	0.80	[0.78 to 0.80]	0.81	0.91	[0.81 to 0.91]	0.28	0.52	[0.28 to 0.52]
Negligible-Main-Critical	1.00	1.00	[1 to 1]	1.00	1.00	[1 to 1]	1.00	1.00	[1 to 1]

Table 7 Consequence impact of failure and uncertainty interval (U)

	Computer	Cushion	PLC
Interval	[0.96 to 0.96]	[0.81 to 0.91]	[0.93 to 1]
U_{1-2}	Negligible-Main	Main-Critical	Negligible-Main

algorithms based on the probability theory are not capable to calculate risk, because there is a lack of data in a real situation. Reliability analyses should necessarily be a risk base linked with the losses from failures, in which decision on allocation reliability or reallocation in uncertainty condition on a basis of unknown data is new challenge. When failure time is unknown, loss of production will be occurred.

In the theoretical concept, we have applied tools, such as Dempster-Shafer theory. Dissimilarity assessment is a central problem in the DST, where the difference information content between several evidence should be quantified (Sarabi-Jamab et al. 2013).

Dempster-Shafer theory provides significant patterns about the risk and reliability, and it can be extracted from data originated in a factory. If we do not have enough available data, we can use qualitative and precise numbers together to calculate the risk. These results in discovering new knowledge about the failure risk to the factory. In this paper, due to the lack of information, we have introduced a method that determines a range for the consequence impact and calculates the probability of failure with relation between the risk and reliability. Decision maker should get to a relative certainty for assigning decisions. The uncertainty is classified under two main groups which are aleatory uncertainty and epistemic uncertainty. The major intention of this study is epistemic uncertainty which is due

to the lack or incompleteness of the correct data. One of the theories which are used for decision making in such conditions is the evidence theory or Dempster-Shafer theory.

Evidence theory provides a proper tool when there is incomplete or imperfect information. In this research, using the evidence theory, the uncertainty range of belief and plausibility has been achieved, and this range provides a measure for decision making in uncertainty condition that is due to incomplete information. Dempster-Shafer theory does not provide the limitations of a model and is a logical expression of ignorance. As in the assessment of real risks in the working environments, we encounter the problem of ignorance; it is possible to reduce the degree of ambiguity and increase the reliability of the results by integrating the qualitative risk assessment diagram and the Dempster-Shafer theory. In this investigation, the Dempster-Shafer theory was used for the identification in uncertainty conditions, and its findings were applied beside a risk diagram for specifying the risk of a production system. Indeed, its application in the production organizations was examined. In the qualitative approach, there is no estimation on the probability of failure but it could rank in the fuzzy logic categories; a computational intelligence technique can provide a convenient way to represent linguistic variables, subjective probability, and ordinal categories. Linguistic variables are designed to describe imprecise facts about a system and project. It is different from the frequency of repeatable events. Hence, subjective probability is a better way to represent risk as compared to quantitative objective probability of failure. Furthermore, fuzzy severity categories are more credible than numeric scores.

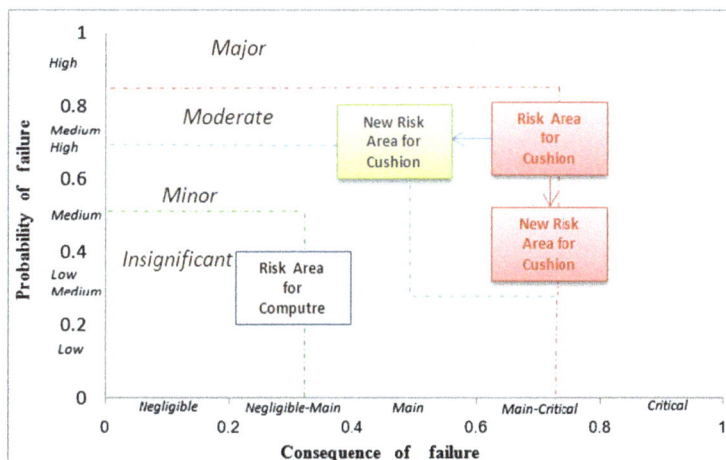

Figure 3 Square diagram of risk assessment for two essential parts in the fault tree.

Endnote
[a]MIL_STD_882 B.

Competing interests
The authors declare that there is no conflict of interests.

Authors' contributions
All authors MK, FK and AK, have made adequate effort on all parts of the work necessary for the development of this manuscript according to his/her expertise. All authors read and approved the final manuscript.

Acknowledgment
The authors would like to appreciate constructive comments and helpful suggestions from the associate editor and the referees which have helped improving the presentation of the paper.

Author details
[1]Department of Industrial Engineering, Robat Karim Branch, Islamic Azad University, Tehran, Iran. [2]Department of Statistics, Science and Research Branch, Islamic Azad University, Tehran, Iran. [3]Department of Management, Naragh Branch, Islamic Azad University, Naragh, Iran.

References
Ariful I, Des T (2012) Risk determinants of small and medium-sized manufacturing enterprises (SMEs) - an exploratory study. J Ind Eng Int 8:12

Baraldi P, Compare M, Zio E (2013) Maintenance policy performance assessment in presence of imprecision based on Dempster–Shafer Theory of Evidence. Inform Sci 245:112–131

Bauer M (1997) Approximation algorithms and decision making in the Dempster-Shafer theory of evidence - an empirical study. Int J Approx Reason 17:217–237

Bayes T (1763) An essay toward solving a problem in the doctrine of chances. Phil Trans Roy Soc (London) 53:370–418

Benferaht S, Saffiotti A, Smets P (2000) Belief functions and default reasoning. Artif Intell 122:1–69

Beynon M, Cosker D, Marshall D (2001) An expert system for multi-criteria decision making using Dempster-Shafer theory. Expert Syst Appl 20:357–367

Binaghi E, Madella P (1999) Fuzzy Dempster–Shafer reasoning for rule-based classifiers. Int J Intell Syst 14:559–583

Binaghi E, Gallo I, Madella P (2000) A neural model for fuzzy Dempster–Shafer classifiers. Int J Approx Reason 25(2):89–121

Biswas G, Oliff M, Sen A (1988) An expert decision support system for production control. Decis Support Syst 4:235–248

Buchanan BJ, Shortcliffe EH (1975) A model of inexact reasoning in medicine. Math Biosci 23:351–379

Buede D, Girardi P (1997) A target identification comparison of Bayesian and Dempster-Shafer multisensor fusion. IEEE Trans Syst Man Cybern A 27(5):569–577

Carazas FG, Souza GFM (2010) Risk-based decision making method for maintenance policy selection of thermal power plant equipment. Energy 35:964–975

Caselton WF, Luo W (1992) Decision making with imprecise probabilities: Dempster & Shafer theory and applications. Water Resour Res 28(12):3071–3078

Chang M-K, Chang R-R, Shu C-M, Lin K-N (2005) Application of risk based inspection in refinery and processing piping. J Loss Prev Process Ind 18:397–402

Dempster AP (1967) Upper and lower probabilities induced by a multi-valued mapping. Ann Math Stat 38:325–339

Denoeux T (1995) A k-nearest neighbour classification rule based on Dempster-Shafer theory. IEEE Trans Syst Man Cybern 25(5):804–813

Dubois D, Prade H (1988) Possibility theory: an approach to computerized processing of uncertainty, 1st edition. Plenum Press, New York

Faghihinia E, Mollaverdi N (2012) Building a maintenance policy through a multi-criterion decision-making model. J Ind Eng Int 8:14

Fedrizzi M, Kacprzyk J (1988) On measuring consensus in the setting of fuzzy preference relations, Non-conventional Preference Relations in Decision Making. 301:129–141

Gui-Wu W, Xiaofei Z, Rui L (2013) Some hybrid aggregating operators in linguistic decision making with Dempster–Shafer belief structure. Comp Ind Eng 65:646–651

Henley EJ, Kumamoto H (1981) Reliability engineering and risk assessment. Prentice-Hall, Englewood Cliffs, NJ

Ip HHS, Ng JMC (1994) Human face recognition using Dempster-Shafer theory. Int Conf Image Process 2:292–295

Khan FI, Haddara M (2003) Risk-based maintenance (RBM): A quantitative approach for maintenance/inspection scheduling and planning. J Loss Prev Process Ind 16:561–573

Klir GJ (1989) Is there more to uncertainty than some probability theorists might have us believe? Int J Gen Syst 15:347–378

Klir GJ, Wierman MJ (1998) Uncertainty based information: elements of generalized information theory. Physica-Verlag, Heidelberg

Knezevic J (1997) System maintainability, analysis, engineering and management. Chapman and Hall, London

Lee NS, Grize YL, Dehnad K (1987) Quantitative models for reasoning under uncertainty in knowledge-based expert systems. Int J Intell Syst 2:15–38

Liu J, Yang JB, Wang J, Sii HS, Wang YM (2004) Fuzzy rule-based evidential reasoning approach for safety analysis. Int J Gen Syst 33(2–3):183–204

Monney PA (2003) Analyzing linear regression models with hints and the Dempster–Shafer theory. Int J Intell Syst 18(1):5–29

Moubray J (1991) Reliability centered maintenance. Butterworth-Heinemann

Oberkampf WL, Helton JC, Sentz K (2001) Mathematical representation of uncertainty. American Institute of Aeronautics and Astronautics Non-Deterministic Approaches Forum, Seattle, WA, Paper No. 2001-1645

Parsons S (2001) Qualitative methods for reasoning under uncertainty. MIT Press, Cambridge, MA

Sarabi-Jamab A, Araabi BN, Augustin T (2013) Information-based dissimilarity assessment in Dempster–Shafer theory. Knowledge-Based Systems 54:114–127

Shafer G (1976) A mathematical theory of evidence. Princeton University Press, Princeton, NJ

Siow CHR, Yang JB, Dale BG (2001) A new modeling framework for organizational self-assessment: development and application. Qual Manag J 8(4):34–47

Sundaram S, Ramya B (2013) Distance-based test for uncertainty hypothesis testing. J Uncertainty Anal Appl 1:4

Tabrizi H, Razmi J (2013) A multi-period distribution network design model under demand uncertainty. J Ind Eng Int 9:13

Telmoudi A, Chakhar S (2004) Data fusion application from evidential databases as a support for decision making. Inf Softw Technol 46(8):547–555

Vose D (2003) Risk analysis: A quantitative guide. John Wiley & Sons, London, UK

Walley P (1987) Belief-function representations of statistical evidence. Ann Statist 15:1439–1465

Wang J, Yang JB (2001) A subjective safety based decision making approach for evaluation of safety requirements specifications in software development. Int J Reliab Qual Saf Eng 8(1):35–57

Wang YM, Yang JB, Xu DL (2006) Environmental impact assessment using the evidential reasoning approach. Eur J Oper Res 174(3):1885–1913

Wu WZ, Zhang M, Li HZ, Mi JS (2005) Knowledge reduction in random information systems via Dempster–Shafer theory of evidence. Inform Sci 174 (3–4):143–164

Yen J (1989) A Dempster-Shafer approach to diagnosing hierarchical hypotheses. Commun ACM 32(5):573–585

Zimmermann HJ (1991) Fuzzy set theory and its applications, 2nd edition. Kluwer, Dordrecht

11

Incorporating location, routing, and inventory decisions in a bi-objective supply chain design problem with risk-pooling

Reza Tavakkoli-Moghaddam[1*], Fateme Forouzanfar[2] and Sadoullah Ebrahimnejad[3]

Abstract

This paper considers a single-sourcing network design problem for a three-level supply chain. For the first time, a novel mathematical model is presented considering risk-pooling, the inventory existence at distribution centers (DCs) under demand uncertainty, the existence of several alternatives to transport the product between facilities, and routing of vehicles from distribution centers to customer in a stochastic supply chain system, simultaneously. This problem is formulated as a bi-objective stochastic mixed-integer nonlinear programming model. The aim of this model is to determine the number of located distribution centers, their locations, and capacity levels, and allocating customers to distribution centers and distribution centers to suppliers. It also determines the inventory control decisions on the amount of ordered products and the amount of safety stocks at each opened DC, selecting a type of vehicle for transportation. Moreover, it determines routing decisions, such as determination of vehicles' routes starting from an opened distribution center to serve its allocated customers and returning to that distribution center. All are done in a way that the total system cost and the total transportation time are minimized. The Lingo software is used to solve the presented model. The computational results are illustrated in this paper.

Keywords: Stochastic supply chain; Inventory control; Risk-pooling; Uncertainty; Capacity levels

Background

Nowadays, rapid economic changes and competitive pressure in the global market make companies pay more attention on supply chain topics. The company whose supply chain network structure is more appropriate has higher competitive advantage. This structure helps to overcome environmental disturbances (Dullaert et al. 2007). Analyzing location issues and decision making about facility location is considered as one of the important issues of decision making in companies. Certainly, appropriate facility location has high effects on economic benefits, appropriate service, and customer's satisfaction. Propounding the supply chain because of its effect on factors of operational efficiency, such as inventory, response, and lead time, specific attention is focused on how to create a distribution network. Facility location and how to relate them with customers are an important factor in designing a distribution network (Francis et al. 1992; Gong et al. 1997).

As nowadays living conditions have changed due to increasing world changes, mutually, situations have changed where supply chains are confronted with and influenced by them. The manager is confronted with more unknown conditions and new risks. Customers' demands have been more uncertain and various, and the lead time on their services is very effective. The demand variety can be recognized as one of the important sources of uncertainty in a supply chain (Gupta et al. 2000). Hence, inventory and product holding in a distribution center are an important issue in the supply chain (Liao et al. 2010). The inventory existence in these centers can lead to a great success in reaching the risk-pooling advantage in order to overcome the variability of customer demands. The proposed risk-pooling strategy and centralizing the inventory at distribution centers are considered as one of the effective ways to manage such a demand uncertainty to achieve appropriate service levels to customers. The lead time is one of the effective

* Correspondence: tavakoli@ut.ac.ir
[1]School of Industrial & Systems Engineering and Department of Engineering Science, College of Engineering, University of Tehran, Tehran, Iran
Full list of author information is available at the end of the article

factors in safety stock levels due to customer demand uncertainty (Park et al. 2010). For sure, whether the amount of the level is low for the product, it is considered an additional value that one can gain a long-term or short-term competitive advantage in the market.

The lead time is dependent on different factors, such as transportation mode. Different modes of transportation include a reverse relation between cost and time. It contains different routes for any type of vehicles (Cardona-Valdés et al. 2011). The implicit assumption is that a faster transportation mode is also the more expensive one, creating a trade-off between cost and time affecting the distribution network configuration. In the recent decades, the topic of multi-depot heterogeneous vehicle routing problem is presented in order to increase the productivity and efficiency of transportation systems, in which this model leads to the least cost function by minimizing the number of vehicles (Bettinelli et al. 2011).

Literature review

One of the important factors of the total productivity and profitability of a supply chain is to consider its distribution network, which can be used to achieve variety of the supply chain objectives. Designing a distribution network consists of three subproblems, namely, location allocation, vehicle routing, and inventory control. In the literature, there are some research studies amalgamating two of the above subproblems, such as location-routing problems, inventory-routing problems, and location-inventory problems. These three subproblems of a distribution network design are considered in few papers simultaneously. Location-routing problems are surveyed and classified by Min et al. (1998) and Nagy and Salhi (2007). Inventory-routing problems are studied in several studies (Zhao et al. 2008; Yu et al. 2008; Oppen and Loketangen 2008; Day et al. 2009). In addition, a number of studies have considered location-inventory problems (Daskin et al. 2002; Shen 2005). Finally, Ahmadi Javid and Azad (2010) presented a new model for a location-routing-inventory problem. They considered one objective for their model and did not consider transportation time and risk-pooling. However, in this paper, we present a multi-objective model to concurrently optimize location, allocation, capacity, inventory, selection of vehicles, and routing decisions with risk-pooling in a stochastic supply chain system for the first time. These decisions are made in a way that the total system cost and the total transportation time are minimized.

Problem formulation
Problem description

The trade-off between cost and time creates a bi-objective problem. One criterion tries to minimize the fixed cost of locating the opened distribution centers, the safety stock costs of distribution center by considering uncertainty in customer's demand, inventory ordering and holding costs, the transportation costs from a plant to its allocated distribution centers, and also vehicle routing costs beginning from a distribution center (DC) with the aim of replying to and covering the devoted customer's demands to the DC by considering risk-pooling. The other criterion looks for the reduction of the time to transport the product along the supply chain. It is desired to minimize the transportation time from a plant to customers. The important assumptions in this paper are as follows:

1. One kind of product is involved (Paksoy and Chang 2010).
2. Each distribution center j is assumed to follow the (Q_i, R_j) inventory policy (Ahmadi Javid and Azad 2010).
3. The inventory control is to be conducted only at DCs in this paper (Park et al. 2010; Ahmadi Javid and Azad 2010).
4. A single-sourcing strategy is considered in the whole supply chain (Park et al. 2010).
5. It is considered that the customers' demands after reaching the retailer are independent and follows a normal distribution (Park et al. 2010; Ahmadi Javid and Azad 2010).
6. Each plant has a limited capacity (Cardona-Valdés et al. 2011).
7. We consider different capacity levels for each distribution center, and finally, one capacity for each of them is selected (Ahmadi Javid and Azad 2010).
8. Each DC with the limited capacity carries on-hand inventory to satisfy demands from customer demand zones as well as safety stock to deal with the mutability of the customer demands at customer demand zones to attain risk-pooling profits (Park et al. 2010).
9. All customers should be served.
10. The number of available vehicles for each type and the number of allowed routes for each DC are limited (Bettinelli et al. 2011).
11. There are several modes of transportation between two consecutive levels.
12. Between two nodes on an echelon, only one type of vehicle is used.
13. A faster transportation mode is the more expensive one (Cardona-Valdés et al. 2011).
14. The amount of products is transported from each plant to each distribution center that is associated with it, and an equal amount of products has been ordered from the desired distribution center to that plant.

15. To determine all feasible routes, the following factors are taken into account:

- Each customer should be visited by only one vehicle.
- Each route begins at a DC and ends at the same DC.
- The sum of the demands of the customers served in each route must not exceed the capacity of the associated vehicle.
- Each of the distribution center and the vehicle have various limited, and determined capacity (Bettinelli et al. 2011; Marinakis and Marinaki 2010).

Model formulation

Following are the notations introduced for the mathematical description of the proposed model.

1. Indices
   (a) I, set of plants indexed by i
   (b) J, set of candidate DC locations indexed by j
   (c) K, set of customer demand zones indexed by k
   (d) N_j, set of capacity levels available to DC_j $(j \in J)$
   (e) Ω_{jl_2}, set of all feasible routes using a vehicle of type l_2 from DC_j $(j \in J)$
   (f) LP_{ij}, set of vehicles l_1 between nodes i and j
   (g) LW_{jk}, set of vehicles l_2 between nodes j and k
2. Parameters
   (a) F_j^n, yearly fixed cost for opening and operating distribution center j with capacity level n $(\forall\, n \in N_j, \forall\, j \in J)$
   (b) CP_{ijl_1}, cost of transporting one unit of product from plant i to distribution center j using vehicle l_1
   (c) $\mathbf{CW_{rl_2}}$, cost of sending one unit of product in route r using vehicle l_2 (These costs include the fixed cost of vehicle plus the transportation cost of each demand unit in route r. The mentioned transportation cost for each demand unit is not related to customer demand zone, and it is considered fixed for all locations in each route r.)
   (d) $\mathbf{TP_{ijl_1}}$, time for transporting any quantity of a product from plant i to DC_j using vehicle l_1
   (e) $\mathbf{TW_{jl_2r}}$, time for transporting any quantity of a product from DC_j on route r using vehicle l_2
   (f) λ_j, safety stock factor of DC_j $(j \in J)$
   (g) h_j, unit inventory holding cast at DC_j $(j \in J)$, (annually)
   (h) μ_k, mean demand at customer demand zone k
   (i) δ_k^2, variance of demand at customer demand zone k
   (j) E_j, fixed inventory ordering cost at DC_j
   (k) b_j^n, capacity with level n for DC_j
   (l) MP_i, capacity of plant i
   (l) ω_{l_2}, number of available vehicles of each type l_2
   (n) g_j, number of routes associated with each distribution center j

3. Binary coefficients
   (a) P_{kr}, 1 if and only if customer k is visited by route r, and 0 otherwise
4. Decision variables
   (a) U_j^n, 1 if distribution center j is opened with capacity level n, and 0 otherwise
   (b) A_{ijl_1}, binary variable equal to 1 if vehicle l_1 connecting plant i and DC_j is used, and equal to 0 otherwise
   (c) B_{jkl_2}, binary variable equal to 1 if vehicle l_2 connecting DC_j and customer k is used, and equal to 0 otherwise
   (d) X_r, 1 if and only if route r is selected, and 0 otherwise
   (e) X_{ijl_1}, quantity transported from plant i to DC_j using vehicle l_1
5. Mathematical model
   (a) The problem formulation is as follows:

$$\min f_1 = \sum_{n \in N_j}\sum_{j \in J} F_j^n U_j^n + \sum_{i \in I}\sum_{j \in J}\sum_{l_1 \in LP_{ij}} CP_{ijl_1}A_{ijl_1}X_{ijl_1}$$
$$+ \sum_{j \in J}\sum_{k \in K}\sum_{l_2 \in LW_{jk}}\sum_{r \in \Omega_{jl_2}} CW_{rl_2}\mu_k P_{kr}X_r$$
$$+ \sum_{i \in I}\sum_{j \in J}\sum_{k \in K}\sum_{l_1 \in LP_{ij}}\sum_{l_2 \in LW_{jk}} \frac{E_j B_{jkl_2}\mu_k}{X_{ijl_1}A_{ijl_1}}$$
$$+ \sum_{i \in I}\sum_{j \in J}\sum_{l_1 \in LP_{ij}} \frac{A_{ijl_1}X_{ijl_1}h_j}{2}$$
$$+ \sum_{j \in J}\lambda_j h_j \sqrt{\sum_{i \in I}\sum_{k \in K}\sum_{l_1 \in LP_{ij}}\sum_{l_2 \in LW_{jk}} \delta_k^2 B_{jkl_2}L_{jil_1}A_{ijl_1}}$$

$$\min f_2 = \max_j\left(\max_{i,l_1}\left(TP_{ijl_1}A_{ijl_1}\right) + \max_{r,l_2}\left(TW_{jl_2r}X_r\right)\right)$$

s.t.

$$\sum_{n \in N_j} U_j^n \leq 1 \quad \forall j \in J \tag{1}$$

$$\sum_{k \in K}\sum_{l_2 \in LW_{jk}} \mu_k B_{jkl_2} \leq \sum_{n \in N_j} b_j^n U_j^n \forall j \in J \tag{2}$$

Table 1 U_j^n is 1 if distribution center j is opened with capacity level n. and 0 otherwise

	DC1		DC2	
	Capacity 1	Capacity 2	Capacity 1	Capacity 2
Example 1	1	0	0	1
Example 2	0	1	1	0
Example 3	0	1	1	0
Example 4	1	0	0	1
Example 5	0	1	1	0

Table 2 X_r is 1 if and only if route r is selected, and 0 otherwise

	Route 1		Route 2		Route 3		Route 4		Route 5		Route 6	
	Truck	Airplane	Truck	Airplane	Truck	Airplane	Truck	Airplane	Truck	Airplane	Truck	Airplane
Example 1	1	0	0	1	0	0	0	0	-	-	-	-
Example 2	0	0	0	0	0	1	0	1	-	-	-	-
Example 3	0	0	0	0	0	0	0	0	0	1	0	1
Example 4	0	0	0	0	1	0	1	0	0	0	0	0
Example 5	1	0	1	0	0	0	0	0	-	-	-	-

$$\sum_{i\in I}\sum_{l_1\in \mathrm{LP}_{ij}} X_{ijl_1} + \lambda_j \sqrt{\sum_{i\in I}\sum_{k\in K}\sum_{l_1\in \mathrm{LP}_{ij}}\sum_{l_2\in \mathrm{LW}_{jk}} \delta_k{}^2 B_{ikl_2} L_{jil_1} A_{ijl_1}}$$
$$\leq \sum_{n\in N_j} b_j{}^n U_j{}^n \forall j \in J \tag{3}$$

$$\sum_{j\in J}\sum_{l_1\in \mathrm{LP}_{ij}} X_{ijl_1} \leq \mathrm{MP}_i \forall i \in I \tag{4}$$

$$\sum_{i\in I}\sum_{l_1\in \mathrm{LP}_{ij}} A_{ijl_1} \geq \sum_{n\in N_j} U_j{}^n \forall j \in J \tag{5}$$

$$\sum_{j\in J}\sum_{l_2\in \mathrm{LW}_{jk}} B_{jkl_2} = 1 \forall k \in K \tag{6}$$

$$\sum_{l_2\in \mathrm{LW}_{jk}} B_{jkl_2} \leq 1 \forall j \in J, \forall k \in K \tag{7}$$

$$\sum_{l_1\in \mathrm{LP}_{ij}} A_{ijl_1} \leq 1 \forall i \in I, \forall j \in J \tag{8}$$

$$\sum_{l_1\in \mathrm{LP}_{ij}}\sum_{i\in I} A_{ijl_1} \geq \sum_{l_2\in \mathrm{LW}_{jk}} B_{jkl_2} \forall j \in J, \forall k \in K \tag{9}$$

$$\sum_{n\in N_j} U_j{}^n \geq \sum_{l_2\in \mathrm{LW}_{jk}} B_{jkl_2} \forall j \in J, \forall k \in K \tag{10}$$

$$\sum_{j\in J}\sum_{l_2\in \mathrm{LW}_{jk}}\sum_{r\in \Omega_{jl_2}} P_{kr} X_r \geq 1 \forall k \in K \tag{11}$$

$$\sum_{j\in J}\sum_{r\in \Omega_{jl_2}} X_r \leq W_{l_2} \forall l_2 \in \mathrm{LW}_{jk} \tag{12}$$

$$\sum_{l_2\in \mathrm{LW}_{jk}}\sum_{r\in \Omega_{jl_2}} X_r \leq g_j \forall j \in J \tag{13}$$

$$X_{ijl_1} - A_{ijl_1} \geq 0 \forall i \in I, \forall j \in J, \forall l_1 \in \mathrm{LP}_{ij} \tag{14}$$

$$\mu_k \geq \sum_{l_2\in \mathrm{LW}_{jk}}\sum_{n\in N_j}\sum_{j\in J} B_{jkl_2} U_j{}^n \forall k \in K \tag{15}$$

$$\mathrm{MP}_i - A_{ijl_1} X_{ijl_1} \geq 0 \forall i \in I, \forall j \in J, \forall l_1 \in \mathrm{LP}_{ij} \tag{16}$$

$$U_j{}^n \in \{0,1\} \forall j \in J, \forall n \in N_j \tag{17}$$

$$X_r \in \{0,1\} \forall r \in \underset{j\in J, l_2\in \mathrm{LW}_{jk}}{\cup} \Omega_{jl_2} \tag{18}$$

$$A_{ijl_1} \in \{0,1\} \forall i \in I, \forall j \in J, \forall l_1 \in \mathrm{LP}_{ij} \tag{19}$$

$$B_{jkl_2} \in \{0,1\} \forall j \in J, \forall k \in K, \forall l_2 \in \mathrm{LW}_{jk} \tag{20}$$

$$X_{ijl_1} > 0 \forall i \in I, \forall j \in J, \forall l_1 \in \mathrm{LP}_{ij} \tag{21}$$

In this model, the first objective function minimizes the total expected costs consisting of the fixed cost for

Table 3 A_{ijl_1} binary variable, where it is 1 if vehicle l_1 connecting plant i and DC_j is used

	Plant 1				Plant 2			
	DC1		DC2		DC1		DC2	
	Train	Airplane	Train	Airplane	Train	Airplane	Train	Airplane
Example 1	1	0	0	0	0	0	1	0
Example 2	0	0	1	0	1	0	0	0
Example 3	0	1	0	0	0	0	0	1
Example 4	1	0	0	1	0	0	0	0
Example 5	0	0	0	0	1	0	0	1

Table 4 B_{jkl_2} binary variable, where it is 1 if vehicle l_2 connecting DC_j and customer k is used

	DC1						DC2					
	Demand 1		Demand 2		Demand 3		Demand 1		Demand 2		Demand 3	
	Truck	Airplane	Truck	Airplane	Truck	Airplane	Truck	Airplane	Truck	Airplane	Truck	Airplane
Example 1	0	0	1	0	1	0	0	1	0	0	0	0
Example 2	0	1	0	1	0	0	0	0	0	0	0	1
Example 3	0	1	0	0	0	0	0	0	0	1	0	1
Example 4	1	0	0	0	1	0	0	0	1	0	0	0
Example 5	1	0	1	0	0	0	0	0	0	0	1	0

opening distribution centers with a certain capacity level, transportation costs from plants to distribution centers, annual routing costs between distribution centers and customer demand zones, and expected annual inventory costs. The second objective function looks for the minimum time to transport the product along any path from the plants to the customers.

Constraint (1) ensures that each distribution center can be assigned to only one capacity level. Constraints (2) and (3) are the capacity constraints associated with the distribution centers, and also, constraint (4) is the capacity constraints associated with the plants. Constraint (5) states that if distribution center j with n capacity is opened, it is serviced by a plant. Constraint (6) represents the single-sourcing constraints for each customer demand zone. Constraints (7) and (8) ensure that if two nodes on an echelon are related to each other, one type of vehicle transports products between them. Constraint (9) makes sure that if the distribution center j gives the service to the customer k, that center must get services at least from a plant. Constraint (10) ensures that if the distribution center j is allocated to customer k by vehicle l_2, that center should certainly be established with a determined capacity level. Constraint (11) is the standard set covering constraints, modeling assumption 9. Constraints (12) and (13) impose limits on the maximum number of available vehicles of each type and the maximum number of allowed routes for each DC, modeling assumption 10. Constraint (14) makes sure that if plant i gives the service to the DC_j, the amount of

transported products from that plant to the desired distribution center would be more than one. Constraint (15) implies that customers' demands of zone k are more than 1. Constraint (16) is the capacity constraint associated with plant i. Constraints (17) to (20) enforce the integrality restrictions on the binary variables. Finally, constraint (21) enforces the non-negativity restrictions on the other decision variables.

Solution method

Optimization is a mathematical procedure to determine devoting the optimal allocation to scarce resources, and it helps to get the best result from the model solution. In this paper, we consider five examples, and then they are solved by the Lingo 9.0 software to show that this model works well. This software is a comprehensive tool designed to make building and solving linear, nonlinear, and integer optimization models faster, easier, and more efficient. It provides a completely integrated package that includes a powerful language for expressing optimization models, a full featured environment for building and editing problems, and a set of fast built-in solvers. Objective functions (f_i) have been normalized between zero and one. In other words, they have been without any dimension (i.e., scaleless). By using the following formula, these objectives are converted to a single objective function, where f_1' and f_2' are the normalized forms of f_1 and f_2 objective functions.

Table 5 X_{ijl_1} quantity transported from plant i to DC_j using vehicle l_1

	Plant 1				Plant 2			
	DC1		DC2		DC1		DC2	
	Train	Airplane	Train	Airplane	Train	Airplane	Train	Airplane
Example 1	4	0	0	0	0	0	1	0
Example 2	0	0	2	0	8	0	0	0
Example 3	0	6	0	0	0	0	0	2
Example 4	5	0	0	6	0	0	0	0
Example 5	0	0	0	0	6	0	0	1

$$\min f = \alpha f_1{}' + (1-\alpha)f_2{}'$$

To minimize deviations from the ideal, this function is reduced. As the first objective function (f_1) is more important than the second objective function (f_2) in the given problem, the coefficients of the above formula are considered in the form of $\alpha = 0.7$ and $1 - \alpha = 0.3$.

This problem is implemented by this software, and a global optimal solution is obtained. The computational results are shown in Tables 1, 2, 3, 4, and 5.

Conclusions

In this paper, a new mathematical model to design a three-level supply chain has been presented by considering the inventory under uncertain demands, risk-pooling, vehicle routing, transportation time, and cost. The decision related to the transportation options has an impact on the transportation time from plants to customers. The trade-off between cost and time is considered in the formulation of a mathematical model that minimizes both criteria. Therefore, this model holding two objectives has been formulated for the first time as a location-inventory-routing problem with a risk-pooling strategy in a three-level supply chain. The Lingo software has been used to solve the given problem. Some future studies are as follows: considering each parameter as a fuzzy, multi-period planning and solving the presented model by using heuristic or meta-heuristic algorithms.

Competing interests
The authors declare that they have no competing interests.

Authors' contributions
RTM managed the study and was responsible for integrating and revising the manuscript. FF contributed to the development of goals, designed the mathematical model, and drafted the manuscript. SE solved the presented model. All authors read and approved the final manuscript.

Acknowledgements
The authors would like to thank autonomous reviewers for their valuable comments that improve the quality and presentation of this paper.

Author details
[1]School of Industrial & Systems Engineering and Department of Engineering Science, College of Engineering, University of Tehran, Tehran, Iran. [2]School of Industrial Engineering, South Tehran Branch, Islamic Azad University, Tehran, Iran. [3]Department of Industrial Engineering, Karaj Branch, Islamic Azad University, Karaj, Iran.

References
Ahmadi Javid A, Azad N (2010) Incorporating location, routing and inventory decisions in supply chain network design. Transport Res E Logist Transport Rev 46:582–597

Bettinelli A, Ceselli A, Righini G (2011) A branch-and-cut-and-price algorithm for the multi-depot heterogeneous vehicle routing problem with time windows. Transport Res C Emerg Tech 19:723–740

Cardona-Valdés Y, Álvarez A, Ozdemir D (2011) A bi-objective supply chain design problem with uncertainty. Transport Res C Emerg Tech 19(5):821–832

Daskin M, Coullard C, Shen ZJ (2002) An inventory–location model: formulation, solution algorithm and computational results. Ann Oper Res 110:83–106

Day JM, Wright PD, Schoenherr T, Venkataramanan M, Gaudette K (2009) Improving routing and scheduling decisions at a distributor of industrial gasses. Omega 37:227–237

Dullaert W, Braysy O, Goetschalckx M, Raa B (2007) Supply chain (re)design: support for managerial and policy decisions. Eur J Transport Infrastruct Res 7(2):73–91

Francis RL Jr, McGinnis LF, White JA (1992) Facility layout and location: an analytical approach, 2nd edn. Prentice-Hall, Englewood Cliffs

Gong D, Gen M, Yamazaki G, Xu W (1997) Hybrid evolutionary method for capacitated location-allocation problem. Comput Ind Eng 33:577–580

Gupta A, Maranas CD, McDonald CM (2000) Mid-term supply chain planning under demand uncertainty: customer demand satisfaction and inventory management. Comput Chem Eng 24:2613–2621

Liao CJ, Lin Y, Shih SC (2010) Vehicle routing with cross-docking in the supply chain. Expert Syst Appl 37:6868–6873

Marinakis Y, Marinaki M (2010) A hybrid genetic-particle swarm optimization algorithm for the vehicle routing problem. Expert Syst Appl 37:1446–1455

Min H, Jayaraman V, Srivastava R (1998) Combined location–routing problems: a synthesis and future research directions. Eur J Oper Res 108:1–15

Nagy G, Salhi S (2007) Location-routing, issues, models, and methods: a review. Eur J Oper Res 117:649–672

Oppen J, Loketangen A (2008) A tabu search approach for the livestock collection problem. Comput Oper Res 35:3213–3229

Paksoy T, Chang C-T (2010) Revised multi-choice goal programming for multi-period, multi-stage inventory controlled supply chain model with popup stores in guerrilla marketing. Appl Math Model 34:3586–3598

Park S, Lee T-E, Sup Sung C (2010) A three-level supply chain network design model with risk-pooling and lead times. Transport Res E Logist Transport Rev 46:563–581

Shen ZJ (2005) A multi-commodity supply chain design problem. IIE Transactions 37:753–762

Yu Y, Chen H, Chu F (2008) A new model and hybrid approach for large scale inventory routing problems. Eur J Oper Res 189:1022–1040

Zhao Q, Chen S, Zhang C (2008) Model and algorithm for inventory/routing decision in a three-echelon logistics system. Eur J Oper Res 191:623–635

12

Integration of QFD, AHP, and LPP methods in supplier development problems under uncertainty

Zahra Shad[1], Emad Roghanian[1] and Fatemeh Mojibian[2*]

Abstract

Quality function deployment (QFD) is a customer-driven approach, widely used to develop or process new product in order to maximize customer satisfaction. Last researches used linear physical programming (LPP) procedure to optimize QFD; however, QFD issue involved uncertainties, or fuzziness, which requires taking them into account for more realistic study. In this paper a set of fuzzy data is used to address linguistic values parameterized by triangular fuzzy numbers. The proposed integrated approach includes analytic hierarchy process (AHP), QFD, and LPP to maximize overall customer satisfaction under uncertain conditions and apply them in the supplier development problem. The fuzzy AHP (FAHP) approach is adopted as a powerful method to obtain the relationship between the customer requirements (CRs) and engineering characteristics (ECs) to construct the house of quality (HOQ) in QFD method. LPP is used to obtain the optimal achievement level of the ECs and subsequently the customer satisfaction level under different degrees of uncertainty. The effectiveness of proposed method will be illustrated by an example.

Keywords: Quality function deployment; Fuzzy; Analytic hierarchy process; Linear physical programming; Supplier development

Introduction

The increasing global competition and cooperation and the vertical disintegration of production activities have created the logistical challenge of coordinating the entire supply chain (SC) effectively, in upstream to downstream activities (Gebennini et al. 2009). Supply chain management (SCM) integrates suppliers, manufacturers, distributors, and customers to meet final consumer needs and expectations efficiently and effectively (Cox 1999).

Quality function deployment (QFD) was developed by Yoji Akao in the 1960s. The basis of QFD is to obtain and translate customer requirements into engineering characteristics and subsequently, into part characteristics, process plans, and production requirements. This paper concentrated on the house of quality (HOQ) which translates customer requirements into the engineering characteristics. By better managing the SC, companies can increase their customers' satisfaction and

achieve sustainable business success. SC has different levels and each level can be considered as a customer of the previous level in which customer satisfaction should be maximized. QFD can be used as a useful method to translate the requirements of each level to the engineering characteristics (ECs) of the previous level. The analytic hierarchy process (AHP) method can be used as a powerful multi-criteria tool to extract the relationships between the requirements of each level and ECs of the previous level. Humans are often uncertain in assigning the evaluation scores in crisp AHP, so fuzzy analytic hierarchy process (FAHP) can capture this difficulty. Although QFD implementation has extended recently, a few researchers focused in the supply chain (e.g., Zarei et al. 2011, Hassanzadeh Amin and Razmi 2009).

Satisfying customer requirement is a multi-objective optimization problem. Different optimization methods have been applied in the field of QFD to maximize customer satisfaction. Mathematical programming is one of these optimization methods. The linear programming model is used to maximize the overall customer satisfaction (e.g., Chen and Ko 2009; Lai et al. 2007). Park

* Correspondence: F.mojibian@modares.ac.ir
[2]Department of Management and Economics, Tarbiat Modares University, Jalal Ale Ahmad Highway, P.O.BOX: 14115-111, Tehran, Iran
Full list of author information is available at the end of the article

and Kim (1998) used integer programming to optimize product design in the QFD. Chen and Weng (2006) used goal programming to determine the fulfillment levels of the design requirements in the QFD. Delice and Güngör (2009) applied mixed integer linear programming (MILP) to acquire the optimized solution of alternative customer requirements (CRs). Chen and Ko (2010) consider the close link between the four phases using the means-end chain (MEC) concept to build up a set of fuzzy linear programming models to determine the contribution levels of each 'how' for customer satisfaction.

Bhattacharya et al. (2010) present a concurrent engineering approach integrating AHP with QFD in combination with cost factor measure (CFM) which has been delineated to rank and subsequently selects candidate-suppliers under multiple conflicting-in-nature criteria environment within a value-chain framework. Raissi et al. (2012) prioritize engineering characteristic in QFD using fuzzy common set of weight. Lai et al. (2006) used linear physical programming (LPP) as an effective multi-objective optimization method to optimize QFD. In this paper we extended Lai et al.'s (2006) approach by using fuzzy numbers instead of the crisp numbers to build HOQ. We used HOQ with triangular fuzzy numbers to extract mathematical model to deal with the fuzziness of the problem to achieve the optimal values of the ECs under different degrees of uncertainty.

Due to the high importance of the SCM, the aim of this paper is to develop a useful approach by integrating fuzzy AHP, fuzzy QFD (FQFD), and LPP to obtain the optimal values of the ECs of the suppliers. Supplier development is an important issue in the context of the SCM. Also, supplier development is a multi-criterion decision making (MCDM) problem which includes both qualitative and quantitative factors (e.g., Xia and Wu 2007; Chan and Kumar 2007).

In this section literature review of QFD, fuzzy AHP, and LPP methods, and applying LPP with QFD and fuzzy linear programming are presented. In Section 'Proposed methodology,' we present the proposed methodology and illustrated it in solving a numerical example in Section 'Numerical example'. In Section 'Discussion of results,' the obtained results are discussed and finally in the last section (Section 'Conclusion'), the conclusion is presented.

Quality function deployment

QFD aims at identifying the customers together with their demands for the product, which are translated into product characteristics. QFD methodology has introduced twofold principles in product development. First, the needs of the customer should be carefully considered during the development process, Secondly,

the importance of the different product characteristics should be analyzed and ranked (Bevilacqua et al. 2006).

Many researchers applied QFD to present new product or to improve product design as follows: Fung et al. (2005) applied an asymmetric fuzzy linear regression approach to estimate the functional relationships for product planning based on QFD. Kahraman et al. (2006) proposed a fuzzy optimization model based on FQFD to determine the product engineering requirements in designing a product. Soota et al. (2011) propose a method to foster product development using combination of QFD and analytic network process (ANP). Sener and Karsak (2011) combined fuzzy linear regression and fuzzy multiple objective programming for setting target levels in the QFD. Based on the Kano's category of design requirements, Chen and Ko (2008) presented a fuzzy nonlinear model to determine the performance level of each design requirements to maximize customer satisfaction. Raharjo et al. (2008) applied AHP to overcome the priorities change over time in the QFD. Sharma and Rawani (2008) develop a post-HoQ model through a well-defined and structured approach to comprehensive matrix and SWOT analysis. Raissi et al. (2011) proposed a novel methodology using common set of weight (CSW) method as a well-known technique in DEA to aggregate each of the requirements expressed by customers and comparisons among the product produced by own company with competitive products.

In the supply chain field, researchers used QFD as an effective decision making tool as follows: Bottani and Rizzi (2006) proposed a FQFD approach to deploy HOQ to efficiently and effectively improve the logistic process. Bottani (2009) presented an original approach to show the applicability of the QFD methodology to enhance the agility of enterprises. Zarei et al. (2011) studied QFD application to identify viable lean enabler for increasing the leanness of food chain. Yousefi et al. (2011) propose an original approach for the management tools selection based on the quality function deployment approach, a methodology which has been successfully adopted in development of new products.

Fuzzy analytic hierarchy process

AHP is a decision support tool that can adequately represent qualitative and subjective assessments under the multiple criteria decision making environment. AHP is strongly connected to human judgment, and pair-wise comparisons in AHP may cause an assessment bias of the evaluator, which makes the comparison judgment matrix inconsistent (Aydogan 2011). Because of this problem, using the fuzzy set theory can solve evaluation bias problem in AHP. Various applications of the FAHP can be found to solve MCDM problems. Kahraman et al. (2004) used FAHP to compare catering firms. Chan and Kumar (2007) applied FAHP for solving the global supplier selection problem. Haghighi et al. (2010) applied FAHP to prioritize factors that impact electronic banking development in Iran. Rung Yu and Shing (2013) propose a two-stage fuzzy logarithmic preference programming with multi-criteria decision making in order to derive the priorities of comparison matrices in the AHP and the ANP.

Different methods of FAHP were employed to extract the weight of criteria based on pair-wise comparison matrices. Extent analysis method proposed by Chang (1992, 1996) is a popular approach to determine the weight of criteria (e.g., Kahraman et al. 2004; Haghighi et al. 2010).

Geometric mean technique proposed by Buckley (1985) also was used to define the fuzzy geometric mean and fuzzy weights of each criterion (e.g., Chen et al. 2008; Güngör et al. 2009). After constructing pair-wise comparison matrices, $(\tilde{\mathbf{D}})$ according to geometric mean technique by using Equations 5 and 6, we can define the fuzzy weights of each criterion as following:

$$\tilde{\mathbf{D}} = \begin{bmatrix} 1 & \tilde{d}_{12} \cdots & \tilde{d}_{1n} \\ \vdots & \ddots & \vdots \\ \tilde{d}_{n1} & \tilde{d}_{n2} \cdots & 1 \end{bmatrix} = \begin{bmatrix} 1 & \tilde{d}_{12} \cdots & \tilde{d}_{1n} \\ \vdots & \ddots & \vdots \\ 1/\tilde{d}_{1n} & 1/\tilde{d}_{2n} \cdots & 1 \end{bmatrix}$$

(1)

where $\tilde{d}_{ij} = \begin{cases} triangular\ fuzzy\ number, & i \neq j \\ 1 & i = j \end{cases}$.

A fuzzy number $\tilde{d}$ on $\mathbb{R}$ to be a triangular fuzzy number if its membership function $\mu_{\tilde{d}}(x) : \mathbb{R} \rightarrow [0,1]$ can be defined by the following equation:

$$\mu_{\tilde{d}}(x) = \begin{cases} \dfrac{x - d^l}{d^m - d^l}, & d^l \leq x \leq d^m \\ \dfrac{d^r - x}{d^r - d^m}, & d^m \leq x \leq d^r \\ 0 & otherwise. \end{cases}$$

(2)

Let $\tilde{a}$ and $\tilde{b}$ be two triangular fuzzy numbers parameterized by the triplet (a_1, a_2, a_3) and (b_1, b_2, b_3), respectively, then the operational laws of these two triangular fuzzy numbers are as follows:

$$\tilde{a} \oplus \tilde{b} = (a_1, a_2, a_3) \oplus (b_1, b_2, b_3) = (a_1 + b_1, a_2 + b_2, a_3 + b_3)$$

(3)

$$\tilde{a} \otimes \tilde{b} = (a_1, a_2, a_3) \otimes (b_1, b_2, b_3) \cong (a_1 \times b_1, a_2 \times b_2, a_3 \times b_3)$$

(4)

$$\tilde{r}_{ij} = \left(\tilde{d}_{i1} \otimes \cdots \otimes \tilde{d}_{ij} \otimes \cdots \otimes \tilde{d}_{in} \right)^{\frac{1}{n}}$$

(5)

and the normalized weight of each criterion is obtained as follows:

$$\tilde{r}_{ij}{}' = \tilde{r}_{ij} \otimes \left(\tilde{r}_{i1} \oplus \cdots \oplus \tilde{r}_{ij} \oplus \cdots \oplus \tilde{r}_{in} \right)^{-1}$$

(6)

In this paper the normalized fuzzy weights are used to construct fuzzy HOQ of the QFD.

Linear physical programming (LPP)

LPP is a multi-objective optimization method that develops an aggregate objective function of the criteria in a piece-wise, Archimedean-goal-programming fashion. The physical programming approach in its nonlinear (general) form was developed by Messac (1996) and in its piece-wise linear form, LPP, provides the means for Decision makers (DMs) to express his/her priority with respect to each criterion using four classes, i.e., the Decision maker (DM) declares each criterion as belonging to one of four distinct classes. Class functions allowed the Decision makers (DMs) to express the ranges of differing levels of preference for each criterion. A criterion falls into one of four classes of penalty functions, hereby called class functions, which are defined as follows:

Class 1S smaller-is-better, i.e., minimization
Class 2S larger-is-better, i.e., maximization
Class 3S value-is-better
Class 4S range-is-better

LPP has been used in several diverse applications. Maria et al. (2003) used LPP in production planning. Melachrinoudis et al. (2005) propose a LPP model which enables a decision maker to consider multiple criteria (i.e., cost, customer service, and intangible benefits) and to express criteria preferences not in a traditional form of weights, but in ranges of different degrees of desirability.

Tian and Zuo (2006) proposed a multi-objective optimization model by using physical programming for redundancy allocation for multi-state series–parallel systems.

Applying LPP with QFD

By applying LPP, the satisfaction level of each customer requirement is classified in one of six different ranges (ideal range, desirable range, tolerable range, undesirable range, highly undesirable range, and unacceptable range). According to the proposed methodology by Lai et al. (2006), each engineering characteristic usually needs cost for improvement. So the last row of the HOQ is the cost index for each engineering characteristic. $X_j = (j = 1, 2,..., q)$ is defined as the value of the engineering characteristic j. The normalized value of engineering characteristic j is defined as follows:

$$x_j = X_j / \max\{X_j\} \quad \text{and} \quad 0 \leq x_j \leq 1.$$

(7)

The proposed algorithm by Messac et al. (1996) to obtain the weights of the different ranges is as follows: The value of a class function z_i at the intersection of given ranges is the same for any customer requirement. The loss function z_i ($i = 1, 2, ..., p$) is defined in LPP and can be viewed as a loss of customer satisfaction. z_s is defined as the value of class function at range intersection s. It can be expressed mathematically as follows:

$$z_s \equiv z_i(t_{is}). \tag{8}$$

t_{is} is the limit of different ranges, and s denotes a range. z_s is a constant for all i and $\tilde{z}^s$ and is defined as follows:

$$\tilde{z}^s \equiv z^s - z^{s-1} \qquad (2 \leq s \leq 5) \tag{9}$$

$$z^1 \equiv 0. \tag{10}$$

According to the LPP method, we can define $\tilde{z}^s$ as follows:

$$\tilde{z}^s = \beta(p-1)\tilde{z}^{s-1} \quad (3 \leq s \leq 5) \tag{11}$$

where p denotes the number of customer requirements, and β is the convexity parameter. t_{is} is defined as follows:

$$\tilde{t}_{is} = t_{i(s-1)} - t_{is} \qquad (2 \leq s \leq 5). \tag{12}$$

The importance weight of each customer satisfaction level is given by

$$w_{is} = \tilde{z}^s / \tilde{t}_{is} \qquad (2 \leq s \leq 5) \tag{13}$$

$$w_{i1} = 0. \tag{14}$$

The importance weight of each range for every customer requirement can be calculated as follows:

$$\tilde{w}_{is} = w_{is} - w_{i(s-1)} \qquad (2 \leq s \leq 5). \tag{15}$$

Finally, by solving the following proposed mathematical model by Lai et al. (2006), the optimal achievement level of the each EC, allocated budget to each EC, and CRs satisfaction level can be determined:

$$\min_{d_{is}^-, x} \sum_{i=1}^{p} \sum_{s=2}^{5} (\tilde{w}_{is} d_{is}^-) \tag{16}$$

subject to

$$\sum_{j=1}^{q} r_{ij} x_j + d_{is}^- \geq t_{i(s-1)} \quad i = 1, ..., p \quad s = 2, ..., 5 \tag{17}$$

$$\sum_{j=1}^{q} c_j x_j \leq B \tag{18}$$

$$d_{is}^- \geq 0 \qquad i = 1, ..., p \qquad s = 2, ..., 5 \tag{19}$$

$$0 \leq x_j \leq 1 \qquad J = 1, ..., q. \tag{20}$$

The deviational variable, denoted by d_{is}^-, can be viewed as the distance from the value of the performance rating of customer requirement i under consideration to $t_{i(s-1)}$, starting from the left-hand side. C_j is the cost of unit improvement of the engineering characteristic, and B is the cost limit for improvement for all of the engineering characteristics.

Fuzzy linear programming

Linear programming (LP) is the optimization technique most frequently applied in real-world problems. Any linear programming model representing real-world situations involves a lot of parameters whose values are assigned by experts, so some of these parameters or whole of them can be fuzzy. In this paper, for solving the fuzzy mathematical model, we use Jiménez's approach. According to Jiménez (1996), the expected interval (EI) of triangular fuzzy number $\tilde{d}$ can be defined as follows:

$$EI(\tilde{d}) = \left[E_1^d, E_2^d\right] = \left[\frac{1}{2}(d^l + d^m), \ \frac{1}{2}(d^m + d^r)\right]. \tag{21}$$

Moreover, according to the ranking method of Jiménez (1996) for any pair of fuzzy numbers $\tilde{a}$ and $\tilde{b}$, the degree in which $\tilde{a}$ is bigger than $\tilde{b}$ is defined as follows:

$$\mu_M(\tilde{a}, \tilde{b}) = \begin{cases} 0 & \text{if } E_2^a - E_1^b < 0 \\ \dfrac{E_2^a - E_1^b}{E_2^a - E_1^b - (E_1^a - E_2^b)} & \text{if } 0 \in \left[E_1^a - E_2^b, E_2^a - E_1^b\right] \\ 1 & \text{if } E_1^a - E_2^b > 0. \end{cases} \tag{22}$$

When $\mu_M(\tilde{a}, \tilde{b})$, it will demonstrate that $\tilde{a}$ is bigger than, or equal, to $\tilde{b}$ at least in a degree α and it will be represented it by $\tilde{a} \geq_\alpha \tilde{b}$ for two types of the constraints as the following:

$$\tilde{a}_i x \geq \tilde{b}_i \qquad i = 1, ..., m \tag{23}$$

$$\tilde{a}_i x \leq \tilde{b}_i \qquad i = m + 1, ..., t \tag{24}$$

Table 1 Triangular fuzzy conversion scale

Linguistic scale	Triangular fuzzy scale	Triangular fuzzy reciprocal scale
Equal	(1,1,1)	(1,1,1)
Weak	(2/3,1,3/2)	(2/3,1,3/2)
Fairly strong	(3/2,2,5/2)	(2/5,1/2,2/3)
Very strong	(5/2,3,7/2)	(2/7,1/3,2/5)
Absolute	(7/2,4,9/2)	(2/9,1/4,2/7)

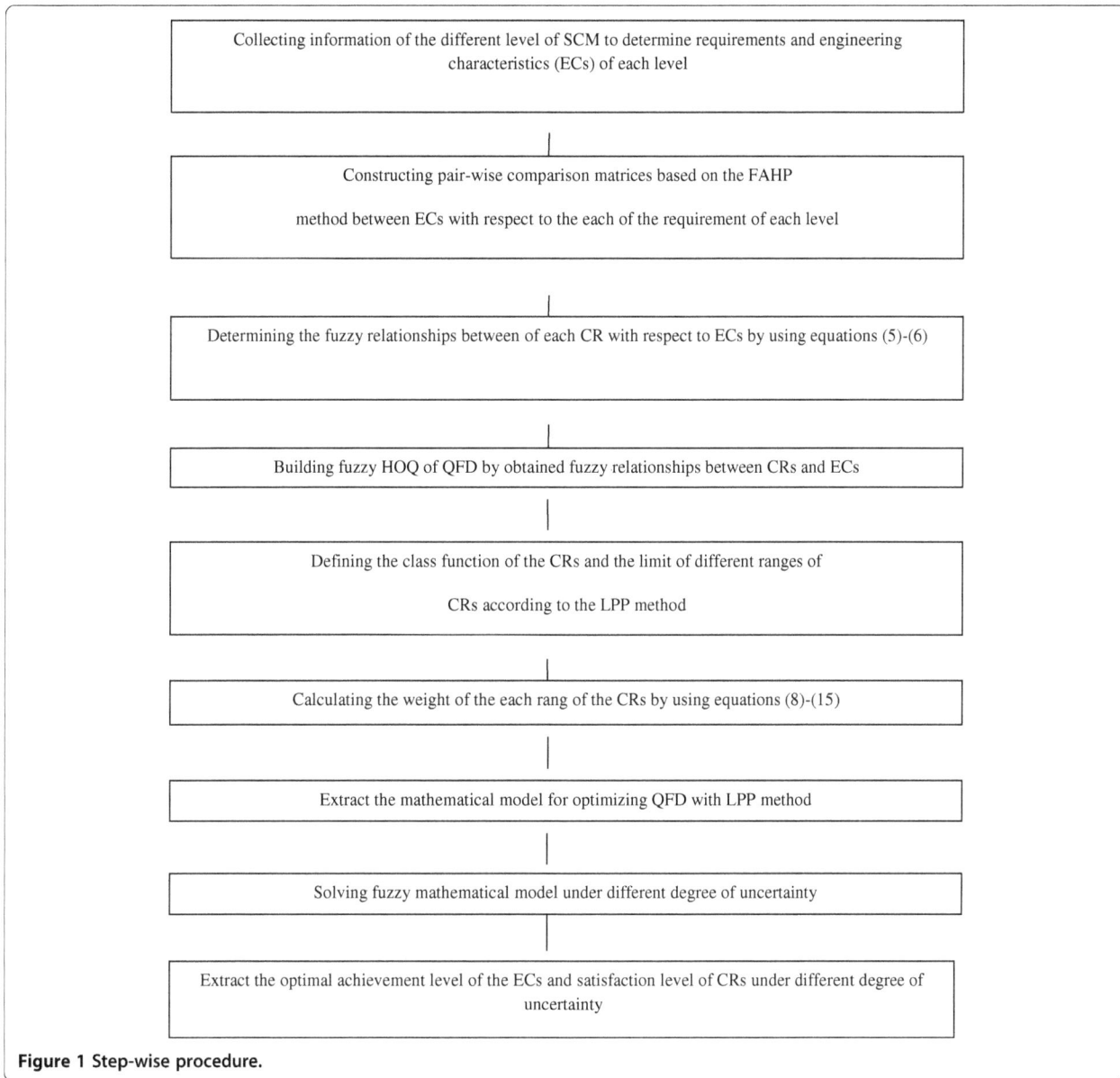

Figure 1 Step-wise procedure.

According to the Jiménez et al. (2007), a decision vector $x \in \mathfrak{R}^n$ is feasible in degree α if $\min_{i=1,\dots,m} = \{\mu_M(\tilde{a}_i x, b_i)\} = \alpha$. According to Equation 20, the equation $\tilde{a}_i x \geq b_i$ is equivalent to the following:

$$\frac{E_2^{a_i x} - E_1^{b_i}}{E_2^{a_i x} - E_1^{a_i x} + E_2^{b_i} - E_1^{b_i}} \geq \alpha \qquad i = 1, \dots, m. \qquad (25)$$

So the equation can be rewritten as follows:

$$\left[(1-\alpha)E_2^{a_i} + \alpha E_1^{a_i}\right] x \geq \alpha E_2^{b_i} + (1-\alpha) E_1^{b_i} \qquad i = 1, \dots, m. \qquad (26)$$

We can do this for $\tilde{a}_i x \leq b_i$, so this is equation equivalent to the following respectively:

Table 2 Important CRs and ECs

Customer requirements	Engineering characteristics
Cost	EF = experience of the sector
Conformity	IN = capacity for innovation to follow up the customer's evolution in terms of changes in its strategy and market
Punctuality	SQ = quality system certification
Efficacy	FL = flexibility of response to the customer's requests
Lead time	RR = ability to manage orders on-line (EDI-system)

Table 3 Pair-wise comparison matrix between the engineering characteristics with respect to the cost

Cost	EF	IN	SQ	FL	RP
EF	(1,1,1)	(1,1,1)	(3/2,2,5/2)	(2/3,1,3/2)	(2/3,1,3/2)
IN	(1,1,1)	(1,1,1)	(2/5,1/2,2/3)	(1,1,1)	(2/3,1,3/2)
SQ	(2/5,1/2,2/3)	(3/2,2,5/2)	(1,1,1)	(3/2,2,5/2)	(1,1,1)
FL	(2/3,1,3/2)	(1,1,1)	(2/5,1/2,2/3)	(1,1,1)	(2/3,1,3/2)
RP	(2/3,1,3/2)	(2/3,1,3/2)	(1,1,1)	(2/3,1,3/2)	(1,1,1)

$$\left[\alpha E_2^{a_i} + (1-\alpha) E_1^{a_i} \right] x \leq \alpha \, E_1^{b_i} + (1-\alpha) \, E_2^{b_i} \quad i = m+1, \ldots, t. \tag{27}$$

In this paper Jiménez's approach is used to solve the mathematical model.

Proposed methodology

Because of the ambiguity and fuzziness of the real-world problems, crisp number cannot deal with the problem carefully. We extended Lai et al.'s (2006) proposed methodology by combining FAHP method to construct HOQ with the fuzzy numbers. Triangular fuzzy number in Table 1 is used for weighting the ECs with respect to the each CR. So Equation 17 is converted to the following equation:

$$\sum_{j=1}^{q} \tilde{r}_{ij}' x_j + d_{is}^- \geq t_{i(s-1)} \quad i = 1, \ldots, p \quad s = 2, \ldots, 5 \tag{28}$$

where $\tilde{r}_{ij}'$ is triangular fuzzy number which is obtained by geometric mean method based on the pair-wise comparison according to FAHP. We use Jiménez's approach to solve the mathematical model. In Figure 1, the step-wise procedure of the proposed methodology is shown.

Numerical example

We illustrate our proposed methodology step by step by solving an example of supplier development:

Step 1. Information about company requirements and characteristics of the suppliers to satisfy these requirements are collected. Important CRs and ECs are shown in Table 2.

Step 2. Pair-wise comparison matrices based on the FAHP method between ECs with respect to the each of the CRs are constructed. For example, the relationship between the engineering characteristics with respect to the cost is shown in Table 3. Similarly, other pair-wise comparison matrices can be obtained.

Step 3. Fuzzy relationships of each CR with respect to ECs by using Equations 3, 4, 5, and 6 according to the geometric mean method are determined. For example, the fuzzy relationships between the first requirement and ECs are determined as follows:

$$\tilde{r}_{11} = \left(\tilde{d}_{11} \otimes \tilde{d}_{12} \otimes \tilde{d}_{13} \otimes \tilde{d}_{14} \otimes \tilde{d}_{15} \right)^{1/5}$$

$$\tilde{r}_{11} = \left((1 \times 1 \times \ldots \times 2/3)^{1/5}, \quad (1 \times 1 \times \ldots \times 1)^{1/5}, \right.$$
$$\left. (1 \times 1 \times \ldots \times 3/2)^{1/5} \right) = (0.922, 1.149, 1.413)$$

Similarly, we can compute the remaining $\tilde{r}_{ij}$, which are the following:

$$\tilde{r}_{12} = (0.708, 0.871, 1.084), \quad \tilde{r}_{13} = (0.979, 1.149, 1.33),$$
$$\tilde{r}_{14} = (0.653, 0.871, 1.176), \quad \tilde{r}_{15} = (0.784, 0.871, 1.275).$$

We normalized the calculated weights as follows:

$$\tilde{r}_{11}' = \tilde{r}_{11} \otimes (\tilde{r}_{11} \oplus \tilde{r}_{12} \oplus \tilde{r}_{13} \oplus \tilde{r}_{14} \oplus \tilde{r}_{15})^{-1}$$
$$\tilde{r}_{11}' = (0.922, \ 1.149, \ 1.413) \otimes \Big((0.922, \ 1.149, \ 1.413)$$
$$\oplus \ldots \oplus 0.784, \ 0.871, \ 1.275) \Big)^{-1}$$
$$= (0.1, \ 0.15, \ 0.23).$$

The remaining $\tilde{r}_{ij}'$ values are as follows:

$$\tilde{r}_{12}' = (0.1, \ 0.15, \ 0.23), \quad \tilde{r}_{13}' = (0.15, \ 0.21, \ 0.28),$$
$$\tilde{r}_{14}' = (0.09, \ 0.15, \ 0.23), \quad \tilde{r}_{15}' = (0.11, \ 0.16, \ 0.29).$$

Step 4. Fuzzy HOQ of QFD is constructed by fuzzy relationships. Table 4 is has shown the fuzzy HOQ which is build by applying FAHP.

Step 5. Table 5 is shown the class function of the CRs and the limit of different ranges of CRs.

Step 6. After determining the limit of different ranges, the weight of the each range of the CRs according to the Messac et al. (1996) $\beta = 1.1$ and $z^2 = 0.1$

Table 4 Fuzzy HOQ of the QFD

	EC$_1$			EC$_2$			EC$_3$			EC$_4$			EC$_5$		
EC$_1$	0.10	0.15	0.23	0.11	0.15	0.22	0.15	0.21	0.28	0.09	0.15	0.23	0.11	0.16	0.29
EC$_2$	0.10	0.15	0.23	0.11	0.16	0.23	0.12	0.17	0.23	0.11	0.15	0.24	0.12	0.18	0.26
EC$_3$	0.12	0.19	0.30	0.11	0.17	0.25	0.11	0.15	0.19	0.14	0.21	0.31	0.09	0.14	0.21
EC$_4$	0.13	0.19	0.27	0.11	0.17	0.27	0.12	0.16	0.21	0.09	0.12	0.15	0.13	0.21	0.31
EC$_5$	0.12	0.17	0.26	0.12	0.20	0.32	0.10	0.14	0.20	0.14	0.19	0.26	0.09	0.14	0.21

Table 5 Class function of the CRs and the limit of different ranges of CRs

Customer requirements	Class function	The limit of different ranges of CRs according to the LPP method				
		t_1	t_2	t_3	t_4	t_5
Cost	1S	0.14	0.36	0.57	0.71	1
Conformity	2S	1	0.89	0.74	0.47	0.32
Punctuality	2S	1	0.7	0.55	0.3	0.1
Efficacy	2S	1	0.75	0.65	0.5	0.2
Lead time	1S	0	0.29	0.57	0.86	1

Table 7 Optimal achievement levels of the CRs under different values of a

a	Satisfaction levels of the CRs under different values of a				
	CR_1	CR_2	CR_3	CR_4	CR_5
0.5	2.17	2.07	2.03	2.29	2.05
0.6	2.18	2.08	2.04	2.29	2.06
0.7	2.25	2.16	2.14	2.35	2.15
0.8	2.18	2.13	2.18	2.3	2.18
0.9	2.11	2.06	2.07	2.24	2.09
1	2.05	1.99	1.99	2.19	2.01

(*small positive number*) is calculated by applying Equations 8, 9, 10, 11, 12, 13, 14, and 15.

The weights of the different ranges of the cost are as following:

$$\tilde{w}_{12} = 0.001, \quad \tilde{w}_{13} = 1.587,$$
$$\tilde{w}_{14} = 11.499, \quad \tilde{w}_{15} = 16.262$$

The weights of the other customer requirement can be defined similarly.

Step 7. Using Equations 16, 17, 18, 19, and 20, we extract the mathematical model of the problem. We exchange the Equation 17 with Equation 28 in our model. Now we have a model with fuzzy constraints. Step 8. Applying the Equations 26 and 27, the fuzzy model is exchanged to the LP model. We solved the model with different degrees of uncertainty. Tables 6 and 7 showed the optimal achievement levels of ECs and CRs under different degrees of uncertainty which are obtained by solving the model.

Discussion of results

The obtained results of this numerical example in Table 6 show that in engineering characteristics, x_3 and x_4 which demonstrate respectively quality system certification and flexibility of response to the customer's requests, have not been fully achieved in some degree

Table 6 Optimal achievement levels of the ECs under different values of a

a	Optimal achievement levels of the ECs under different values of a				
	x_1	x_2	x_3	x_4	x_5
0.5	1	1	1	0	1
0.6	1	1	1	0.02	1
0.7	1	1	1	0.17	1
0.8	1	1	0.7	0.44	1
0.9	1	1	0.69	0.29	1
1	1	1	0.69	0.16	1

of uncertainty, while the other three characters have been obtained completely in all calculated degree of uncertainty.

The results of Table 7 indicate that the satisfaction level of CR_4 is rather higher than the other four requirements, so in this example, efficacy is more important than cost, conformity, punctuality, and lead time. Unlike the existing literature, this method integrates three different concepts such as AHP, QFD, and LPP to achieve the optimal values of the ECs and CRs under different degrees of uncertainty. So with respect to the company strategy, managers can use the results of proposed method to improve and develop engineering characteristics of suppliers in order to meet their requirements.

Conclusions

In this paper we proposed a simple and useful methodology by integrating AHP, QFD, and LPP for supplier development problems under uncertainty conditions. We used fuzzy AHP to determine the relationships between customer's requirements and engineering characteristics for building the relation matrix in the QFD method. Then, applying LPP, we formulated the mathematical model to optimize QFD. Proposed methodology helps decision makers to deal with the vagueness and impreciseness involved in the real problems. In addition, it helps them to maximize overall customer satisfaction in supplier development. Also, the proposed methodology can be used in the product design, product development, process development, and other decision making problems.

For the future work, we suggest to consider the correlation between engineering characteristics to increase the reliability of the obtained solutions or use the other type of fuzzy programming to obtain optimal achievement level of engineering characteristics and customer satisfaction level.

Competing interests
The authors declare that there is no conflict of interests.

Authors' contributions
All authors ZS, ER and FM, have made adequate effort on all parts of the work necessary for the development of this manuscript according to his/her expertise. All authors read and approved the final manuscript.

Author details
[1]Department of Industrial Engineering, Khaje Nasir Toosi University of Technology, Tehran, Iran. [2]Department of Management and Economics, Tarbiat Modares University, Jalal Ale Ahmad Highway, P.O.BOX: 14115-111, Tehran, Iran.

References
Aydogan EK (2011) Performance measurement model for Turkish aviation firms using the rough-AHP and TOPSIS methods under fuzzy environment. Expert Syst Appl 38:3992–3998

Bevilacqua M, Ciarapica FE, Giacchetta G (2006) A fuzzy-QFD approach to supplier selection. J Purchas Suppl Manag 12:14–27

Bhattacharya A, Geraghty J, Young P (2010) Supplier selection paradigm: an integrated hierarchical QFD methodology under multiple-criteria environment. Appl Soft Comput 10:1013–1027

Bottani E (2009) A fuzzy QFD approach to achieve agility. Int J Prod Econ 119 (2):380–391

Bottani E, Rizzi A (2006) Strategic management of logistics service: a fuzzy-QFD approach. Int J Prod Econ 103(2):585–599

Buckley JJ (1985) Fuzzy hierarchical analysis. Fuzzy Set Syst 17(1):233–247

Chan FTS, Kumar N (2007) Global supplier development considering risk factors using fuzzy extended AHP-based approach. Omega, Int J Manag Sci 35:417–431

Chang DY (1992) Extent analysis and synthetic decision, optimization techniques and applications, vol 1. World Scientific, Singapore, 352

Chang DY (1996) Applications of the extent analysis method on fuzzy AHP. Eu J Operat Res 95:649–655

Chen LH, Ko WC (2008) A fuzzy nonlinear model for quality function deployment considering Kano's concept. Math Comput Model 48:581–593

Chen LH, Ko WC (2009) Fuzzy linear programming models for new product design using QFD with FMEA. Appl Math Model 33:633–647

Chen LH, Ko WC (2010) Fuzzy linear programming models for NPD using a four-phase QFD activity process based on the means-end chain concept. Eur J Operat Res 201:619–632

Chen LH, Weng MC (2006) An evaluation approach to engineering design in QFD processes using fuzzy goal programming models. Eur J Operat Res 172:230–248

Chen MF, Tseng GH, Ding CG (2008) Combining fuzzy AHP with MDS in identifying the preference similarity of alternatives. Appl Soft Comput 8:110–117

Cox A (1999) Power value and supply chain management. Supply Chain Management. Int J 4(4):167–175

Delice EK, Güngör Z (2009) A new mixed integer linear programming model for product development using quality function deployment. Comput Indust Eng 57:906–912

Fung RYK, Chen Y, Chen L, Tang J (2005) A fuzzy expected value-based goal programming model for product planning using quality function deployment. Eng Optimization 37(6):633–647

Gebennini E, Gamberinni R, Manzini R (2009) An integrated production-distribution model for the dynamic location and allocation problem with safety stock optimization. Int J Prod Econ 122:286–304

Güngör Z, Serhadlıoğlu G, Kesen SE (2009) A fuzzy AHP approach to personnel selection problem. Appl Soft Comput 9:641–646

Haghighi M, Divandari A, Keimasi M (2010) The impact of 3D e-readiness on e-banking development in Iran: a fuzzy AHP analysis. Expert Syst Appl 37:4084–4093

Hassanzadeh Amin S, Razmi J (2009) An integrated fuzzy model for supplier management: a case study of ISP selection and evaluation. Expert Syst Appl 36:8639–8648

Jiménez M (1996) Ranking fuzzy numbers through the comparison of its expected intervals. Int J Uncertainty, Fuzziness Knowledge-Based Syst 4 (4):379–388

Jiménez M, Arenas M, Bilbao A, Rodriguez MV (2007) Linear programming with fuzzy parameters: an interactive method resolution. Eur J Oper Res 177:1599–1609

Kahraman C, Cebeci U, Ruan D (2004) Multi-attribute comparison of catering service companies using fuzzy AHP: the case of Turkey. Int J Prod Econ 87:171–184

Kahraman C, Ertay T, Büyüközkan G (2006) A fuzzy optimization model for QFD planning process using analytic network process. Eur J Oper Res 171:390–411

Lai X, Xie M, Tan KC (2006) QFD optimization using linear physical programming. Eng Optimization 38(5):593–607

Lai X, Xie M, Tan KC (2007) Optimizing product design using quantitative quality function deployment: a case study. Qual Reliability Eng Int 23:45–572

Maria A, Mattson CA, Ismail-Yahaya A, Messac A (2003) Linear physical programming for production optimization. Eng Optimization 35(1):19–37

Melachrinoudis E, Messac A, Min H (2005) Consolidating a warehouse network: a physical programming approach. Int J Prod Econ 97:1–17

Messac A (1996) Physical programming: effective optimization for computational design. AIAA J 34(1):149–158

Messac A, Gupta SM, Akbulut B (1996) Linear physical programming: a new approach to multiple objective optimization. Trans Oper Res 8:39–59

Park T, Kim K (1998) Determination of an optimal set of design requirements using house of quality. J Oper Manage 16:569–581

Raharjo H, Brombacher AC, Xie M (2008) Dealing with subjectivity in early product design phase: a systematic approach to exploit Quality Function Deployment potentials. Comp Ind Eng 55:253–278

Raissi S, Izadi M, Saati S (2011) A novel method on customer requirements preferences based on common set of weight. Aust J Basic Appl Sci 5 (6):1544–1552

Raissi S, Izadi M, Saati S (2012) Prioritizing engineering characteristic in QFD using fuzzy common set of weight. Am J Sci Res 49:34–49

Rung Yu J, Shing WY (2013) Fuzzy analytic hierarchy process and analytic network process: an integrated fuzzy logarithmic preference programming. Appl Soft Comput 13:1792–1799

Sener Z, Karsak EE (2011) A combined fuzzy linear regression and fuzzy multiple objective programming approach for setting target levels in quality function deployment. Expert Syst Appl 38:3015–3022

Sharma JR, Rawani AM (2008) Quality function deployment: integrating comprehensive matrix and SWOT analysis for effective decision making. J Ind Eng Int 4(6):19–31

Soota t, Singh H, Mishra RC (2011) Fostering product development using combination of QFD and ANP: a case study. J Ind Eng Int 7(14):29–40

Tian Z, Zuo MJ (2006) Redundancy allocation for multi-state systems using physical programming and genetic algorithms. Reliability Eng Syst Saf 91:1049–1056

Xia W, Wu Z (2007) Supplier selection with multiple criteria in volume discount environments. Omega 35(5):494–504

Yousefi S, Mohammadi M, Haghighat Monfared J (2011) Selection effective management tools on setting European Foundation for Quality Management (EFQM) model by a quality function deployment (QFD) approach. Expert Syst Appl 38:9633–9647

Zarei M, Fakhrzad MB, Jamali Paghaleh M (2011) Food supply chain leanness using a developed QFD model. J Food Eng 102:25–33

13

Developing and solving two-echelon inventory system for perishable items in a supply chain: case study (Mashhad Behrouz Company)

Mirbahador Gholi AriaNezhad[1], Ahmad Makuie[2] and Saeed Khayatmoghadam[3*]

Abstract

In this research, a new two-echelon model has been presented to control the inventory of perishable goods. The performance of the model lies in a supply chain and is based on real conditions and data. The main purpose of the model is to minimize the maintenance cost of the entire chain. However, if the good is perished before reaching the customer (the expiration date is over), the cost would be added to other costs such as transportation, production, and maintenance costs in the target function. As real conditions are required, some limitations such as production time, storage capacity, inventory level, transportation methods, and sustainability time are considered in the model. Also, due to the complexity of the model, the solution approach is based on genetic algorithm under MATLAB to solve and confirm the accuracy of the model's performance. As can be noted, the manipulation of parametric figures can solve the problem of reaching the optimum point. Using real data from a food production facility, the model was utilized with the same approach and the obtained results confirm the accuracy of the model.

Keywords: Two-echelon inventory control; Genetic algorithm; Supply chain; Perishable good

Introduction

Most manufacturing systems inevitably maintain some quantities of their products under their inventories in order to respond to customers' needs appropriately and to prevent extra costs. Thus, inventory control and maintenance is a common problem for most factories, especially for those organizations that are involved in a supply chain.

There are many differences among inventory control and maintenance systems due to quantity and complexity of items, type and nature of items, costs of operating system, multi-echelonment of system, probability degree of system, and even competitors status. It is obvious that all existing cases should be considered to plan an inventory control system properly.

This study intends to present a model for managing short-lived products requiring a two-echelon inventory control. In the following section, we are presenting some of the more significant and recent studies that have been done in this aspect.

Background studies

Kyung and Dae (1989) have offered an innovative approach for the probable inventory control model in a way that a two-echelon distribution system has a central warehouse and stores items in the local warehouses for distribution. The offered algorithm was a step-by-step algorithm designed to obtain the optimum or nearly optimum result and also to minimize the sum of system cost variables in a year.

Have investigated an economic ordering policy for the deterministic two-echelon distribution systems. In this article, an algorithm is recommended to determine the economic ordering policy in order to provide producer's items centrally and distribute these items from the central warehouse to some local warehouses. Also, the products are to be distributed to customers through local warehouses. In this case, the purpose is to minimize the producers' overall costs that are results of order costs, distribution costs, and related inventory

---
* Correspondence: skhayatmoghadam@gmail.com
[3]Dep. Of Management, Mashhad Branch, Islamic Azad University, Mashhad, Iran
Full list of author information is available at the end of the article

shipment costs to central and local warehouses. Dada (1992) has analyzed the two-echelon system for spare parts. This system allows a rapid dispatch at the time of inventory shortage.

Bookbinder and Chen (1992) have studied the two-echelon inventory system based on a multi-criterion point of view. The definition of the developed model was dealt with, specifying the optimum quantity of economic ordering in the central and local warehouses; it designed and resolved a two-criterion system to minimize the related costs of inventory system and dis-placement systems.

Shtub and Simon (1994) have discussed about the determination of order points in the two-echelon in-ventory system of spare parts. This system includes a central warehouse in which all service centers are supported. Each service center meets a random de-mand. The related costs of inventory transferred to both echelons are identical, but the probability of shortage occurrence among several maintenance ser-vice centers is different. The inventory management policy specifies the amount of order point for each service center.

Bertrand and Bookbinder (1998) have assessed the two-echelon system with the possibility of redistribu-tion. They developed an algorithm to perform this redistribu-tion. In this study, the developed model has been evaluated.

In another research, Miner (2005) has designed an in-ventory control model with several suppliers. At the same year, a mathematical model was designed in order to reduce warehouse-kept quantity by lowering the pur-chase batch.

Moon et al. (2005) have expanded the model of eco-nomic ordering quantity for perishable and improvable goods by considering the time value of money. Yang and Wee 2002 have presented a model for the inte-grated planning of production and inventory of perish-able goods; this model allocated to study a single product status and a system that consisted of one pro-ducer and a few retailers. This model was introduced supposing the limited production rate, the demand without waiting time, and the integrated production and inventory for perishable goods (Moon et al. 2005). Rau et al. (2003) extended a multi-echelon model among suppliers, producers, and customers for perish-able goods in a way that a numerical example has shown that after specifying the overall cost function, the integration approach in comparison with decision making resulted in the reduction of overall cost. Chen and Lee (2004) proposed the multi-objective synchro-nized optimization opposite to indefinite prices. They were the first researchers who raised multi-objective optimization in supply chain networks.

Based on the supply chain approach and considering the required service levels, Hwang (2002) has designed a logistic system that includes some manufacturing cen-ters, warehouses or distribution center, and customers with indefinite demands in which distances are distributed randomly. In order to solve this problem, initially, random overall coverage was used to establish warehouses, and the objective function was stated to minimize the logistic costs and the number of warehouses which can be estab-lished. To decide about finding directions and specifying the ordering quantity of warehouses to production cen-ters, an object-oriented planning approach based on gen-etic algorithm had been taken so that the total logistic costs could be minimized. In this case, all demands should be satisfied; however, there are some limitations on travel time, capacity, speed, type, and number of transportation means (Hwang 2002).

Bollapragada et al. (1998) surveyed the distribution system including one depot and some warehouses. In this system, the demand is created at random and at warehouse level, and the process is as follows: at the be-ginning of every depot, the order is presented to a sup-plier who is not from the system, and this order will be received by a depot after a fixed waiting time. The depot will then forward the received orders to the warehouses. The fixed waiting time was considered between the depot and the warehouses, and the shortage was sup-posed as the delayed orders; the warehouses have been evaluated at random.

Erenguc et al. (1999) have investigated on inventory decisions in the supply chain and they proposed a math-ematical model based on the following assumptions in-cluding: the delayed orders are not allowed. Waiting times among factories and the distribution centers as well as waiting times among the distribution centers and customers are zero. In this model, each distribution cen-ter decides about inventory, and each customer focuses on determining the order quantities in order to balance between maintenance and ordering costs.

Hoque and Goyal (2000) have studied on specifying optimization policies for the integrated system of pro-duction and inventory, which comprised a single buyer and single vendor, and have considered the following as-sumptions in developing this model: First, the demand rate is definite and fixed. Second, the total accumulative production can be transferred to identical or different batches, but in any case, the fixed cost will be calculated for each dispatch. Third, shortage is not allowed, and transportation time is too slight, so it will not be accounted. Also, all values were supposed fixed and def-inite. Finally, the time horizon under this study has been regarded indefinite. The particular feature of this model is that it is studied under the condition of the limited capacity of transportation means.

Zhou and Min (2002) have designed a supply chain network which balanced the transportation cost and service level in the best manner in which the working load given to all distribution centers are identical. Accordingly, this caused the decrease of shortage in warehouse inventories, the postponed orders, and the delay in responding to the customers' needs; at the same time, the loading and usage rates of distribution centers are increased. For this aim, they considered an objective function to minimize the maximum transportation distance related to distribution centers and used the formula of balanced star comprehensive tree; finally, they used the genetic algorithm for solving.

Clarifying the research question

In this research, a two-echelon warehousing system was observed for short-lived goods. Following the required surveys and in view of different limitations covering the research question, a program was written on MATLAB environment, and the genetic algorithm based on the written program was used for testing the accuracy of the presented model as well for solving the model in order to obtain the required values that can be noticed in the objective function which is minimizing the costs. In addition, as this research was done based on a real situation, after the model was analyzed, this model was resolved based on real numbers, in which the related result also was obtained, and solved using the genetic algorithm. The obtained results have been confirmed and used by related factories. The presented model includes the following assumptions:

1. The system has one producer and several consumers. Products are produced in the original factory and then transferred to the given warehouses of other factories (this means a two-echelon nature).
2. This model has been considered for different transportation capacities (truck, trailer, etc.).

Table 1 Decisions and variables

Variable	Definition of decision
X_{kLm}^t	The quantity of goods L which were transferred from the original factory to the factory k via transportation model m in a period t
PZ^t	This denotes whether the original factory was active in period t or not (variables of 0 and 1)
B_{kL}^t	Accumulative quantity of delayed demand of goods L till the period t in the applicant factory k
In_{kL}^t	Accumulative quantity of goods inventory L till the period t in the applicant factory k
v_{kL}^t	This denotes the delay or maintenance of goods L till the period t in the applicant factory k (variables of 0 and 1)

Table 2 Parameters and variables

Variable	Definition of parameters
Pc_{kLm}^t	Transportation cost for each unit of goods L from the original factory to factory k through the model m in the period t
PF^t	Fixed cost of the original factory in the period t
Pb_L^t	Production cost of goods L in the original factory in the period t
dh_L^t	Maintenance cost for each unit of goods L in the original factory in the period t
dk_L^t	Demand of product L by the factory k in the period t
ps_L^t	Required time for production of each unit of goods L in the original factory in the period t
pu^t	Total production time which is available for the original factory in the period t
Du^t	Total maintenance capacity of warehouse in the original factory in the period t
Pv_L^t	Quantity (volume) of each unit of goods L in the period t
Pv_m^t	Total volume capacity of forwarded goods based on the transportation model m from the original factory in the period t
bl_k^t	Maximum delayed order of product L in the original factory in the period t
π_{kL}^t	Cost coefficient related to daily penalty of early delivery of goods L in the factory k in the period t (earlier delivery than due time)
δ_{kL}^t	Cost coefficient related to daily penalty of delay delivery of goods L in the factory k in the period t
W	Cost of each unit of perished goods
EX_L	Consumption date of the goods L (month)
Q_{KL}^t	Maintenance capacity of product (goods) L in the factory k

Table 3 Variable and corresponding L_1 and L_2

Variable	L_1	L_2
Pc_{kLm}^t	6,000 Rial (for each unit)	6,000 Rial (for each unit)
PF^t	20,000,000 Rial	20,000,000 Rial
Pb_L^t	300,000 Rial	300,000 Rial
dh_L^t	100 Rial	100 Rial
dk_L^t	150 units	150 Unit
ps_L^t	1/12,500	1/6,250
pu^t	0.5 t	0.5 t
Du^t	8,000 m^2	8,000 m^2
Pv_L^t	0.03 m^3	0.02 m^3
Pv_m^t	3 to 20 m^3	3 to 20 m^3
bl_k^t	20 units	20 units
π_{kL}^t	500 Rial	500 Rial
δ_{kL}^t	100 Rial	100 Rial
W	10,000 Rial	10,000 Rial
EX_L	52 weeks	26 weeks
Q_{KL}^t	100 units	100 unit

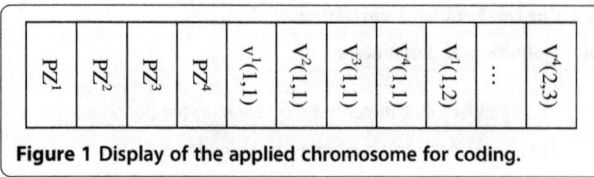

Figure 1 Display of the applied chromosome for coding.

3. The system has been considered for some very significant goods which are highly consumed (at least two highly consumed goods).
4. Demand is fixed and definite in each period.
5. Shortage (lack of inventory) is allowed.
6. Goods that are short-lived have expiration dates (perishable).
7. Expiration date is an integral multiple of the period length.
8. The goods, in some cases, might have been produced before they were ordered by the ordering party, so on their dispatching time, they might have an older manufacturing date than expected.
9. Warehousing system of the original factory (producer) is in the form of FIFO.
10. As the goods' transportation time from the producer factory to other factories is short, the transportation time can be ignored.
11. The producer factory has some limitations for goods maintenance.
12. The customers' demands can be predetermined in each period.
13. The overall capacity of the original producer's warehouse is considered.

Specification and applied indices of the model

The following are the specifications and applied indices of the model (for more details, see Table 1):

- Total number of customers (factories) $K = 1, ..., k$
- Total time periods $T = 1, ..., L$
- Transportation models $M = 1, ..., m$

Proposed model

The proposed mathematical model is as follows:

$$
\begin{aligned}
\min z = &\left(\sum_t \sum_m \sum_L \sum_k PC_{KLm}^t . X_{KLm}^t \right) \\
&+ \left(\sum_t PF^t . Pz^t \right) + \left(\sum_t \sum_L Pb_L^t . \left(\sum_m \sum_k X_{KLm}^t \right) \right) \\
&+ \left(\sum_L \sum_{t \neq T} dh_L^t . \sum_k \left(\sum_m \sum_{t'=1}^t X_{KLm}^{t'} - \sum_{t'}^t d_{KL}^{t'} \right) \right) \\
&+ \left(\sum_l \sum_k \sum_t \delta_{KL}^t . B_{KL}^t + \sum_l \sum_k \sum_t \pi_{kL}^t . In_{kL}^t \right) \\
&+ \left(\sum_t \sum_m \sum_L \sum_k X_{KLm}^t - \sum_{t'}^T \sum_k \sum_L d_{KL}^{t'} \right) . W
\end{aligned}
$$

(1)

Describing the relations

The first relation is the objective function of the proposed model which includes the total transportation costs of goods from the original factory to the applicant factory, the production cost of goods in the original factory including fixed and variable costs of factory for each production unit, the maintenance costs of inventory in the original factory warehouse, the resulted costs from delayed delivery or earlier delivery of order to the applicant factories, and the resulted costs from perishing the goods in the original factory warehouse.

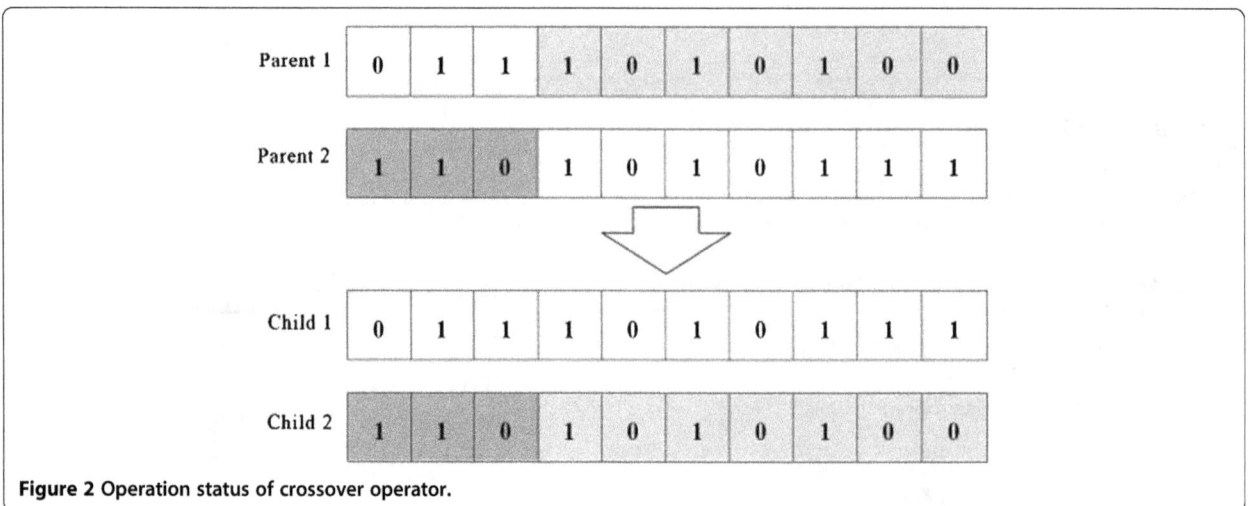

Figure 2 Operation status of crossover operator.

Table 4 Descending trend of reaching the lowest cost based on 30-time repetition and 30 chromosomes

Repetition	Amount function (cost)
1	34,450,000
2	34,450,000
3	30,450,000
4	30,450,000
5	30,450,000
6	30,450,000
7	30,450,000
8	30,450,000
9	30,450,000
10	28,450,000
11	28,450,000
12	24,450,000
13	24,450,000
14	22,450,000
15	22,450,000
16	22,450,000
17	22,450,000
18	22,450,000
19	22,450,000
20	20,450,000
21	20,450,000
22	20,450,000
23	20,450,000
24	20,450,000
25	20,450,000
26	20,450,000
27	20,450,000
28	20,450,000
29	18,450,000
30	18,450,000

Table 5 The model's suggested times for the main factory activity or non-activity in the t time period

Period	Activity/non-activity
1	1
2	1
3	0
4	1
5	0
6	1
7	1
8	1
9	1
10	0
11	1
12	0
13	1
14	1
15	1
16	1
17	0
18	1
19	1
20	0
21	1
22	1
23	1
24	0
25	1
26	0

Limitations

The following are the limitations:

Equation 2 shows the related limitation to the factory production time:

$$\text{St}: \sum_{L} \text{PS}_L^t \left(\sum_k \sum_m X_{KLm}^t \right) \leq \text{Pu}^t . \text{PZ}^t \quad \forall_t \tag{2}$$

Equation 3 shows the related limitation to the capacity of original factory warehouses:

$$\sum_{L} \text{PV}_L^t \left(\sum_k \sum_m X_{KLm}^t \right) \leq \text{Du}^t . \text{PZ}^t \quad \forall_t \tag{3}$$

Equation 4 shows the related limitation to the capacity of transportation means in the specified period:

$$\sum_k \sum_L \text{PV}_L^t . X_{KLm}^t \leq \text{PV}_m^t \quad \forall_{t,m} \tag{4}$$

Equation 5 shows that the limitation specifies the inventory level in the applicants' factories:

$$\sum_m \sum_{t'} X_{KLm}^{t'} - \text{In}_{kL}^t + B_{KL}^t = \sum_{t'} d_{KL}^{t'} \quad \forall_{k,L,t \neq T} \tag{5}$$

These two limitations (Equations 6 and 7) represent how delay or early delivery will occur in each the planning period:

$$\text{In}_{kL}^t \leq Q_{kL}^t . V_{kL}^t \quad \forall_{k,L,t} \tag{6}$$

$$B_{kL}^t \leq bL_{kL}^t . \left(1 - V_{kL}^t \right) \quad \forall_{k,L,t} \tag{7}$$

Table 6 Maintaining or not maintaining sent goods in the first applying factory (k1) in the *t* period

Period	k1	k2
1	1	1
2	1	1
3	1	1
4	1	1
5	0	1
6	1	1
7	1	1
8	0	1
9	1	1
10	1	1
11	1	1
12	1	1
13	1	1
14	1	1
15	0	1
16	1	1
17	1	1
18	1	1
19	1	1
20	1	1
21	1	1
22	1	1
23	1	1
24	1	1
25	1	1
26	1	0

Table 7 Maintaining or not maintaining sent goods in the first applying factory (k2) in the *t* period

Period	k1	k2
1	1	1
2	1	1
3	1	1
4	1	1
5	1	1
6	1	1
7	0	1
8	0	1
9	1	1
10	0	1
11	1	1
12	0	1
13	0	1
14	1	1
15	1	1
16	1	1
17	0	1
18	0	1
19	1	1
20	1	1
21	1	1
22	1	1
23	1	1
24	1	1
25	1	1
26	1	1

This limitation (Equation 8) is related to the factory capacity:

$$\sum_m \sum_{t'}^{t} X_{KLm}^{t'} - \sum_{t'}^{t} d_{KL}^{t'} \leq Q_{kL}^t \quad \forall_{k,L,t \neq T} \tag{8}$$

Equation 9 shows the related limitation to perishing the goods:

$$\left(\sum_m \sum_{t'}^{t} X_{KLm}^{t} - \sum_{t''=1}^{min(t+EX_L,T)} d_{KL}^{t''} \right) > 0 \quad \forall_{k,t \neq T} \tag{9}$$

$$PZ^t, V_{kL}^t = 0 \ or \ 1 \quad \forall_{k,t} \tag{10}$$

$$X_{KLm}, In_{kL}^t, B_{kL}^t \geq 0 \quad \forall_{k,L,m,t} \tag{11}$$

Case study (Mashhad Behrooz Company)

Mashhad Behrooz Company is a manufacturing company of all kinds of compotes, conserves, jams, pickles, and other food products; it has been active in this industry for more than three decades. At present, this company is involved in a supply chain under the brands of Yek & Yek, Pardis, and Bartar; the variety of goods, limited warehousing space in this company, and short durability of products are major factors which cause some problems in the process of planning and controlling the inventory.

Thus, in view of the above-mentioned issues and with regard to the type of products and their real status, the following values are considered as parameters (Table 2). It should be noted that the proposed model is in a form that the model output also can be evaluated easily by changing indices (for instance, number of customers or number of factories).

Applied indices and their specifications based on the case study

The following are the applied indices and their specifications based on the case study:

- $K = 3$ indicates the number of customers (Golestan Company (k1), Yek & Yek (k2), Bartar (k3)).
- $L = 2$ indicates the number of goods: (L_1) conserved wax bean and (L_2) jam.
- The measurement unit of goods in each carton is 48 pieces.
- $M = 4$ indicates the capacities of transportation models (10- to 7-ton truck, trailer, pickup truck).
- $T = 52$ indicates the 1-year planning which includes 52 weeks; $t = 1$ indicates 1 week, and each week covers six working days.

Ranges of parameters

Based on the case study, the ranges of parameters are as follows (Table 3):

The daily production capacity of L_1 is about 40,000 to 100,000 pieces, and the daily production capacity of L_2 is about 20,000 to 50,000 pieces.

Applied optimization approach using genetic algorithm

The genetic algorithm was combined with linear programming (LP) and was used in solving the problem of optimization which is a mixed integer linear problem. In this way, the discontinuous variables of the problem are modeled in the genetic algorithm, and in the section of fitness of the genetic algorithm, a LP is recalled.

Coding the problem

In the genetic algorithm of each generation, there are a few chromosomes (Figure 1) which act as the feasible responses. In order to code the discontinuous variables, the binary alphabet was applied. The discontinuous variables of the problem include PZ^t and V_{kl}^t. Thus, the number of bits of a chromosome is equal to

$$Nb = T + K \times L \times T.$$

For example, if the values of these three parameters are supposed as $T = 4$, $K = 2$, and $L = 3$, we will have

$$Nb = 4 + 2 \times 3 \times 4 = 28$$

Fitness

To evaluate the fitness of chromosomes, the value of the proposed objective function is calculated. In such a way that the value of one chromosome in the genetic algorithm is assumed, the values of discontinuous variables of the problem are known. Thus, by replacing these values in the objective function and considering the limitations of the problem, a linear programming problem is formed. By solving this linear problem, the total value of the objective function will be defined and attributed to that chromosome. In this section, if the LP has no feasible response (as this is a minimizing problem), the given chromosome would be penalized with a big value.

Operators of the designed genetic algorithm

In this problem, selection operator, elitism, crossover operator, mutation operator, and recent reduction operator are determined in such a form that the algorithm

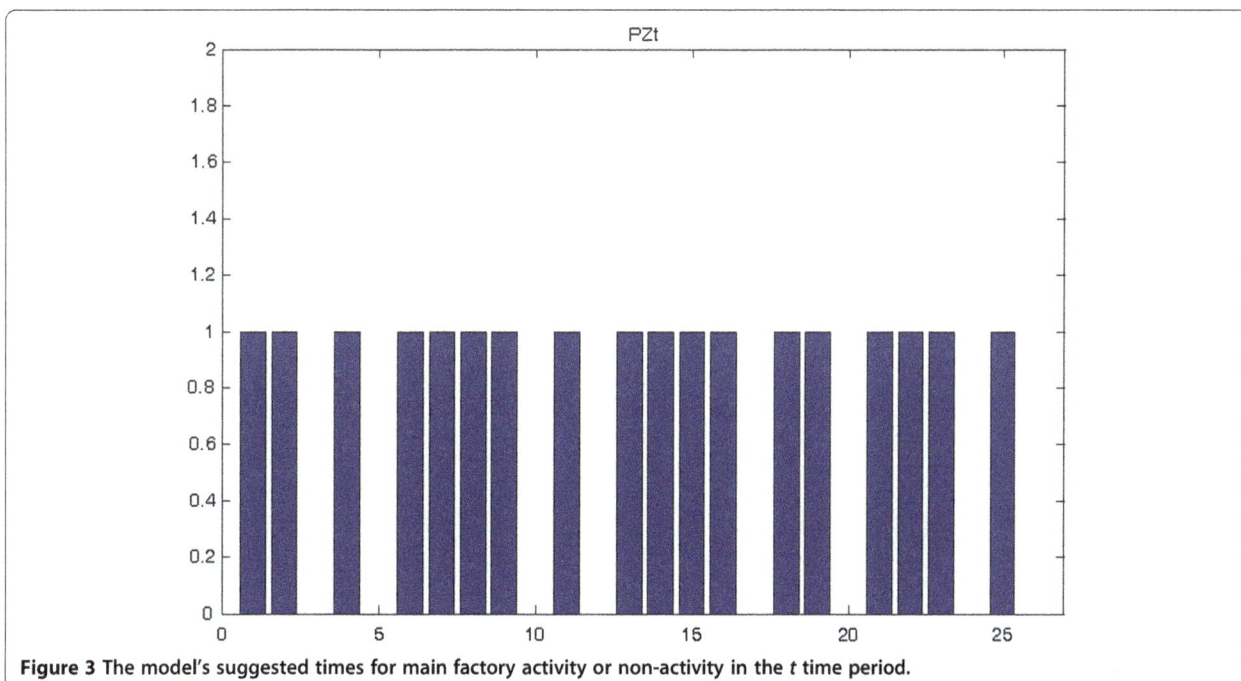

Figure 3 The model's suggested times for main factory activity or non-activity in the t time period.

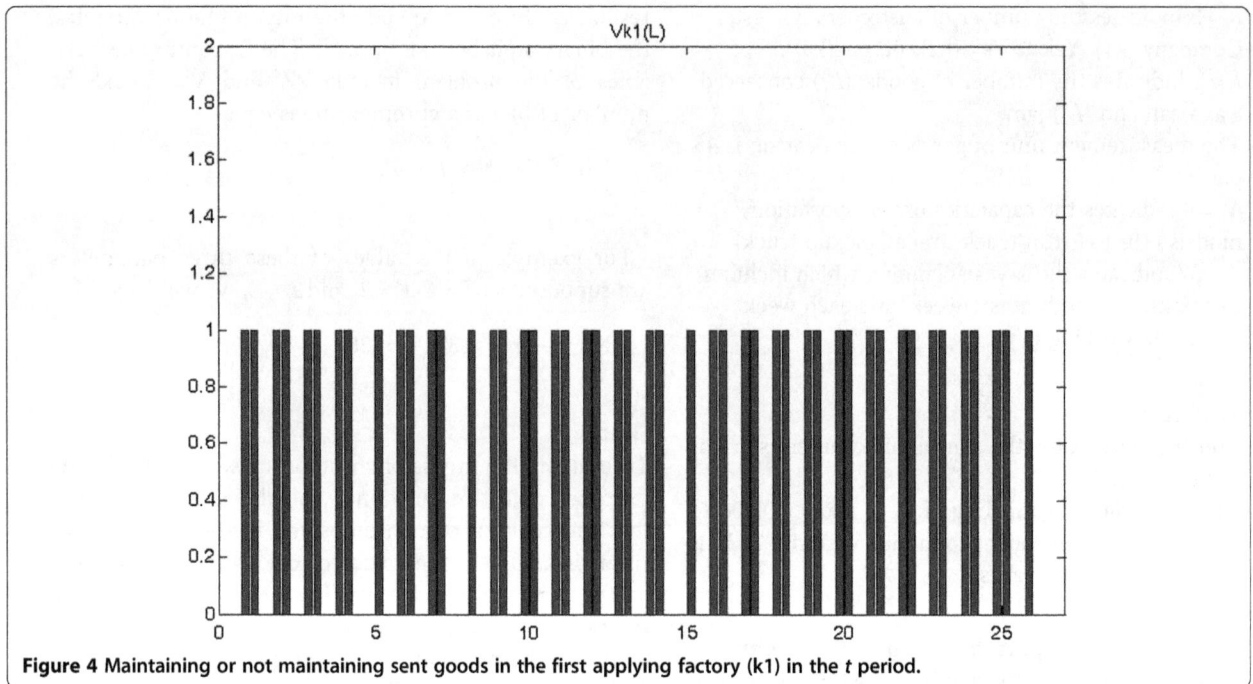

Figure 4 Maintaining or not maintaining sent goods in the first applying factory (k1) in the *t* period.

reaches a proper result; for example, in a crossover operator (Figure 2), the crossover is done in the form of a single point with the probability of 0.8.

Sample problems and solving the proposed model
Now, we solve the problem based on real data in order to analyze the application of the proposed model. A

genetic algorithm is designed and proposed using MATLAB software for solving the considered case study. Initially, the program commenced with 15 people for each generation, and there were 30 frequencies for the number of people in each generation, and the frequencies have been increased for the purpose of further research. In addition, this process has been repeated several times

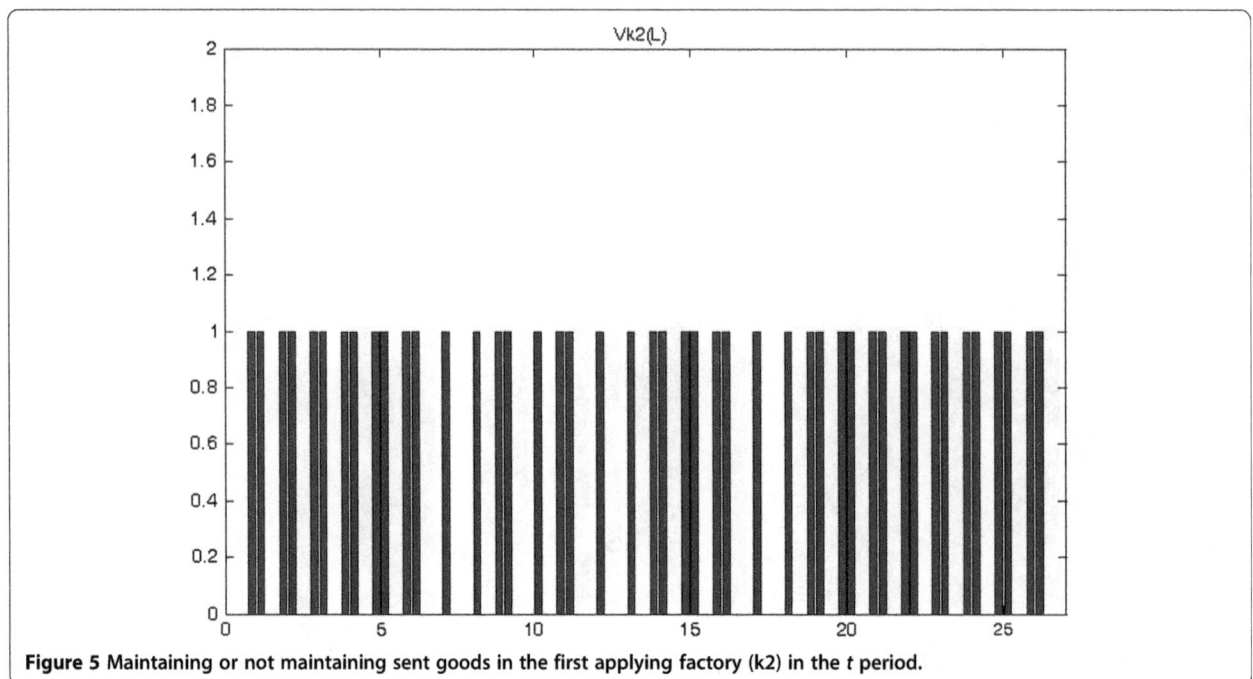

Figure 5 Maintaining or not maintaining sent goods in the first applying factory (k2) in the *t* period.

Table 8 Solution and repetition of the presented model considering the change in related parameters

	Repetition	Max-Gen.	Num-pop	Amount function (cost)	Time
1	1	30	15	26,450,000	27
2	2	30	15	26,450,000	26
3	3	30	15	26,450,000	25
4	1	40	15	24,450,000	37
5	2	40	15	24,450,000	36
6	3	40	15	24,450,000	23
7	1	50	15	20,450,000	45
8	2	50	15	20,450,000	45
9	3	50	15	20,450,000	46
10	1	30	20	26,450,000	37
11	2	30	20	2,650,000	33
12	3	30	20	2,650,000	32
13	1	40	20	26,450,000	47
14	2	40	20	26,450,000	45
15	3	40	20	26,450,000	45
16	1	50	20	26,450,000	60
17	2	50	20	26,450,000	59
18	3	50	20	26,450,000	59
19	1	30	30	18,450,000	60
20	2	30	30	18,450,000	64
21	3	30	30	18,450,000	61
22	1	40	30	18,450,000	80
23	2	40	30	18,450,000	81
24	3	40	30	18,450,000	80
25	1	50	30	16,450,000	100
26	2	50	30	16,450,000	100
27	3	50	30	16,450,000	100
28	1	70	15	18,450,000	66
29	2	70	20	24,450,000	86
30	3	70	30	12,450,000	136
31	1	100	30	12,450,000	188
32	2	100	20	24,450,000	122
33	3	100	15	18,450,000	96
34	1	150	15	18,450,000	153
35	1	150	30	12,450,000	138
36	1	200	30	12,450,000	370

with similar numbers to verify its accuracy. The parameter of time was noticed as a part of this problem.

Conclusions

Based on the obtained results, the following consequences can be referred:

1. By solving the model using genetic algorithm, it is obvious that the operation of the model is quite accurate.
2. Based on testing with real numbers, the practicality of this model can be emphasized.
3. This model was designed for those systems which work in a real setting because in real settings, there may be some goods which are sold late, past their consumption date, and they would not be usable anymore; this model, in addition to minimizing the related costs, minimizes the costs of perishable goods as well.
4. Another output of this model is proposing the time of production, stop, dispatch, and other issues which are completely suitable for real systems.
5. It can be planned easily for a wider range of products by changing the number and specifications of different products.

Suggestions for further research
The following are suggested for further research:

1. Dependence of sales price on durability of goods in a way that the price of goods will be decreased through passing time from production date of the goods
2. Defining the scrapped price for those goods which are on their due expiry date
3. The possibility of returning those goods which are not confirmed in quality by applicant
4. Defining the neural network as a secondary solution of these models in a way that the output of the genetic algorithm was defined as the input to the neural network; thus, through training, with any change in number of people in each generation and other parameters, in this network, we can find the related response with higher speed
5. Extending the two-echelon model to models with three or more echelons

Table 4 shows two examples to solve the model using a software and using different data in which descending cost trend can be observed.

Of course, other information that were results of solving the model include the amount of goods sent (L_1, L_2) by different transportation models (M1, M2, M3, and M4) from the main factory to the applying factories (k1, k2) in the t time period. The total amount of delayed goods (L_1, L_2) in applying factories in the t time period, the model's suggested time for the main factory activity or non-activity in the t time period, and maintaining or not maintaining the sent goods in applying factories (k1, k2) in the t time period are shown in Tables 5, 6, and 7 and Figures 3, 4, and 5, wherein we show two recent examples based on 30 chromosomes and 30-time repetitions.

Finally, the Table 8 is the overall result in solving the model using the software, and different data for several repetitions.

As we can see, with this number of experiments, the best obtained result is the amount 12,450,000 which is shown in rows 35, 31, 30, and 36 of Table 8. Thus, considering the obtained result, we do not continue the solving process.

Competing interests

The authors declare that they have no competing interests.

Authors' contributions

Late Dr. MGA was a distinguished full professor of Industrial Engineering in Iran. He was an excellent expert in the field of operational research and modeling. Dr. AM, Associate Professor of school of Industrial Engineering at University of Science and Technology, Tehran, Iran. An expert in the field of research, production planning, Supply chain, Decision making and modeling techniques. SK, assistant professor of Dep. of management at Islamic Azad University. Mashhad Branch, most of his time is devoted to and occupied by doing research on the field of production planning and inventory control.

Acknowledgements

authors are grateful to the editor of Journal of Industrial Engineering International as well the anonymous referees for valuable and constructive comments to enhance quality of the paper.

Author details

[1]Dep. Of Industrial Management, Science and Research Branch, Islamic Azad University, Tehran, Iran. [2]Dep. Of Industrial Engineering, University of Science and Technology, Tehran, Iran. [3]Dep. Of Management, Mashhad Branch, Islamic Azad University, Mashhad, Iran.

References

Bertrand IP, Bookbinder JH (1998) Stock redistribution in two-echelon logistic systems. J Oper Res Soc 49(9):966–975

Bollapragada S, Akella R, Srinivasan R (1998) Centralized ordering and allocation policies in a two-echelon system with non-identical warehouses. Eur J Oper Res 106:74–81

Bookbinder JH, Chen VYX (1992) Multi criteria trade-offs in a warehouse-retailer system. J Oper Res Soc 43(7):707–720

Chen C, Lee W (2004) Multi-objective optimization of multi-echelon supply chain networks with uncertain product demands and prices. Comput Chem Eng 28:1131–1144

Dada M (1992) A two-echelon inventory system with priority shipments. Manag Sci 38(8):1140–1153

Dai T, Qi X (2007) An acquisition policy for a multi-supplier system with a finite-time horizon. Comp Oper Res 34(9):2758–2773

Erenguc SS, Simpson NC, Vakharia AL (1999) Integrated production /distribution in supply chain, an invited review. Euro J Oper Res 115:219–236

Hoque MA, Goyal SK (2000) An optimal policy for a single-vendor single-buyer integrated production-inventory system with capacity constraint of the transport equipment. Int J Prod Econ 65:305–315

Hwang HS (2002) Design of supply chain logistics system considering service level. Comput Ind Eng 43:283–297

Kyung PS, Dae KH (1989) Stochastic inventory model for two-echelon distribution system. Comp Ind Eng 16(2):245–255

Miner S (2005) Multiple-Supplier inventory models in supply chain management a review. Int J Prod Eco

Moon, et al. (2005) Economic order quantity models for ameliorating /deteriorating items under inflation and tied discounting. Eur J Oper Res 162:773–785

Rau HMY, Wu H, Wee M (2003) Integrated inventory model for determining items under a multi-echelon supply chain environment. Int J Prod Eco 86:155–168

Shtub A, Simon M (1994) Determination of reorder points for spare parts in a two-echelon inventory system: the case of non-identical maintenance facilities. Eur J Oper Res 73(3):458–464

Wang W, Fung RYK, Chai Y (2003) Approach of just-in-time distribution requirements planning for supply chain management. Int J Prod Econ 91:101–107

Yang PC, Wee HM (2002) A Single vendor and multiple buyers production inventory policy for a deteriorating item. Eur J Oper Res 145:570–581

Zhou G, Min H (2002) The balanced allocation of customers to multiple distribution centers in the supply chain network: a genetic algorithm approach. Comput Ind Eng 43:251–261

14

A customer oriented systematic framework to extract business strategy in Indian electricity services

Suchismita Satapathy[1*] and Pravudatta Mishra[2]

Abstract

Competition in the electric service industry is highlighting the importance of a number of issues affecting the nature and quality of customer service. The quality of service(s) provided to electricity customers may be enhanced by competition, if doing so offers service suppliers a competitive advantage. On the other hand, service quality offered to some consumers could decline if utilities focus their attention on those customers most likely to exercise choice, while reducing effort and investment to serve customers less likely to choose alternatives. Service quality is defined as the way in which the utility interacts with and responds to the needs of its customers. To achieve maximum consumer satisfaction in electricity service, This paper has designed a framework by QFD by measuring service quality of electricity utility sector in ANN and also find interrelationship between these design requirements by ISM.

Keywords: Service quality; ANN; QFD; ISM; Electricity utility; Consumer satisfaction

Introduction

In a competitive market for the provision of goods and services, there will be many different suppliers, each of whom may choose to offer a product with a particular combination of price and quality attributes. Financial risk management is often a high priority for participants in deregulated electricity markets due to the substantial price and volume risks that the markets can exhibit. A consequence of the complexity of a wholesale electricity market can be extremely high price volatility at times of peak demand and supply shortages. Consumers will choose the combination of price and quality that best meets their individual preferences. Suppliers will compete to improve their market share by striving to offer better quality products for the lowest price. In economic terms, electricity (both power and energy) is a commodity capable of being bought, sold, and traded. An electricity market is a system for effecting purchases, through bids to buy; sales, through offers to sell; and short-term trades,

generally in the form of financial or obligation swaps. Bids and offers use supply and demand principles to set the price.

Many infrastructure industries, including electricity distribution, are natural monopolies where one supplier can supply the whole market cheaper than two or more suppliers. In these markets, consumers have limited choice and limited bargaining power and are typically presented with only one quality of the product by the infrastructure provider. In the absence of economic regulation, the infrastructure provider will have an incentive to exploit its monopoly power and charge a high price for the product. In electricity systems that were government-owned, the service provided was often 'gold-plated' as suppliers concentrated on achieving a high level of engineering excellence. The service was, thus, expensive but often of a high standard.

When consumers purchase electricity they are purchasing a service with a number of different attributes. The most obvious of these is having electricity supplied at the place and time they want to use it. However, there are many other attribute dimensions that form the 'product' purchased and make up the level of service quality received. These include the reliability of the supply available (determined by the number of interruptions suffered and

* Correspondence: suchismitasatapathy9@gmail.com
[1]School of Mechanical Engineering, KIIT University, Bhubaneswar, 751024 Odisha, India
Full list of author information is available at the end of the article

the duration of any interruptions), the technical characteristics of the supply and their variability (voltage levels, frequency, and harmonics), and customer service (e.g., the timeliness and responsiveness of the supplier to requests for telephone assistance and the accuracy of billing). In addition to these direct attributes affecting their own consumption, some consumers may also be willing to pay a contribution towards societal goals such as having a high-quality electricity supply generally available, achieving environmental objectives and ensuring public safety.

Which attributes of service quality regulators concentrate on in incentive schemes should be determined by those that consumers value the most highly. For instance, it is not productive to 'incentivize' the distributor to minimize phone waiting times while ignoring the reliability of the electricity supply received when consumers are not particularly worried about phone response times but desperately need improvements in reliability. Also, consumers' priorities will change over time. As people become more affluent, they are generally prepared to pay more for a reliable electricity supply. But as reliability improves, consumers will generally be prepared to pay less for additional improvements. Changes in technology have also changed consumers' demand for service quality attributes. Greater use of computers and sophisticated electrical equipment has reduced preparedness to put up with poor-quality supply, particularly in rural areas. Changes in lifestyle choices have also changed service quality demand patterns. Regulators and distributors have used a range of techniques to ascertain and attempt to quantify consumers' preferences for service quality. Customers normally prefer better quality service to inferior quality service and are prepared to pay a premium for better service. However, the size of the premium they are prepared to pay will depend on their individual preferences and the amount of quality involved. Consumers typically exhibit reduced marginal willingness to pay as the amount of quality increases. That is, as they attain higher quality levels, consumers value additional improvements in quality less so they are prepared to pay less to go from a very good service to an excellent service than they were to go from a poor service into a mediocre service. Distributors, on the other hand, face increasing marginal costs of improving quality. For instance, improved maintenance practices and some basic strengthening of the network may improve service quality from poor to medium at modest cost. However, to go from medium to high service quality levels is likely to require major capital expenditure to strengthen and possibly duplicate parts of the network and make greater use of undergrounding. For this, a best system must be designed to satisfy customer's needs. Rabbani et al. (2012) have applied a new bi-objective fuzzy linear regression model in order to fill the gap in the field of forecasting using possibilistic programming.

Additionally, the proposed model is compared with three promising fuzzy linear regression models from the literature in order to forecast the energy consumption in the USA, Japan, Canada, and Australia during 2010 to 2015. Tinnilä (2012) has proposed a framework by catching the essential dimensions of the wide range of service facilities. Khare (2011) has attempted to understand the Indian customers' perceptions towards the service quality of multinational banks. Hajeeh (2010) has applied the analytical hierarchy process (AHP) to assess the different policies that could be used to bring about electricity conservation.

Even though on service quality of electricity utility industries few researches have been done, still it is an important matter for research. By improving service quality, the development of utility industry will be possible. For the improvement of service quality in electricity utility industry, the most important thing is required to find the points on which the service quality failure occurs in Indian electricity utility industry. Thus, artificial neural network (ANN) is implemented to get those important points for improvement. After getting the cause of failure, the rectification is done by designing the system for the improvement of electricity utility industry. For designing/framing the new electricity industry, quality function deployment (QFD) is used. Then, interpretive structural modeling (ISM) method is used to find the correlation between the design requirements of electricity utility industry. Thus, in this paper, three techniques/tools QFD, ISM, and ANN are used.

Literature review

Improving service quality is a top priority for firms that aim to differentiate their services in today's highly competitive business environment. Wyk et al. (1992) have briefly explored the advent of quality management systems, the rapid growth in popularity, and the benefits of service industries such as electricity utilities. Lamedica et al. (2001) have explained regarding the problem of the power quality (PQ) cost of industrial customers which is particularly important at present in the new scenery of competitive electricity markets. Frazier (1999) has described about the purpose of a power system which is to provide economical and reliable electric service to the end user. Hamoud and El-Nahas (2003) have described a probabilistic method for evaluating the level of supply reliability to a customer entering into a performance-based contract with a transmission provider. Carter (1989) has discussed about the most common power problems plaguing industries in this era. Chan (1993) has described in his paper about the increasing demand of good services from customers of utility industries. Mamlook et al. (2009) have described a fuzzy inference model for short-term load forecasting in power system operation. Rekettye and Pinter (2010) have established relationship between satisfaction and price acceptance in the context of the Hungarian electricity utility sector. In business markets, technical support and product customization have been shown to be additional drivers of customers' satisfaction levels (Kim et al. 2008; Kumar 2002).

An important theoretical approach for investigating the service quality is SERVQUAL analysis (Wisher and Corney 2001). Wisher and Corney defined service quality as a 'global judgment or attitude, relating to the superiority of the service', and explicated it as involving of the outcome and process of service act. Ulkuniemi and Pekkarinen (2011) have studied how modularity makes services visible and how it enables the customers to participate in service co-creation. Ozcelik (2010) has provided compelling evidence that the expansion of the scope of the Six Sigma methodology from the manufacturing to the service sector had been quite successful. Nakano (2011) have described, based on an ethnographic study, how, in expertise and experience-based practices, professionals follow a search/assemble process to deliver their services. Dahiyat et al. (2011) have examined the mediating roles of customer satisfaction and customer trust on the relationship between service quality and customer loyalty in Jordan's mobile service operators.

Azadeh and Movagha (2010) have suggested an integrated approach based on data envelopment analysis (DEA) and principal component analysis (PCA) for efficiency assessment and optimization of transmission systems. Rengarajan and Loganathan (2012) explained fuzzy logic-based power theft prevention and PQ improvement by comparing between the total load supplied by the distribution transformer and the total load used by the consumer, and the error signal is used to identify the power theft. Shaikh et al. (2012) focused on the right initiation at the right time to ease control action to enhance stability, reliability, and security of the power system so as to provide a preventive plan to minimize the chances of failure in power system as possible. Chau (2009) reviewed the evolution and development of customer service performance measures in the electricity sector since privatization in 1989 and then examine the impact of a specific recent energy regulatory requirement (known as information and incentives project (IIP)) on the organizational management of an exemplar electricity distribution company. Wattana and Sharma (2011) examined the veracity of this argument by analyzing both the technical and environmental performance of the Thai electricity industry. Parida et al. (2011) explore the methods already proposed in various literatures to overcome the issues associated with VAr management in a competitive environment. Managing reactive power support service in competitive electricity market environment has become an important constituent of ancillary services. Srivastava et al. (2012) have factorized the environmental and social pillars of sustainable development when defining access to modern energy forms which significantly inform the level of effort involved in meeting the goal of energy access to all.

Sahney (2008) has applied the quality function deployment technique to identify performance indicators critical to the success of online retailing. These were prioritized qualitatively through interpretive structural modeling. Chung et al. (2008) have applied discriminant analysis and artificial neural network which are utilized in this study to create an insolvency predictive model that could effectively predict any future failure of a finance company and validated in New Zealand.

Ikiz and Masoudi (2008) have described the development of a conceptual framework to measure the hotel service quality using the SERVQUAL model as a starting point and then identify service design and hotel guests' requirements using a QFD approach.

Bouchereau and Rowlands (2000) have applied fuzzy logic, artificial neural networks, and the Taguchi method combinedly with QFD to resolve some of its drawbacks and propose a synergy between QFD and the three methods. Sherwood (1993) has identified few problems of service industries by the ANN method.

Methodology

The customer is satisfied when he/she feels that the service performance fits well with his/her personal framework (confirming). If it remains below expectations, then the customer will be dissatisfied (disconfirming). To determine the customer satisfaction with service quality of electricity utility industry, a standard questionnaire is designed for all the customers of different categories like domestic consumer. A questionnaire survey was conducted for agricultural customer, industrial and commercial customer, and others. The questionnaire consists of 26 items to investigate the respondent's perception about the service quality of electricity utility industry. Five hundred questionnaires are circulated to different consumers by Internet, phone, and personal contacts. Among them, 293 responses (150 from domestic customer, 43 from industrial and commercial customer, 54 from agricultural and 46 from public organizational consumers) are obtained. The response rate is 78%, which is good for this type of survey. The respondents are advised to respond each item of the questionnaire in a 7-point Likert-type scale (1 = totally disagree, 2 = partially disagree, 3 = somewhat disagree, 4 = no opinion, 5 = somewhat agree, 6 = partially agree, and 7 = totally agree). The details of items in the questionnaire are given in Table 1. After collecting consumer perception, the analysis was done.

Data analysis
Factor analysis
A series of statistical analyses are required before the steps followed for quantification of service quality. These

Table 1 Questionnaire for survey

SI number	Items	Rank						
1	Advance information is received about power shut downs, and notices and regulations are provided	1	2	3	4	5	6	7
2	Load shedding is effected according to the statutory guidelines as per the OERC norms and policy of government	1	2	3	4	5	6	7
3	The quality of power supplied is maintained regularly without interruption	1	2	3	4	5	6	7
4	The correct energy bill is served regularly and in complete shape	1	2	3	4	5	6	7
5	For the ease of customer the modes of payment of electric bills are through cash collecting counter, e_payment mobile vans and through customer care centre	1	2	3	4	5	6	7
6	Fines are imposed for delayed payment and rebates for punctual payment	1	2	3	4	5	6	7
7	Utility staffs are available for registering complaint, enquiry and maintenance work	1	2	3	4	5	6	7
8	For old citizens providing bills, collecting money and receipts are done by the utility organizations	1	2	3	4	5	6	7
9	The meter checking is done by trained employees	1	2	3	4	5	6	7
10	Meters and sub-meters are provided to all consumers	1	2	3	4	5	6	7
11	A complaint number is received at the time of registering, and the complains are rectified in time	1	2	3	4	5	6	7
12	In case of any problem with transformer it is immediately replaced	1	2	3	4	5	6	7
13	In necessary condition load enhancement is done	1	2	3	4	5	6	7
14	They are well equipped and try to maintain safety and security of customers	1	2	3	4	5	6	7
15	The redressal forum has been approached for justice when dissatisfied with the response of the distribution company	1	2	3	4	5	6	7
16	In case of any maintenance work the electricians are available and the rectification is done in time	1	2	3	4	5	6	7
17	No bribes, humiliations or extra charges are demanded in the utility organization	1	2	3	4	5	6	7
18	Applying for a new electricity supply is easy and it is provided in regimal time	1	2	3	4	5	6	7
19	The basic electricity safety lessons including ghost mark are displayed at the required places like electric poles, succession light, no entry areas	1	2	3	4	5	6	7
20	Due to absence of voltage fluctuation the whole system works properly	1	2	3	4	5	6	7
21	The utility service work force is knowledgeable, self-confident, skilled and reliable	1	2	3	4	5	6	7
22	The electricity charges, penalties are imposed as per the OERC guidelines and as per Indian electricity act 2003 for disobeying electricity rule	1	2	3	4	5	6	7
23	Supply is according to demand	1	2	3	4	5	6	7
24	There is no queue or delay at the mobile vans or cash counter during bill payment	1	2	3	4	5	6	7
25	Authentic receipt is provided to consumer against bill payment	1	2	3	4	5	6	7
26	The justice, accidental benefits and subsidies are provided to consumers categorically	1	2	3	4	5	6	7

analyses are conducted using SPSS which is one of the most widely used statistical software packages. Two such techniques such as reliability test and exploratory factor analysis (EFA) were conducted on all the 393 valid responses using SPSS 17.0 software. Reliability test is the most widely used method to measure internal consistency of variables, and exploratory factor analysis is a technique whose major goal is to identify the underlying relationships between the measured variables and their corresponding dimension under which they are grouped. The internal consistency of the survey data are tested by computing Cronbach's alpha (α). From the test results shown in Table 2, it is observed that out of them only 22 items have α value of 0.6 and above. Hence, these items are categorized under seven dimensions such as reliability, tangibility, empathy, responsiveness, assurance, security, and

stability which are selected on the basis of their relevance to the service quality in electricity utility sector. Since the four items of the questionnaire with serial numbers 6, 8, 15, and 17 have α value less than 0.6, they are not useful and hence discarded from further analysis. Moreover, the overall value of α taking all seven dimensions into consideration is found to be 0.7493 which is above the acceptable value of 0.7 for demonstrating internal consistency of the established scale. From EFA, the total percentage of variance explained has been obtained as 79%, and this is an acceptable value for the principal component varimax rotated factor loading procedure. The value of Kaiser-Meyer-Olkin (KMO), which is a measure of sampling adequacy, is found to be 0.79, indicating that the factor analysis test has proceeded correctly and the sample size of 393 used in the data survey is adequate as the minimum acceptable value

Table 2 Factorial analysis

Dimension	Item number	Factor 1	Factor 2	Factor 3	Factor 4	Factor 5	Factor 6	Factor 7	Cronbanch's alpha (0.7493)
Reliability	11	0.770							0.736
	12	0.648							0.711
	16	0.717							0.703
	18	0.672							0.729
	23	0.725							0.739
Tangibility	9		0.674						0.734
	10		0.796						0.729
	13		0.602						0.747
	20		0.672						0.753
	25		0.675						0.769
Assurance	21			0.608					0.738
	26			0.755					0.755
Empathy	5				0.697				0.740
	7				0.795				0.746
	24				0.680				0.732
Responsiveness	2					0.663			0.755
	3					0.761			0.740
	22					0.785			0.739
Security	14						0.736		0.741
	19						0.735		0.722
Stability	1							0.849	0.759
	4							0.695	0.764

of KMO is 0.5. Therefore, it can be concluded that the matrix did not suffer from singularity or multi-collinearity. The result of Bartlett test of sphericity shows that it is highly significant (significance = 0.000), which indicates that the factor analysis is correct and suitable for testing multi-dimensionality. Therefore, the above statistical tests have confirmed that the proposed 22 items and 7 dimensions of instruments as listed in Table 3 are sound enough for the further analysis. In Table 4, the dimensions are ranked as per their cumulative variance.

Linear discriminant analysis (LDA) is a pattern recognition method providing a classification model based on the combination of variables that best predicts the category or group to which a given object belongs. In this study, the independent variable is the service quality provided by electricity utility industry as per customer's perception, and group variables are the four sectors of electricity utility consumers such as domestic consumer, agricultural consumer, industrial and commercial consumer, and public organizational consumer. Stepwise discriminant analysis is performed on the data related to 22 items belonging to 7 dimensions selected in the factor analysis done earlier using the same SPSS 17.0 statistical software. This analysis focused on finding out all

those items which are the most significant in any of the four types of consumer sectors. All the responses received are divided sector-wise into four groups. In LDA, the minimum F entry (F_{min}) and maximum F removal (F_{max}) are set to their default setting values of 3.84 and 2.71, respectively. Wilk's lambda (λ) value of 1.00 occurs when observed group means are equal (i.e., all the variances are explained by factors other than the difference between those group means), while a λ value less than 1.00 occurs when within-group variability is small compared to the total variability. Thus, a small λ indicates that group means appear to differ. The results of LDA are listed in Tables 5, 6, 7, 8, 9, 10, and 11. In each of these tables, λ value is less than 1. Thus, the result is fit for the analysis. The differences in mean response values are compared between the groups taking two at a time, and the minimum difference value is listed in the last column of the table. The corresponding group numbers for which this minimum difference squared value occurs has also been shown in all these tables.

Table 5 shows the results of discriminant analysis for all the five items belonging to the first-ranked dimension, i.e., reliability. Items 11 and 16 have the significance values as 0.000 and lambda value less compared

Table 3 Naming of constructs

Constructs	Item number	Items
Reliability	11	A complaint number is received at the time of registering, and the complains are rectified in time
	12	In case of any problem with transformer it is immediately replaced
	16	In case of any maintenance work the electricians are available and the rectification is done in time
	18	Applying for a new electricity supply is easy and it is provided in regimal time
	23	Supply is according to demand
Tangibility	9	The meter checking is done by trained employees
	10	Meters and sub-meters are provided to all consumers
	13	In necessary condition load enhancement is done
	20	Due to absence of voltage fluctuation the whole system works properly
	25	Authentic receipt is provided to consumer against bill payment
Assurance	21	The utility service work force is knowledgeable, self-confident, skilled and reliable
	26	The justice, accidental benefits and subsidies are provided to consumers categorically
Empathy	5	For the ease of customer the modes of payment of electric bills are through cash collecting counter, e_payment mobile vans and through customer care centre
	7	Utility staffs are available for registering complaint, enquiry and maintenance work
	24	There is no queue or delay at the mobile vans or cash counter during bill payment
Responsiveness	2	Load shedding is effected according to the statutory guidelines as per the OERC norms and policy of government
	3	The quality of power supplied is maintained regularly without interruption
	22	The electricity charges, penalties are imposed as per the OERC guidelines and as per Indian electricity act 2003 for disobeying electricity rule
Security	14	They are well equipped and try to maintain safety and security of customers
	19	The basic electricity safety lessons including ghost mark are displayed at the required places like electric poles, succession light, no entry areas
Stability	1	Advance information is received about power shut downs, and notices and regulations are provided
	4	The correct energy bill is served regularly and in complete shape

to other items in reliability dimension (i.e., item 12, item 16, item 23).

There are four groups of consumers, i.e., domestic, agricultural, industrial, and public organization which are also referred to as groups 1 to 4, respectively. The stepwise discriminant analysis is done by SPSS 17 software for all these four groups/sectors of consumers for each of the seven dimensions separately. This analysis focused on finding out the sector-wise difference for of each of the item relevant to a given dimension.

Table 4 Ranking of constructs

Dimensions	Percentage of variance explained	Ranking
Reliability	0.863	1
Tangibility	0.836	2
Assurance	0.804	3
Empathy	0.770	4
Responsiveness	0.729	5
Stability	0.685	6
Security	0.678	7

The Wilks' lambda (λ) value of 1.00 occurs when observed group mean responses are equal (i.e., all the variances are explained by factors other than difference between those group means), while a λ value less than 1.00 occurs when within-group variability is small compared to the total variability. Thus, a small λ indicates that group means appear to differ. For example, in Table 3 among five items (i.e., 11, 12, 16, 18, and 23) under dimension reliability, items 11 and 16 have the significance values as 0.000. Further, these two items have Wilks' lambda values (i.e., 0.763, 0.787) less than that of all other items (i.e., items 12, 18, and 23). Then, the mean value difference is calculated for all the four groups for item 11.

For example, the mean difference between domestic (group 1) and agriculture (group 2) groups for item 11 is found as $3.4783 - 2.0000 = 1.4783$. Similarly, for the same item 11, the mean differences are calculated for all other possible pairs of groups, i.e., groups 1 and 3, groups 1 and 4, groups 2 and 3, groups 2 and 4, and groups 3 and 4, respectively. Among all these six pairs of groups of consumers, it is found that the mean difference value is minimum (i.e., $3.8000 - 3.4783 = 0.3217$) in the case of

Table 5 Sector-wise difference for reliability

Item number	Mean response				Group	Wilks' lambda	F value	Significance	Min. D^2
	Domestic (group 1)	Agricultural (group 2)	Industrial (group 3)	Public organization (group 4)					
11	3.4783	2.0000	3.8000	4.0357	1 and 3	0.763	7.852	0.000	0.1089
12	3.5652	3.8750	4.6000	3.3214	1 and 4	0.961	1.018	0.390	0.0959
16	2.9130	3.3333	3.6000	4.9286	2 and 3	0.787	6.869	0.000	0.0711
18	4.8261	4.2083	4.6000	4.9286	1 and 4	0.980	0.504	0.681	0.0105
23	5.0435	3.9583	3.8000	3.8571	3 and 4	0.925	2.045	0.115	0.0032

groups 1 and 3. Therefore, '1 and 3' have been shown under the column 'Group' against item 11 in Table 5.

Thus, comparison has been done among all possible groups against a particular item, and only the pairs of groups having the minimum difference in mean response values are listed in the sixth column (i.e., under the column heading 'Group') of Tables 5, 6, 7, 8, 9, 10, and 11.

Similarly, Table 6 presents the sector-wise results for all the items under the second-ranked dimension, i.e., tangibility. Out of all items, the significant values of three items, i.e., with serial numbers 9, 10, and 13, are observed to be close to 0.000. Items 9 and 10 significantly differ due to minimum mean difference when groups 1 and 2 (i.e., domestic and agricultural sectors) are compared. Similarly, groups 3 and 4 (i.e., industrial and public organization sectors) significantly differ for item 13. As far as tangibility dimension is concerned, no significant difference is found in any other item.

Table 7 shows the sector-wise results for all the items under third-ranked dimension, i.e., assurance. Among all the items, the significant value of only item 26 is found to be close to 0.000, and mean difference is less when groups 1 and 4 (i.e., domestic and public organization sectors) are compared. Under assurance dimension, no other item is observed to be significant.

Table 8 shows the sector-wise results for all the items under fourth-ranked dimension, i.e., empathy. Here, out of all the items, the significant value of only item 24 is found to be close to 0.000, and mean difference is minimum when groups 2 and 4 (i.e., agricultural and public

organization sectors) are compared. Under dimension empathy, no other item is observed to be significant.

Table 9 shows the sector-wise results for all the items under fifth-ranked dimension, i.e., responsiveness. Among all the items, item 3 has the significant value close to 0.000 when groups 2 and 3 (i.e., agricultural and industrial sectors) are compared and less mean difference value is found. Under dimension responsiveness, no other item is found to be significant.

Table 10 shows the sector-wise results for all the items under sixth-ranked dimension, i.e., stability. Among all the items, the significant value of only one item, i.e., item 4, is found to be close to 0.000 when groups 2 and 4 (i.e., agricultural and public organizational sectors) are compared and minimum mean difference value is found. Under dimension stability, no other item has any significance.

Table 11 shows the sector-wise results for all the items under seventh-ranked dimension, i.e., security, where no other items except item 14 have significant value close to 0.000. Item 14 is found to be significantly different when groups 3 and 4 (i.e., industrial and public organization sectors) are compared and less mean difference value is found.

Network parameters

The back propagation module of a neural network package NeuNet Pro version 2.3 is used for the training and testing of survey data. The network parameters are taken as follows: Input layer with 22 nodes, one hidden layer with 11 nodes, and an output layer with a single node. A

Table 6 Sector-wise difference for tangibility

Item number	Mean response				Group	Wilks' lambda	F value	Significance	Min. D^2
	Domestic (group 1)	Agricultural (group 2)	Industrial (group 3)	Public organization (group 4)					
9	3.5217	3.8333	6.4000	4.5357	1 and 2	0.759	8.03	0.000	0.0961
10	2.6522	2.3333	5.2000	3.8214	1 and 2	0.809	6.000	0.001	0.1018
13	2.0870	2.9167	3.6000	4.0000	3 and 4	0.608	16.35	0.000	01600
20	2.8000	2.8200	3.2000	4.000	1 and 2	0.967	4.567	0.280	0.0004
25	5.3913	5.5417	4.6000	6.0714	1 and 2	0.945	1.461	0.232	0.0226

Table 7 Sector-wise difference for assurance

Item number	Mean response				Group	Wilks' lambda	F value	Significance	Min. D^2
	Domestic (group 1)	Agricultural (group2)	Industrial (group 3)	Public organization (group 4)					
21	4.6522	2.9167	1.6000	4.8214	1 and 4	0.786	6.902	0.000	0.0286
26	6.3043	4.5417	4.6000	4.7857	2 and 3	0.813	5.821	0.008	0.0344

single question regarding overall customer evaluation of the service quality is considered as the output.

As per the software recommendations, the number of nodes in the hidden layer (H) is decided by the relation below:

$$H = 2\sqrt{I+1}. \tag{1}$$

Normalization of raw data was carried to obtain values between 0 and 1 for expressing all data in a common scale. Learning rate and momentum parameter are set at 25% and 20%, respectively, during the training phase. The numbers of correct outputs were noted until the root mean square error (RMSE) is minimized to a reasonable value.

Sensitivity analysis

The model is run varying the learning parameter, momentum parameter, and number of cycles until RMSE is minimized. Both the learning parameter and momentum parameter are set at 0.25 and 0.20 to obtain the best results. The model is run for 294,455 number of cycles to obtain RMSE of 0.01 for two outputs, respectively. Training of the network is stopped at this point. In order to find the robustness of the proposed model, sensitivity analysis was carried out. Sensitivity analysis is used to study the impact of service quality evaluation output on the various items (inputs) in electricity utility constraints. The inputs in the test samples are varied one at a time systematically, up and down 10% (±10%) from its base value holding other items at their original values. The scaled change in output is calculated with the current input increased by 10% and the current input decreased by 10%. The scaled change in output is given by the following:

Thus, the results obtained are the scaled output change per 10% change in input. The calculation is repeated for every input and every fact and then averaged across all the facts, yielding a single mean scaled change in output for each input criterion. Increasing input (gap) from its base value causes decrease in service quality due to the widening of the gap, whereas reduction of gap indicates an increased service quality evaluation. Logically, net effect of change in input (gap) results in negative score for average scaled change in output. However, positive or increased service quality is also obtained in all the cases. This irregularity may be attributed to the noisiness of the survey data. Noisy data exists when customers responding to survey have similar evaluation on individual question but different evaluation of the overall service quality. This results in similar input data for the neural network with very different corresponding outputs.

It can be observed from Tables 12 and 13 that there are seven common items rated negative score by all stakeholders. The seven common items include the following: the quality of power supplied (item 3), for the ease of customer, the mode of payment of electric bills (item 5), utility staffs are available for registering complaint (item 7), applying for a new electricity supply is easy (item 18), the basic electricity safety lessons are displayed at the required places like electric poles (item 19), due to no fluctuation of voltage, the whole system works properly (item 20), and no queues and no delay at the time of bill payment (item 24). From Table 12, it is found that the other items, apart from the items mentioned above as common items, also contribute for service quality of electricity utility sector. For example in the domestic sector, two items, item 21 and item 23,

$$\frac{\text{Scaled output for 10\% increase in output–Scaled output for 10\% decrease in output}}{2} = \text{Scaled change in output.}$$

Table 8 Sector-wise difference for empathy

Item number	Mean response				Group	Wilks' lambda	F value	Significance	Min. D^2
	Domestic (group 1)	Agricultural (group 2)	Industrial (group 3)	Public organization (group 4)					
2	3.5652	3.8750	4.6000	3.3214	2 and 4	0.961	1.018	0.390	0.0594
3	2.9130	3.3333	3.6000	4.9286	2 and 3	0.787	6.869	0.000	0.0711
22	4.8261	4.2083	4.6000	4.9286	1 and 4	0.980	0.504	0.681	00105

Table 9 Sector-wise difference for responsiveness

Item number	Mean response				Group	Wilks' lambda	F value	Significance	Min. D^2
	Domestic (group 1)	Agricultural (group2)	Industrial (group 3)	Public organization (group 4)					
5	2.3913	1.7917	3.4000	2.2143	1 and 4	0.965	0.920	0.435	0.0313
7	3.1739	4.3333	2.6000	4.1071	2 and 4	0.899	2.846	0.043	0.1649
24	4.4348	6.3333	3.2000	5.9286	2 and 4	0.790	6.718	0.000	0.1636

contribute to the electricity utility sector. Similarly, two items, item 2 and item 21, contribute towards improvement of agricultural sector in electricity service. Item 25 and item 2 of the industrial sector and public organization sector, respectively, are essential for the improvement of electricity utility sector. It implies that these seven items have strong effect on service quality, and the policy makers of the electricity utility must focus on these areas for improving satisfaction level of different types of consumers. It is found that after the above seven items, the rest of the items must be taken cared according to their ranks and according to their priorities the policies must be formed. After getting seven important parameters/items for the improvement of electricity utility sectors, QFD is implemented for these seven items, and a frame work is designed.

QFD methodology

The QFD transforms all the information on a graphical display known as 'House of Quality'. This house provides a framework that guides the team through the QFD process. It is a matrix that identifies the 'whats' and 'hows', the relationship between them, and the criteria for deciding which of the hows will provide the greatest customer satisfaction. The top of the house identifies the interrelationships between the hows. Enhancements to the QFD process include adding important measures to the customer requirements including target values for product design features and relating product design features to part and mechanism characteristics.

Application of QFD to electricity utility service

After getting seven important parameters/items for the improvement of electricity utility sectors by interview with experts, QFD was applied to identify system design requirements in electricity utility industry for customer

satisfaction. QFD uses a series of matrices to document information collected and represent the QFD team's plan for a customer need and system design requirements. Seven customer requirements and eight design characteristics are identified in an electricity utility industry. Tables 14 and 15 show the details of the customer needs and system design needs.

The house of quality (HoQ) (matrix) is the mainly acknowledged form of QFD. HoQ is constructed from these major components as explained below:

➤ Customer needs (*whats*): a structured list of requirements derived from experts feedback.
➤ Design requirements (*hows*): a structured set of relevant and measurable services/characteristics which are required for fulfilling whats.
➤ Planning matrix (left matrix): gives customer/expert perceptions observed in surveys. It includes the relative importance of requirements.
➤ Interrelationship matrix (center matrix): gives the expert's perceptions of interrelationships between design requirements and customer needs. An appropriate scale is applied and illustrated using symbols or figures. Filling this portion of the matrix involves discussions.
➤ Design correlation (top) matrix: used to identify where design requirements support or impede each other in the system or product design.

After interview, the seven common items selected like the quality of power supplied (item 3), for the ease of customer, the mode of payment of electric bills (item 5), utility staffs are available for registering complaint (item 7), applying for a new electricity supply is easy (item 18), the basic electricity safety lessons are displayed at the required places like electric poles (item 19), due to no fluctuation of voltage, the whole system works properly

Table 10 Sector-wise difference for stability

Item number	Mean response				Group	Wilks' lambda	F value	Significance	Min. D^2
	Domestic (group 1)	Agricultural (group2)	Industrial (group 3)	Public organization (group 4)					
1	5.0435	3.9583	3.8000	3.8571	2 and 4	0.925	2.045	0.115	0.0032
4	2.4348	4.4583	3.8000	4.7500	2 and 4	0.783	7.028	0.000	0.0850

Table 11 Sector-wise difference for security

Item number	Mean response				Group	Wilks' lambda	F value	Significance	Min. D^2
	Domestic (group 1)	Agricultural (group 2)	Industrial (group 3)	Public organization (group 4)					
14	3.4783	2.0000	3.8000	4.0357	3 and 4	0.763	7.852	0.000	0.0555
19	3.3043	3.6250	5.0000	4.5000	1 and 2	0.905	2.667	0.054	0.1028

(item 20), and no queues and no delay at the time of bill payment (item 24) are used for customer needs. By observing these customer needs, design requirements are located at the top of the HoQ. These are very important for improvement of electricity service in India.

Revised customer rating for the needs was determined from the correction matrix:

$$Z_i + \frac{1}{n-1}\sum_{j\neq1}^{n}\left(B_{ij}\cdot Z_j\right),$$

where B_{ij} is the ith initial customer rating, Z_j denotes the relationship between the customer need and customer rating, n is the number of customer needs, and $B_{ij} = \begin{bmatrix} B1_{ij}, & B2_{ij} \end{bmatrix}$.

The individual rating for each design requirement is obtained from the center matrix using the following relation:

$$\text{Design requirement} = \sum_{j}^{n} A_{ij}\cdot X_j,$$

where A_{ij} and X_j denote the relative importance of the ith characteristics with respect to the jth customer need in the relationship matrix and the importance of the jth customer needs (customer ratings) and n is the number of customer needs.

A QFD matrix was prepared, and the respondents were asked to assign numeric relative importance scores to the various customer requirements, from the lowest to the highest, on a scale from 1 to 9. They were also

Table 12 Result of neural network model in output 1 (service quality satisfaction)

Dimension	Item number	Learning parameter	Momentum parameter	Number of cycles	RMS error	Agricultural	Industrial	Domestic	Public organization
Reliability	11	25	20	294,455	0.01	−0.0856	0.0580	−0.0870	−0.0870
	12					−0.06	0.0784	0.073	−0.35
	16					−0.099	−0.136	0.0526	0.071
	18					−0.311	−0.094	−0.156	−0.381
	23					−0.387	0.051	−0.27	0.079
Tangibility	9	25	20	294,455	0.01	−0.0861	−0.059	−0.071	−0.45
	10					−0.095	0.096	−0.018	−0.08
	13					−0.067	0.074	−0.06	−0.16
	20					−0.41	−0.32	−0.29	−0.35
	25					−0.54	−0.33	−0.41	0.09
Assurance	21	25	20	294,455	0.01	−0.45	−0.051	−0.44	0.07
	26					−0.31	−0.16	−0.17	−0.05
Empathy	5	25	20	294,455	0.01	−0.49	−0.57	−0.48	−0.40
	7					−0.24	−0.01	−0.015	−0.45
	24					−0.44	−0.44	−0.35	−0.17
Responsiveness	2	25	20	294,455	0.01	−0.37	−0.06	−0.38	−0.306
	3					−0.21	−0.28	−0.27	−0.28
	22					−0.12	0.21	0.17	0.38
Security	14	25	20	294,455	0.01	−0.07	0.031	−0.024	0.10
	19					−0.33	−0.041	−0.416	−0.09
Stability	1	25	20	294,455	0.01	0.07	0.067	0.07	0.10
	4					0.03	−0.05	0.04	−0.16

Table 13 Rank of items

Agricultural		Industrial		Domestic		Public organization	
Item 5	−0.4901 a	Item 5	−0.5637 a	Item 5	−0.487 a	Item 5	−0.4015 a
Item 24	−0.4481 b	Item 24	−0.4405 b	Item 21	−0.448 b	Item 7	−0.4528 b
Item 21	−0.4503	Item 19	−0.4167 c	Item 19	−0.416 c	Item 18	−0.3811 c
Item 20	−0.4157 c	Item 25	−0.3387	Item 23	−0.3886	Item 20	−0.359 d
Item 2	−0.378	Item 20	−0.3282 d	Item 24	−0.3580 d	Item 2	−0.3062
Item 19	−0.3356 d	Item 3	−0.28 e	Item 20	−0.2916 e	Item 3	0.28 e
Item 18	−0.3114 e	Item 5	−0.2721 f	Item 3	−0.2748 f	Item 19	−0.1618 f
Item 7	−0.2467 f	Item 26	−0.1637 g	Item 7	−0.1579 g	Item 24	−0.1713 g
Item 3	−0.21 g	Item 16	−0.136 h	Item 18	−0.1563 h		

The negative score for average scaled change in output scores per 10% variation in inputs is the norm. Different letters indicate common items.

asked to express, in numeric values, the strong-moderate-weak relationship between what's and how's as .2, .4, .6, and .8 with symbols. After the responses were collected, scores for each of the columns was computed. The absolute values were computed for each column, and the respective what's and how's were then ranked. In Figure 1, all the revised costumer ratings as per the correlation of customer needs are calculated using Equation 1. Then, correlation is calculated using the specified symbols (present in Table 16), and Equation 2 is used for calculating the design requirement. The top matrix correlation is shown in Figure 1. The final rating of design requirements is normalized by dividing each rating with the maximum available rating. The final ratings are shown in Figure 1. Using the normalized ratings, the design requirements are prioritized as per their importance of design requirements. It is shown that design requirement 7 (customer care centers must be open day and night) is the most prioritized design requirement, design requirement 5 (electricity policies must be framed as per the customer requirement) is the second prioritized design requirement, and design requirement 6 (advanced technology transformers, good equipments, and proper power demand side management) is ranked as the third most prioritized design requirement. Then, ISM is implemented to find the relation between these design requirements.

Interpretive structural modeling and interrelationship of design requirements

After prioritizing the design requirements by QFD, ISM methodology is implemented to find hierarchy of interrelations among the identified enablers. ISM methodology helps to impose order and direction on the complexity of relationships among the elements of a system. It is interpretive as the judgment of the group decides whether and how the variables are related. It is structural as on the basis of relationship; an overall structure is extracted from the complex set of variables. It is a modeling technique as the specific relationships and overall structure are portrayed in a graphical model. The various steps involved in the ISM technique are as follows:

➤ Identifying elements which are relevant to the problem or issues - this could be done by survey.

Table 14 Customer requirements

	Items
1	The quality of power supplied
2	The mode of payment of electric bills
3	Utility staffs are available for registering complaint
5	Applying for a new electricity supply is easy
6	The basic electricity safety lessons are displayed at the required places like electric poles
7	Due to no fluctuation of voltage, the whole system works properly
8	No queues and no delay at the time of bill payment

Table 15 Design according to technical requirement

	Requirements
1	Regular power supply without any interruption
2	Availability of mobile vans, customer care center, and online payment system for easy payment of electricity bills
3	For listening, registering complaints using toll-free numbers, customer care people, and websites must be available
4	Quick procedures for new supply of electricity
5	Electricity policies must be framed as per the customer requirement
6	Advanced technology transformers, good equipments, and proper power demand side management
7	Customer care centers must be open day and night
8	Safeties of the customers are taken cared by showing safety lessons in the website, regarding electric poles
9	Regular maintenance work must be done

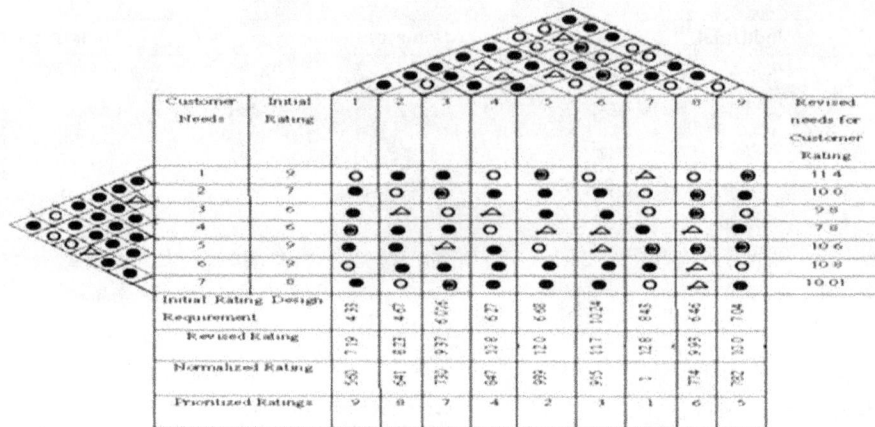

Figure 1 House of quality.

> Establishing a contextual relationship between elements with respect to which pairs of elements would be examined.
> Developing a structural self-interaction matrix (SSIM) of elements which indicates pair-wise relationship between elements of the system.
> Developing a reachability matrix from the SSIM and checking the matrix for transitivity - transitivity of the contextual relation is a basic assumption in ISM which states that if element A is related to B and B is related to C, then A is related to C.
> Partitioning of the reachability matrix into different levels.
> Based on the relationships given above in the reachability matrix, drawing a directed graph (digraph) and removing the transitive links.
> Converting the resultant digraph into an ISM-based model by replacing element nodes with the statements.
> Reviewing the model to check for conceptual inconsistency and making the necessary modifications.

The various steps, which lead to the development of ISM model, are illustrated in Figure 2 by flow chart for ISM.

Application of interpretive structural modeling

Stuart and Tax (1996) discussed that interpretive structural modeling was first projected by Warfield to analyze

Table 16 Symbols for QFD

Relationship		Symbol
.8	Strong	O
.6	Moderate	⊖
.4	Week	Δ
.2	Very week	●

the complex socioeconomic systems. Kumar et al. (2009) have described that ISM is a computer-assisted learning process that enables individuals or groups to develop a map of the complex relationships between the many elements involved in a complex situation. Its basic idea is to use experts' practical experience and knowledge to decompose a complicated system into several sub-systems (elements) and construct a multilevel structural model. ISM is often used to provide fundamental understanding of complex situations as well as to put together a course of action for solving a problem. The ISM approach is useful when a multilevel research design is required where the outcome of the research cannot be predicted based on available research. Anantatmula and Kanungo (2005) have described that ISM approach is different in that it is relatively more efficient (in some cases) and lends itself to being replicated more effectively.

Nelson et al. (2000) have shown that interpretive structural modeling results in a directed graphic representation of a particular relationship among all pairs of elements in a set to aid in structuring a complex issue area. Porter et al. (1980) have given three broad steps for developing an interpretive structural model. These are step 1: ISM begins with an issue or problem, step 2: the next step is to identify the elements that comprise the issue context which are listed, and step 3: in the third step, pairs of elements are compared graphically or in a relation matrix, using a contextual relationship, which is mostly a verb or a verb phrase. Hansen et al. (1979) have elaborated ISM as the representation of a problematique because it captures the richness and the variety of complex phenomena. Anderson et al. (1994) have discussed a framework for business performance improvement and leadership to test its 'fitness' through interpretive structural equation modeling to be used for understanding the predictive behavior of the acquisition success. Mishra (2006) has used ISM to analyze the relationships among

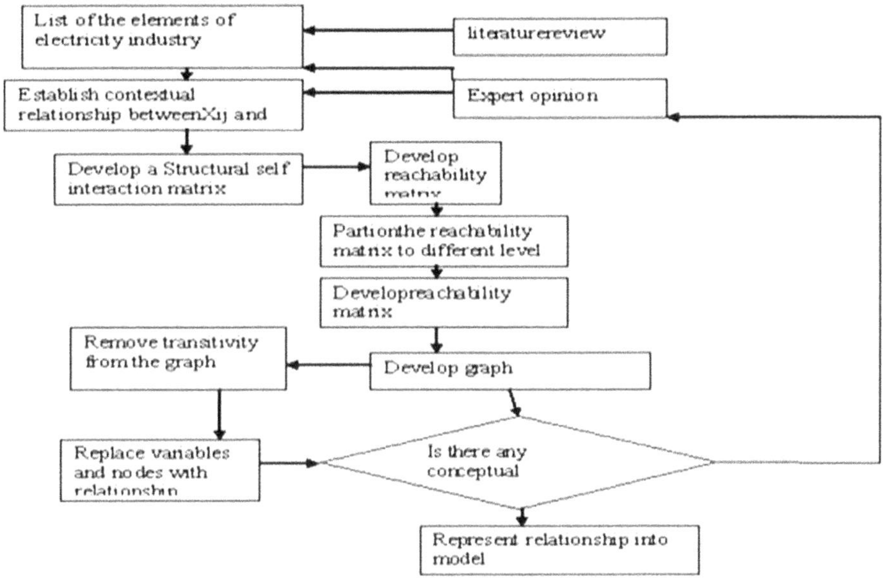

Figure 2 Flow chart of ISM.

innovativeness, learning orientation, market orientation, entrepreneurial orientation, organizational structure, and policies and business performance. The critical success factors of electricity policies and its impact on overall business performance have been analyzed using the ISM methodology, and the relation between the variables is found.

Application of ISM in electricity utility service

Keeping in view the contextual relationships in nine elements (design requirements), the existence of a relationship between two elements (i and j) and the associated direction of relationship have been identified (Table 17).

The procedural steps for ISM are listed as follows:

1. Identification of variables: On the basis of prioritization of nine design requirements from the house of quality, the variables are identified.
2. Contextual relationship: From the identified elements in step 1, a contextual relationship is identified among them with respect to whom pairs of variables would be examined. After resolving the system design requirement set under consideration and the contextual relation, SSIM is prepared. Four symbols are used to denote the direction of relationship between the criterion (i and j):
   (a) V - barrier i will help to achieve barrier j
   (b) A - barrier j will help to achieve barrier i
   (c) X - barriers i and j will help to achieve each other
   (d) O - barriers i and j are unrelated.

Table 17 Victograph on the correlation of factors

Elements/variable	9	8	7	6	5	4	3	2
1	V	A	A	A	A	A	A	A
2	V	A	X	A	X	V	V	
3	V	X	A	X	A	X		
4	V	X	A	X	A			
5	V	O	V	X				
6	V	O	V					
7	V	X						
8	V							
9								

Table 18 Level partition

Elements	1	2	3	4	5	6	7	8	9
1	1	0	0	0	0	0	0	0	1
2	1	1	0	0	1	1	1	1	1
3	1	1	1	1	0	1	0	1	1
4	1	1	1`	1	0	1	0	1	1
5	1	1	1	1	1	1	1	0	1
6	1	1	1	1	1	1	1	0	1
7	1	1	1	1	0	0	1	1	1
8	1	1	1	1	0	0	1	1	1
9	0	0	0	0	0	0	0	0	1

Table 19 Dependency and driver power

Elements	1	2	3	4	5	6	7	8	9	Driver power
1	1	0	0	0	0	0	0	0	1	2
2	1	1	0	0	1	1	1	1	1	7
3	1	1	1	1	0	1	0	1	1	7
4	1	1	1	1	0	1	0	1	1	7
5	1	1	1	1	1	1	1	0	1	8
6	1	1	1	1	1	1	1	0	1	8
7	1	1	1	1	0	0	1	1	1	7
8	1	1	1	1	0	0	1	1	1	7
9	0	0	0	0	0	0	0	0	1	1
Dependence	8	7	6	6	3	5	5	5	9	

Table 21 Iteration 1

Elements	Reachability set	Antecedent set	Intersection set	Level
1	1, 9	1, 2, 3, 4, 5, 6, 7, 8	1	
2	1, 2, 5, 6, 7, 8, 9	2, 3, 4, 5, 6, 7, 8	2, 5, 6, 7, 8	
3	1, 2, 3, 4, 6, 8, 9	3, 4, 5, 6, 7, 8	3, 4, 6, 8	
4	1, 2, 3, 4, 6, 8, 9	3, 4, 5, 6, 7, 8	3, 4, 6, 8	
5	1, 2, 3, 4, 5, 6, 7	2, 5, 6	2, 5, 6	
6	1, 2, 3, 4, 5, 6, 7	2, 3, 4, 5, 6	2, 3, 4, 5, 6	
7	1, 2, 3, 4, 7, 8, 9	2, 5, 6, 7, 8	2, 7, 8, 9	
8	1, 2, 3, 4, 7, 8, 9	2, 3, 4, 7, 8	2, 3, 4, 7, 8	
9	9	1, 2, 3, 4, 5, 6, 7, 8, 9	9	I

Initial reachability matrix

The design requirements of the system in the SSIM are transformed into a binary matrix called the initial reachability matrix by substituting V, A, X, and O with 1 and 0 (Table 18) using the following four substitutions of 1's and 0's as follows:

- If the (i, j) entry in the SSIM is V, then the (i, j) entry in the reachability matrix becomes 1 and the (j, i) entry becomes 0.
- If the (i, j) entry in the SSIM is A, then the (i, j) entry in the reachability matrix becomes 0 and the (j, i) entry becomes 1.
- If the (i, j) entry in the SSIM is X, then the (i, j) entry in the reachability matrix becomes 1 and the (j, i) entry also becomes 1.
- If the (i, j) entry in the SSIM is O, then the (i, j) entry in the reachability matrix becomes 0 and the (j, i) entry also becomes 0.

Final reachability matrix

The reachability matrix obtained in step 3 is converted into the final reachability matrix by scrutinizing it for transitivity. If the transitivity rule is not found to be satisfied, the SSIM is reviewed and modified by the specific feedback about transitive relationship from the experts. From the revised SSIM, the reachability matrix is again worked out and tested for the transitivity rule. The transitivity of the contextual relation is a basic assumption in ISM which states that if element A is related to B and B is related to C, then A is related to C. After checking transitivity, the final reachability matrix is shown in Table 19.

Level partition reachability matrix

The reachability and antecedent sets for each element are found out from the final reachability matrix. The reachability set includes criteria itself and others which it may help to achieve, and the antecedent set consists of itself and other criterion which helps in achieving it. Subsequently, the intersection set is derived, and the variable having the same reachability and intersection sets is given the top level in the ISM hierarchy. Then, the same process is repeated to find out the elements in the next level. This process is continued until the level of each element is found. These levels help in building the diagraph and the final model (Tables 20, 21, 22, 23, 24, 25, 26, and 27).

Table 20 Transitivity relation

Elements	1	2	3	4	5	6	7	8	9	Driver power
1	1	0	0	0	0	0	0	0	1	2
2	1	1	0	0	1	1	1	1	1	7
3	1	1	1	1	0	1	0	1	1	7
4	1	1	1`	1	0	1	0	1	1	7
5	1`	1	1	1	1	1	1	0	1	8
6	1	1	1	1	1	1	1	0	1	8
7	1	1	1	1	0	0	1	1	1	7
8	1	1	1	1	0	0	1	1	1	7
9	0	0	0	0	0	0	0	0	1	1
Dependence	8	7	6	6	3	5	5	5	9	

Table 22 Iteration 2

Elements	Reachability set	Antecedent set	Intersection set	Level
1	1	1, 2, 3, 4, 5, 6, 7, 8	1	II
2	1, 2, 5, 6, 7, 8	2, 3, 4, 5, 6, 7, 8	2, 6, 7, 8	
3	1, 2, 3, 4, 6, 8	3, 4, 5, 6, 7, 8	3, 4, 6, 8	
4	1, 2, 3, 4, 6, 8	3, 4, 5, 6, 7, 8	3, 4, 6, 8	
5	1, 2, 3, 4, 5, 6, 7	2, 5, 6	2, 5, 6	
6	1, 2, 3, 4, 5, 6, 7	2, 3, 4, 5	2, 3, 4, 5	
7	1, 2, 3, 4, 7, 8	2, 5, 6, 7, 8	2, 7, 8	
8	1, 2, 3, 4, 7, 8	2, 3, 4, 7, 8	2, 3, 4, 7, 8	

Table 23 Iteration 3

Elements	Reachability set	Antecedent set	Intersection set	Level
2	2, 5, 6, 7, 8	2, 3, 4, 5, 6, 7, 8	2, 6, 7, 8	
3	2, 3, 4, 6, 8	3, 4, 5, 6, 7, 8	3, 4, 6, 8	
4	2, 3, 4, 6, 8	3, 4, 5, 6, 7, 8	3, 4, 6, 8	
5	2, 3, 4, 5, 6, 7	2, 5, 6	2, 5, 6	
6	2, 3, 4, 5, 6, 7	2, 3, 4, 5, 6	2, 3, 4, 5, 6	
7	2, 3, 4, 7, 8	2, 5, 6, 7, 8	2, 7, 8	
8	2, 3, 4, 7, 8	2, 3, 4, 7, 8	2, 3, 4, 7, 8	III

After the iteration process, few elements are eliminated, and using these variables, an ISM model is drawn as shown in Figure 4. After that, the MICMAC analysis is done, and Figure 3 is plotted by taking the driver power in the x-axis and dependency in the y-axis. In this method, the design requirements are classified into four clusters (Figure 4). The objective behind the behavioral classification of these design requirements is to analyze the driving power and dependence powers that influence the design requirements of electricity utility service.

Classification of enablers

Cluster I: weak driving power and weak dependence
This group is called autonomous or excluded enablers. These enablers have only a few links with the system. They appear quite out of line with the system. However, a distinction may be drawn within this group between the disconnected enablers situated near the axis's origin, whose evolution therefore seems to be rather excluded from the system's global dynamics and secondary levers which, although quite autonomous, are more influent than dependent. These enablers concerned are located in the south-west frame, quite above the diagonal, and can be used as secondary acting enablers or as application points for possible accompanying measures. In the present study, none of the barriers is coming under this category.

Cluster II: weak driving power and strong dependence
These enablers are called depending enablers or rather result enablers. These enablers are little influent and

Table 24 Iteration 4

Elements	Reachability set	Antecedent set	Intersection set	Level
2	2, 5, 6, 7	2, 3, 4, 5, 6, 7	2, 5, 6, 7	IV
3	2, 3, 4, 6	3, 4, 5, 6, 7	3, 4, 6	
4	2, 3, 4, 6	3, 4, 5, 6, 7	3, 4, 6	
5	2, 3, 4, 5, 6, 7	2, 5, 6	2, 5, 6	
6	2, 3, 4, 5, 6, 7	2, 3, 4, 5, 6	2, 3, 4, 5, 6	
7	2, 3, 4, 7	2, 5, 6, 7	2, 7	

Table 25 Iteration 5

Elements	Reachability set	Antecedent set	Intersection set	Level
3	3, 4, 6	3, 4, 5, 6, 7	3, 4, 6	V
4	3, 4, 6	3, 4, 5, 6, 7	3, 4, 6	V
5	3, 4, 5, 6, 7	5, 6	5, 6	
6	3, 4, 5, 6, 7	3, 4, 5, 6	3, 4, 5, 6	
7	3, 4, 7	5, 6, 7	7	

very dependent. This indicates that design requirements 1 (regular power supply without any interruption) and 9 (regular maintenance work must be done) come under this category.

Cluster III: strong driving power and strong dependence
These enablers are at the same time very influent and very dependent. They are also called relay enablers. These enablers are unstable. Any action on these indicators will have impact on others and feedback effect on themselves which may amplify or support the initial pulse. Design requirements 2 (availability of mobile vans, customer care center, and online payment system for easy payment of electricity bills), 3 (for listening, registering complaints using toll-free numbers, customer care people, and websites must be available), and 4 (quick procedures for new supply of electricity) also come in this category. These enablers should be studied even more carefully than the others.

Cluster IV: strong driving power and weak dependence
These enablers are altogether very influent and little dependent. They condition the rest of the system and are also called independent or determinant enablers. The influent enablers are its most crucial elements since they can act on the system depending on how much we can control them as a key factor. They are also considered as entry enablers in the system. The analysis reveals that four design requirements 5 (electricity policies must be framed as per the customer requirement), 6 (advanced technology transformers, good equipments, and proper power demand side management), 7 (customer care centers must be open day and night), 8 (safeties of the customers are taken cared by showing safety lessons in the website, regarding electric poles) are ranked as independent enablers as they are having the maximum driver power. This implies that these variables are key barriers.

Table 26 Iteration 6

Elements	Reachability set	Antecedent set	Intersection set	Level
5	5, 6, 7	5, 6	5, 6	
6	5, 6, 7	5, 6	5, 6	
7	7	5, 6, 7	7	VI

Table 27 Iteration 7

Elements	Reachability set	Antecedent set	Intersection set	Level
5	5, 6	5, 6	5, 6	VI
6	5, 6	5, 6	5, 6	VI

Conclusion, future scope, and limitations

This study provides a framework for system design requirements in electricity utility service as per the customer requirements and then a dependent and independent relationship is established in between the design requirements (see Additional file 1). It is found that few elements are linked with each other and lies in the same cluster. Thus, electricity utility industries must take necessary steps to take care of these design requirements depending on their characteristics to get maximum customer satisfaction for improving the business strategy.

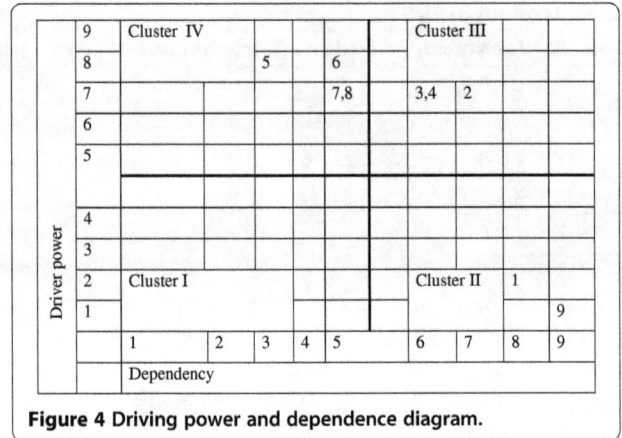

Figure 4 Driving power and dependence diagram.

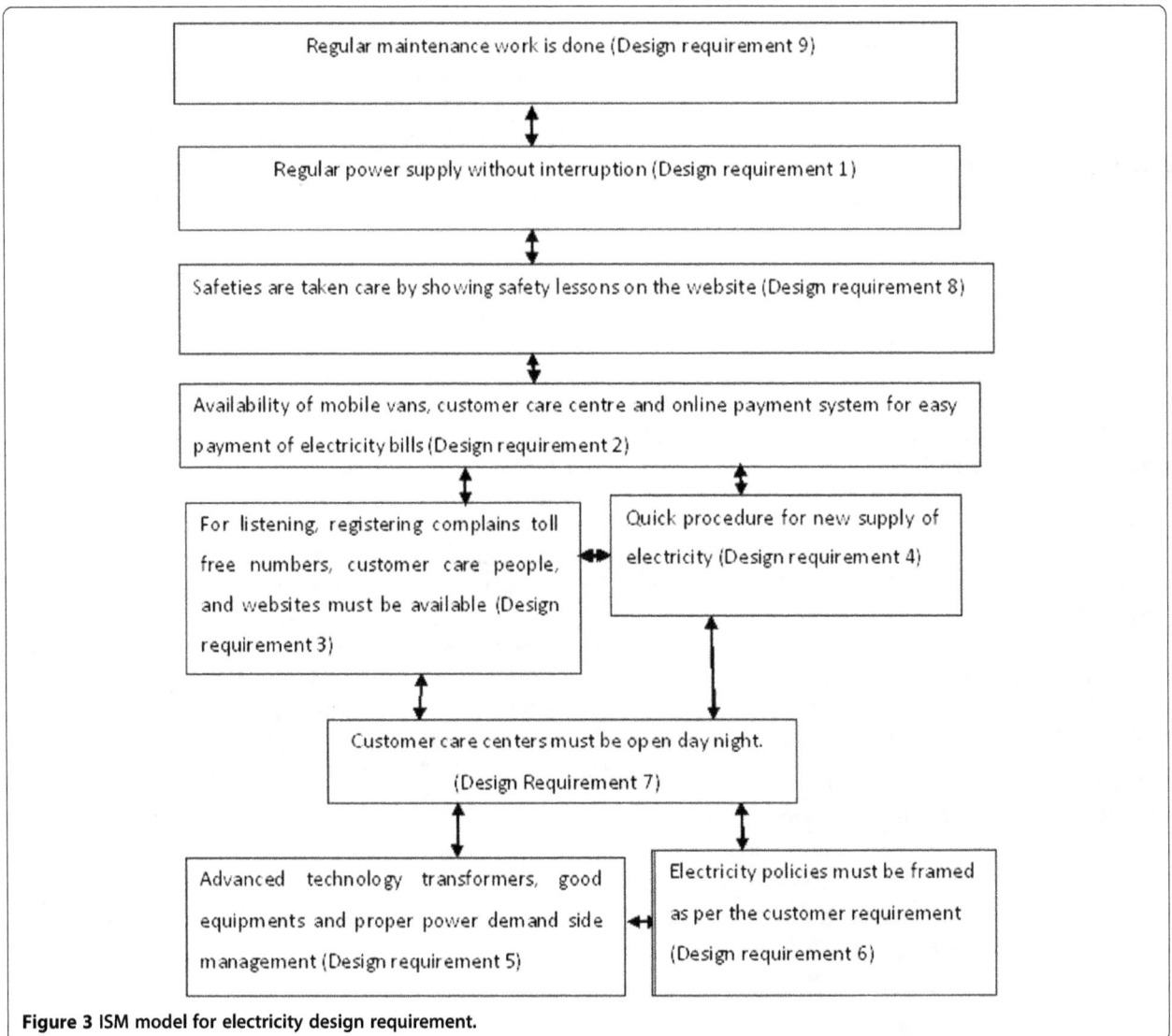

Driver power	1	2	3	4	5	6	7	8	9
9	Cluster IV					Cluster III			
8				5	6				
7					7,8	3,4	2		
6									
5									
4									
3									
2	Cluster I					Cluster II			1
1									9
	1	2	3	4	5	6	7	8	9
	Dependency								

Figure 3 ISM model for electricity design requirement.

- Regular maintenance work is done (Design requirement 9)
- Regular power supply without interruption (Design requirement 1)
- Safeties are taken care by showing safety lessons on the website (Design requirement 8)
- Availability of mobile vans, customer care centre and online payment system for easy payment of electricity bills (Design requirement 2)
- For listening, registering complains toll free numbers, customer care people, and websites must be available (Design requirement 3)
- Quick procedure for new supply of electricity (Design requirement 4)
- Customer care centers must be open day night. (Design Requirement 7)
- Advanced technology transformers, good equipments and proper power demand side management (Design requirement 5)
- Electricity policies must be framed as per the customer requirement (Design requirement 6)

The results of modeling customer evaluation of service quality using neural network reveal important items affecting the service quality. These seven common deficient items of electricity supply common to all the four stakeholders (i.e., domestic, agricultural, industrial, and public organization customers) need substantial improvement for providing service quality in Indian electricity industry. They are identified as item 3 (regarding the quality of power supplied), item 5 (regarding mode of payment of electric bills), item 7 (regarding registration of complaint), item 18 (regarding new electricity connection), item 19 (regarding display of electricity safety lessons), item 20 (regarding fluctuation of voltage), and item 24 (regarding waiting time in queue for bill payment). Thus, the government agencies and administrators must formulate policies and take appropriate measures to improve upon these seven items for enhancing service quality of electricity supply in India.

Based on the seven deficient items, a set of nine system design requirements has been suggested in this work. With the help of quality function deployment, the design requirements were ranked based on their importance in affecting the service quality in the electricity sector. The design requirements with the highest ranking are found to be related with the working hour of customer care center (i.e., customer care centers must be open day and night), whereas the item related to the interruption in power supply (i.e., regular power supply without any interruption) has the lower ranking. These design requirements are proposed to improve the electricity industry for successful service delivery by the Indian electricity industry. The relation between these design requirements is also found.

Service quality measurement of different service industries is a vast area, and as the discipline of service industries change, the approach of research will also change. Thus, in the future, the following studies can be extended to other service industries. A large number of samples from the customers may be collected to have better understanding of quality characteristics in the service sectors. Applying the same methodologies effectively in other service sectors such as healthcare, tourism, hotels and restaurants, transportation, repair and maintenance shop, information service, and recreational services may carry out extension of this research. This research can be extended for other utility sectors like water, telecom, gas etc. Electricity/energy sector covers a large area, and policies vary from consumer to consumer, depending upon the category (agricultural, domestic/industrial). Thus, this research can be done in all the sectors separately, as the service specification varies from country to country and place to place. For different countries, this research can be extended globally. Some other techniques like benchmarking can be implemented to find the variation of service in different sectors/different zones.

The electricity survey is done by collecting the responses from different parts of India. Moreover, the service delivery of electricity does not follow uniform policy throughout India. Secondly, rural problems are not considered, and QFD is not implemented by considering rural culture. Thus, some additional questions must be added for validating the model in rural electrification also. No separate study has been done for different types of consumers like LT and HT consumers.

Additional file

Additional file 1: Research highlights.

Authors' contributions

S.Satapathy is serving as Asst Prof. in KIITs University, Bhubaneswar. Her area of research is service quality management. Her area of interest includes total quality management, statistical process control, and service quality management. She has published more than five research paper in various international and national conferences and journals. P. D. Mishra is a Finance manager in the Southern Electricity Supply of Orissa. He served as a Lecturer in Finance and marketing in the I.C.W.A chapter in Bhubaneswar, India for two years. His area of research is Service Quality Management. His area of interest includes Total Quality Management, Statistical Process Control, Service Quality Management, Strategic Planning and Analyzing Consumer Behavior, Finance and stock market. He has published more than three research paper in various international and national conference and journals.

1. Suchismita Satapathy • Saroj K. Patel • Amitabha Biswas • Pravudatta Mishra, "Interpretive structural modeling for E-electricity utility service, Serv Bus, An International Journal, ISSN 1862-8516, Volume 6, Number 3 Serv Bus (2012) 6:349-367. DOI 10.1007/s11628-012-0139-9.
2. S. Satapathy1, S. K. Patel1, P.D. Mishra, "Discriminate analysis and neural network approach in water utility service" Int. J. Services and Operations Management, Vol. 12, No. 4, 2012.
3. Pravudatta Mishra, S. Satapathy1, S. K. Patel1, "A methodology for evaluation of e-electricity service quality using neural networks", Int. J. Indian Culture and Business Management, Vol. X, No. Y, xxxx. Accepted
4. S. Satapathy1, S. K. Patel1, P.D. Mishra, "A methodology to measure the service quality of online shopping of electronic goods in India" Int. J. Indian Culture and Business Management, Vol. x, No. x, xxxx. Accepted.

Acknowledgements

The authors would like to thank the Editor of Journal of Industrial Engineering International and anonymous referees for their valuable comments on the earlier version of the manuscript, which helped improve the overall contents and presentation of the paper considerably.

Author details

[1]School of Mechanical Engineering, KIIT University, Bhubaneswar, 751024 Odisha, India. [2]Cental electricity supply Utility Dept, Bhubaneswar, 751024 Odisa, India.

References

Anantatmula V, Kanungo S (2007) Establishing and structuring criteria for measuring knowledge management. In: Proceedings of the 38th Hawaii international conference on system sciences, 3–6 January 2005, IEEE Computer Society, New York, pp 192a

Anderson JC, Rungtusanthanam M, Schroeder RG (1994) A theory of quality management underlying the Deming management method. Acad Manag Rev 19(3):472–509

Azadeh A, Movagha SA (2010) An integrated multivariate approach for performance assessment and optimization of electricity transmission systems. Int J Ind Syst Eng 5(2):226–248

Bouchereau V, Rowlands H (2000) Methods and techniques to help quality function deployment (QFD). Benchmarking: Int J 7(1):8–19

Carter WW (1989) Control of power quality in modern industry. In: IEEE textile industry technical conference. Southampton University, Charlotte. 3–4 May 1989

Chan CKS (1993) Customer service calls handling system. In: IEEE 2nd international conference on advances in power system control, operation and management, APSCOM-93, 7–10 December 1993, vol. 2, Hongkong. pp 505–508

Chau VS (2009) Benchmarking service quality in UK electricity distribution networks. Benchmarking: Int J 16(1):47–69

Chung KC, Tan SS, Holdsworth D (2008) Insolvency prediction model using multivariate discriminant analysis and artificial neural network for the finance industry in New Zealand. Int J Bus Manage 3(1):19–29

Dahiyat SE, Akroush MN, Abu-Lail BN (2011) An integrated model of perceived service quality and customer loyalty: an empirical examination of the mediation effects of customer satisfaction and customer trust. Int J Serv Oper Manage 9(4):453–490

Frazier SD (1999) Animals, power systems, and reliability in a deregulated environment. IEEE Power Energy Soc Summer Meet 1:302–307. doi:10.1109/PESS.1999.784364

Hajeeh M (2010) Multicriteria decision making in electricity demand management: the case of Kuwait. Int J of Serv Oper Manage 6(4):423–442. doi:10.1504/IJSOM.2010.032917

Hamoud G, El-Nahas I (2003) Assessment of customer supply reliability in performance-based contracts. IEEE T Power Syst 18(4):1587–1593

Hansen JV, Mckell LJ, Heitger LE (1979) ISMS: computer-aided analysis for design of decision-support systems. Manag Sci 25(11):1069–1081

Ikiz AK, Masoudi A (2008) A QFD and SERVQUAL approach to hotel service design. İsletme Fakültesi Dergisi 9(1):17–31

Khare A (2011) Customers' perception and attitude towards service quality in multinational banks in India. Int J Services Oper Manage 10(2):199–215

Kim H, Jeon S, Kim J (2008) ASP effects in the small-sized enterprise: the case of the Bizmeka service from Korea Telecom. Serv Bus 2(4):287–301

Kumar P (2002) The impact of performance, cost, and competitive considerations on the relationship between satisfaction and repurchase intent in business markets. J Serv Res 5(1):55–68

Kumar N, Prasad R, Shankar R, Lyer KC (2009) Technology transfer for rural housing: an interpretive structural modelling approach. J Adv Manag Res 6(2):188–205

Lamedica R, Esposito G, Tironi E, Zaninelli D, Prudenzi A (2001) A survey on power quality cost in industrial customers. IEEE Conf Power Eng Soc Winter Meeting 2:938–943. doi:10.1109/PESW.2001.916999

Mishra H (2006) Managing leadership in a systems acquisition life cycle: a strategic framework. In: Proceedings of the IEEE Engineering Management Society, Bahia, 17–20 September, pp 84–88

Nakano D (2011) Modular service design in professional services: a qualitative study. Int J Services Oper Manage 9(1):1–17

Nelson K, Nadkarni S, Narayanan VK, Ghods M (2000) Understanding software operations support expertise: a revealed causal mapping approach. MIS Quart J 24(3):475–508

Ozcelik O (2010) Six Sigma implementation in the service sector: notable experiences of major firms in the USA. Int J Services Oper Manage 7(4):401–418

Parida SK, Srivastava SC, Singh SN (2011) A review on reactive power management in electricity markets. Int J Energy Sector Manage 5(2):201–214

Porter AL, Rossini FA, Carpenter SR, Roper AT (1980) A guidebook for technology assessment and impact analysis, vol 46. North Holland series in system science and engineering. , North Holland, New York

Rabbani M, Ghoreyshi SM, Rafiei H, Ghazanfari M (2012) Energy consumption forecasting using a bi-objective fuzzy linear regression model. Int J Services Oper Manage 13(1):1–18

Rengarajan S, Loganathan S (2012) Power theft prevention and power quality improvement using fuzzy logic. Int J Electrical Electron Eng 1(3):2231–5284

Sahney S (2008) Critical success factors in online retail—an application of quality function deployment and interpretive structural modeling. Int J Bus Inf 3 (1):144–162

Shaikh FA, Jain R, Kotnala M, Agarwal N (2012) New techniques for the prevention of power system collapse. Int J Electrical Electron Eng 1(3):2231–5284

Srivastava L, Goswami A, Diljun GM, Chaudhury S (2012) Energy access: revelations from energy consumption patterns in rural India. Energy Policy 47(Supplement 1):11–20

Stuart FI, Tax SS (1996) Planning for service quality: an integrative approach. Int J Serv Ind Manag 7(4):58–77

Tinnilä M (2012) A classification of service facilities, servicescapes and service factories. Int J Serv Oper Manage 11(3):267–291

Ulkuniemi P, Pekkarinen S (2011) Creating value for the business service buyer through modularity. Int J Serv Oper Manage 8(2):127–141

Wattana S, Sharma D (2011) Electricity industry reforms in Thailand: an analysis of productivity. Int J Energy Sector Manage 5(4):494–521

Wisher JD, Corney WJ (2001) Comparing practices for capturing bank customer feedback - Internet versus traditional banking. Benchmarking: Int J 8(3):240–250. doi:10.1108/14635770110396647

Wyk V, Louw E, Inc M (1992) A quality management system for electricity utilities. In: IEEE AFRICON'92 Proceedings, Ezulwini Valley, 22–24 September 1992, pp 584–587. doi:10.1109/AFRCON.1992.624552

15

A neuro-data envelopment analysis approach for optimization of uncorrelated multiple response problems with smaller the better type controllable factors

Mahdi Bashiri[1*], Amir Farshbaf-Geranmayeh[2] and Hamed Mogouie[1]

Abstract

In this paper, a new method is proposed to optimize a multi-response optimization problem based on the Taguchi method for the processes where controllable factors are the smaller-the-better (STB)-type variables and the analyzer desires to find an optimal solution with smaller amount of controllable factors. In such processes, the overall output quality of the product should be maximized while the usage of the process inputs, the controllable factors, should be minimized. Since all possible combinations of factors' levels, are not considered in the Taguchi method, the response values of the possible unpracticed treatments are estimated using the artificial neural network (ANN). The neural network is tuned by the central composite design (CCD) and the genetic algorithm (GA). Then data envelopment analysis (DEA) is applied for determining the efficiency of each treatment. Although the important issue for implementation of DEA is its philosophy, which is maximization of outputs versus minimization of inputs, this important issue has been neglected in previous similar studies in multi-response problems. Finally, the most efficient treatment is determined using the maximin weight model approach. The performance of the proposed method is verified in a plastic molding process. Moreover a sensitivity analysis has been done by an efficiency estimator neural network. The results show efficiency of the proposed approach.

Keywords: Multiple response optimization; Artificial neural networks; Data envelopment analysis; Smaller-the-better-type controllable factors

Introduction

Today's competitive environment impels companies to improve the quality of their products proactively, so the design of experiments (DOEs) can be one of the most efficient methods for this purpose. The experimental design helps us find the effects of the controllable and nuisance factors on one or more responses. Finding the combination of controllable factor levels, namely treatment, which leads to the most appropriate process outputs is one of the common challenges in quality engineering researches.

The Taguchi method is a common strategy in the robust design and involves designing experiments with the use of orthogonal arrays for finding the treatment which optimizes a given performance measure, typically a signal-to-noise (SN) ratio.

The Taguchi method was initially proposed for single response problems (Taguchi and Chowdhury 2000; Maghsoodloo et al. 2004; Robinson et al. 2004; Zang et al. 2005). However, in real industrial problems, there are more than one responses and their simultaneous optimization is of the interest.

Multiple-response optimization (MRO) problems have been studied by several researchers. Existing methods in this field are classified into three basic categories by Ortiz et al. (2004). The performance of each category depends on the complexity of the problem. In the first category, overlying contour plots of each response is applied for finding the space where each response satisfies its specification limits. Myers and Montgomery (2002) expressed that this method is applicable only in the

* Correspondence: bashiri@shahed.ac.ir
[1]Department of industrial Engineering, Faculty of Engineering, Shahed University, Khalij Fars Highway, Tehran P.O. BOX: 3319118651, Iran
Full list of author information is available at the end of the article

situations where the numbers of controllable factors are very few. In the second category, the most important response for the decision maker (DM) is used as the main objective and the rest of responses are considered as constraints. However the approaches in this category do not conform to the basic idea of multiple-response optimization which is the simultaneous optimization of all responses.

The third category, which contains a major proportion of studies in MRO, consists of three main steps to find the optimal treatment. In the first step, a model for describing the relation between controllable factors and responses is built. In the second step, an aggregation approach of responses is applied. Finally, in the third step, an optimization method is used for optimizing the single response which is obtained from the last step. In the mentioned categories, Ordinary Least Square (OLS) is one of the common methods for building the model of relation between controllable factors and responses (first step). Although in some researches such as Bashiri and Moslemi (2013), robust optimization methods are applied; one of the major shortcomings of the studies in this category is that when the mean square error (MSE) of the regression model is a high value, the ability of the model to describe the relationship of the response variables and the controllable factors would be poor (Erzurumlu and Oktem 2007). For overcoming this problem, ANN can be used as a proper substitute method for response estimation. Some authors have compared response surface and regression models with ANN in model building and the preciseness of ANN has been verified in their results (Tsao 2008; Desai et al. 2008; Namvar-Asl et al. 2008; Gauri and Pal 2010).

Furthermore, Niaki and Hoseinzade (2013) used ANN for forecasting S&P indices where the experimental design was used for tuning the parameters of the neural network.

A numerous number of researchers studied about using artificial neural networks for solving MRO problems. The summary of existing approaches is illustrated in Figure 1, where the studies are classified according to the techniques each of them has used.

In MRO problems, different multi-criteria decision making methods have been used to determine the optimum treatment (for a review see Amiri et al. (2012)). Data envelopment analysis (DEA) is one of these techniques which have been used in several researches. Caporaletti et al. (1999) proposed a pure input DEA model for the nominal-the-best (NTB)-type responses using $\left(\bar{y}_{ij} - y_i\right)^2$ and S_{ij}^2 as input variable. In this study, just the experimented treatments have been evaluated. Liao and Chen (2002) proposed an input-oriented basic DEA ratio known as the Charnes, Cooper and Rhodes (CCR) model introduced by Charnes et al. (1978) that uses the normalized mean responses as input variables when the responses are the NTB or the smaller-the-better (STB) type; also, Goel et al. (2007) proposed a new method in multiple-response optimization using the Pareto optimal solution.

In the model of Liao and Chen (2002), when the responses are LTB type, the normalized mean responses are considered as the output variable. Herein again, only the real experimented treatments and their corresponding responses are considered. Liao (Goel et al. 2007) also used a back propagation (BP) neural network (trained with the data of the actual treatments) to estimate the SN ratio of responses for all treatments and then efficient treatments are determined by the CCR DEA model, considering normalized SN ratio as outputs.

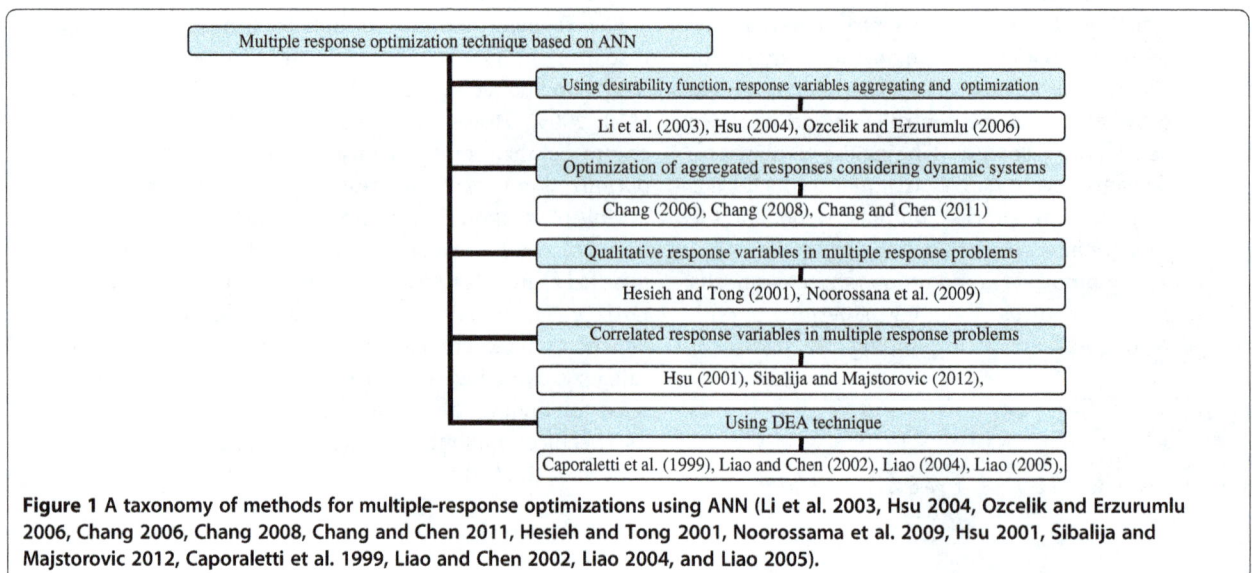

Figure 1 A taxonomy of methods for multiple-response optimizations using ANN (Li et al. 2003, Hsu 2004, Ozcelik and Erzurumlu 2006, Chang 2006, Chang 2008, Chang and Chen 2011, Hesieh and Tong 2001, Noorossama et al. 2009, Hsu 2001, Sibalija and Majstorovic 2012, Caporaletti et al. 1999, Liao and Chen 2002, Liao 2004, and Liao 2005).

In Liao (Goel et al. 2007), the same DEA model is used but all possible treatments are estimated using a BP neural network. In their proposed approach, the most treatment is not selected.

Gutierrez and Lozano (2010) used a similar approach to find the efficient treatment and then sieve the most efficient among the efficient ones.

In the mentioned literature of the studies which have used DEA for the determination of the optimum treatment, only the response variables have been focused on, while the main philosophy of DEA is maximization of the overall process outputs versus the minimization of the total consumed inputs.

Many real world processes can be exampled where the controllable factors are STB type. For instance, in a plastic molding process, one of the factors is the barrel temperature, of which the less value is more preferred. The higher temperature impels more electricity consumption and equivalently more costs. By considering other similar STB controllable factors as inputs and the SN ratio of response variables as the outputs of the process, the main philosophy of DEA would be realized. The final results of DEA for such a problem would assure us that the most appropriate quality of the process is obtained by the least consumption of input variables.

In this paper, a new approach is developed based on the neural network and data envelopment analysis where controllable factors are the STB type and the responses are uncorrelated. Besides, it is. In this method, the ANN is used to estimate the response values for unpracticed treatments in a way that the multiple responses are obtained simultaneously. In this paper, the used neural network is tuned by the method proposed by Bashiri and Geranmayeh (2011). In their method, the Central Composite Design (CCD) and the Genetic Algorithm (GA) have been used to tune and determine the optimum parameters of the neural network, specifically the number of the layers and the number of neurons in each layer.

After tuning the ANN, the unpracticed treatments with smaller intervals of factor levels are generated and their corresponding responses are estimated by the ANN.

In the next step, by using DEA, the efficiency value of each treatment is computed. Finally, the most efficient treatment is determined using the maximin weight model approach.

The remainder of the article is organized as follows: in the next section, the steps of the proposed approach are described. In section 3, the proposed method is implemented in a real world case study and the steps of the proposed method are explained thoroughly. Finally, section 4 summarizes and draws the conclusion.

Proposed approach

This section describes the proposed methodology which is illustrated in Figure 2.

Experimentation phase

In the first step of any design of experiment problems, the responses and their descriptions such as their types

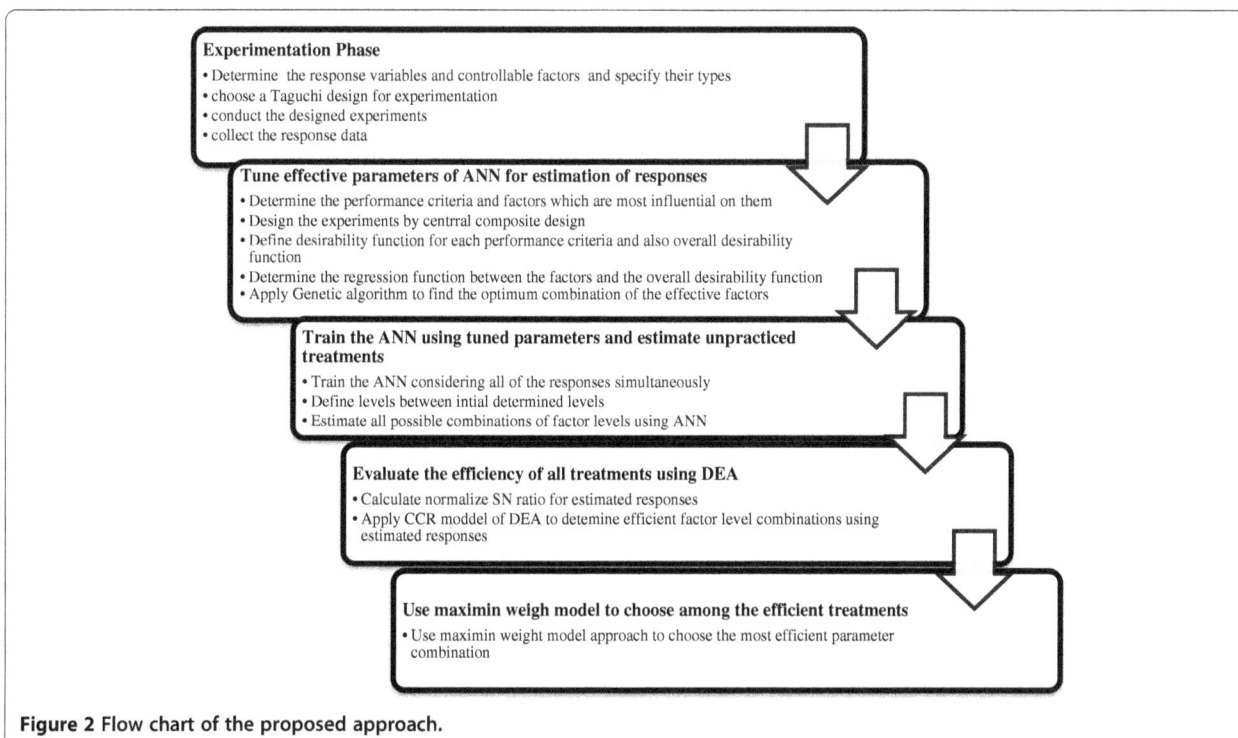

Experimentation Phase
- Determine the response variables and controllable factors and specify their types
- choose a Taguchi design for experimentation
- conduct the designed experiments
- collect the response data

Tune effective parameters of ANN for estimation of responses
- Determine the performance criteria and factors which are most influential on them
- Design the experiments by centrral composite design
- Define desirability function for each performance criteria and also overall desirability function
- Determine the regression function between the factors and the overall desirability function
- Apply Genetic algorithm to find the optimum combination of the effective factors

Train the ANN using tuned parameters and estimate unpracticed treatments
- Train the ANN considering all of the responses simultaneously
- Define levels between intial determined levels
- Estimate all possible combinations of factor levels using ANN

Evaluate the efficiency of all treatments using DEA
- Calculate normalize SN ratio for estimated responses
- Apply CCR moddel of DEA to detemine efficient factor level combinations using estimated responses

Use maximin weigh model to choose among the efficient treatments
- Use maximin weight model approach to choose the most efficient parameter combination

Figure 2 Flow chart of the proposed approach.

should be determined. Also, the controllable factors and their corresponding levels should be specified according to the technical knowledge of the process and the execution limitations. By knowing the required information of responses and the controllable factors, a Taguchi design is chosen and conducted and the response data are collected. The outputs of this step are the SN ratio values computed for each response in each treatment. Note that the responses are assumed to be uncorrelated.

Tune-effective parameters of ANN for estimation of responses

For obtaining desirable results by using Artificial Neural Networks, tuning the parameters of ANN seems to be necessary. For example, the number of layers and the number of neurons in each layer are effective parameters in the performance of neural networks. Some authors have used the design of experiments (DOE) for determining the best combination of effective parameters of ANN. Khaw et al. (1995) used the Taguchi method as well as two simulated data collections to determine the effective parameters of ANN, which caused to increase the velocity and convergence of the Back Propagation (BP) algorithm. Also, other similar researches have been proposed in this field, such as the studies of Kim and Yum (2004), Sukthomya and Tannock (2005), Tortum et al. (2007), Packianather et al. (2000), and Peterson et al. (1995). Bashiri and Geranmayeh (2011) proposed a method for tuning the parameters of the artificial neural network based on CCD and genetic algorithm. Because of the accuracy and generalization capability of this approach, in this study, we applied this method for tuning the parameters of the neural network with some necessary changes in determining the neural network's performance criteria. The condition of the problem determines the proper performance criteria of the neural network.

For training ANNs, the data are divided into three subsets: training, validation, and testing sets. In this study, root mean square error (RMSE) of the test and validation data are considered as ANN's performance criteria in estimation of responses in the multiple response optimization problem. In this step, the optimum number of hidden layers and the number of neurons in hidden layers are obtained.

Train ANN using tuned parameters and estimate the unpracticed treatments

In the previous step, optimal parameters of ANN for this problem are obtained. So, the neural network is ready to be trained and estimate the response values of unpracticed treatments.

After the training phase, neural network builds a model and can estimate other treatments which are not experimented. Since the trained neural network's response estimation is not affected by the number of

factor levels, new levels are defined between the initial factor levels. This procedure improves the accuracy of the solution.

In mentioned studies, the neural network is used to estimate the SN ratio or the mean square deviation (MSD) of responses. But in the cases where a nuisance factor exists, the effect of the nuisance factor is neglected and the nuisance factor is treated as a replicate. In this study, for solving this problem, ANN is applied for estimating responses, not the SN ratios or MSDs of responses. As Chang (Ozcelik and Erzurumlu 2006) applied ANN in dynamic multiple response experiments, factors and nuisance factors are considered as inputs for training of the neural network.

Evaluate the efficiency of all treatments using DEA

DEA is used to compute the relative efficiency of a group of competing decision-making units (DMUs), while there are several inputs and outputs for each DMU (Tbanassoulis 2001). The relative efficiency is the ratio of the weighted sum of outputs to the weighted sum of inputs.

Thus, if a DMU wants to have an upper efficiency, the input data must be minimized and the output data must be maximized. If it is assumed that n available DMUs should be studied, and each of them has m inputs and s outputs, the efficiency of the jth DMU will be computed by Equation 1:

$$
\begin{aligned}
&\text{Maximize} \ \ E_0 = \sum_{r=1}^{s} u_r y_{r0} \\
&\text{s.t.} \\
&\sum_{i=1}^{m} v_i x_{i0} = 1, \\
&\sum_{r=1}^{s} u_r y_{rj} - \sum_{i=1}^{m} v_i x_{ij} \le 0, \\
&u_r \ge 0, \quad r = 1, \dots, s, \\
&v_i \ge 0, \quad i = 1, \dots, m,
\end{aligned}
\tag{1}
$$

where u_r is the weight of output r, v_i is the weight of input i, y_{rj} is the value of output r from DMU j, and x_{ij} is the value of input i from DMU j and where DMU_0 is the DMU under study.

In this paper, each treatment is considered as a DMU, each controllable factor is considered as an input variable and finally each response variable is considered as an output.

For using DEA in MRO problems, input variables should be STB type and response variables should be LTB type; however this condition does not necessarily hold in all processes. For this reason, before applying DEA, it should be checked that the controllable factors are STB type variables and the responses are LTB-type ones.

Although there might be different types of responses in an MRO problem, since we used the Taguchi method, the responses are transformed into the SN ratio values using Equations 2, 3, and 4 and we know that the SN ratio is a LTB-type variable. For the controllable factors, according to their properties, we can assure that whether a controllable factor is STB type or not.

$$x_{ij} = -10 \log_{10} \left[\frac{1}{l} \sum_{k=1}^{i} y_{ijk}^2 \right] \qquad 0 \le y_{ijk} \le \infty \tag{2}$$

(for the smaller-the-better response)

$$x_{ij} = -10 \log_{10} \left[\frac{1}{l} \sum_{k=1}^{i} \frac{1}{y_{ijk}^2} \right] \qquad 0 \le y_{ijk} \le \infty \tag{3}$$

(for the larger-the-better response)

$$x_{ij} = 10 \log_{10} \left[\frac{\bar{y}_{ij}^2}{S_{ij}^2} \right] \qquad 0 \le y_{ijk} \le \infty \tag{4}$$

(for the nominal-the-better response)

Let the SN ratio be x_{ij} for the jth response at the ith trial, for $i = 1, ..., m$, $j = 1, ..., n$. y_{ijk} is the observed data for the jth response at the ith trial, in the kth repetition, $\bar{y}_{ij} = \frac{1}{l} \sum_{k=1}^{l} y_{ijk}$ (the average observed data for the jth response at the ith trial), $S_{ij}^2 = \frac{1}{l-1} \sum_{k=1}^{l} \left(y_{ijk} - \bar{y}_{ij} \right)^2$ (the variation of the observed data for the jth response at the ith trial) for, $i = 1, ..., m$ and $j = 1, ..., n$ and $k = 1, ..., l$. Now, x_{ij} is normalized as $Z_{ij} (0 \le Z_{ij} \le 1)$ by Equation 5 to avoid the effect of adopting different units.

$$Z_{ij} = \frac{X_{ij} - \min \{ X_{ij}, j = 1, 2, ..., n \}}{\max \{ X_{ij}, j = 1, 2, ..., n \} - \min \{ X_{ij}, j = 1, 2, ..., n \}} \tag{5}$$

At the end of this step, an efficiency value is computed for each treatment which represents that how well the input variables, controllable factors, have been minimized and how well the output variables, SN ratios of responses, have been maximized. The efficiency value is a measure ranging from 0 to 1 and sometimes more than one treatment would have the efficiency equal to 1.

Table 1 Response definitions for the injection molding process example

Response	Description	Specification limit (mm)	Type
Y_1	The size of the upper side	483.5 + 0.3	NTB
Y_2	The size of the lower side	483.6 + 0.3	NTB

Table 2 Factors and their levels for the injection molding process example

Row	Factors	Levels	Type
A	Injection pressure (percent of machine max pressure)	40, 50	STB
B	Injection speed (percent of machine max speed)	55, 60, 65	STB
C	Holding Pressure1 (percent of machine max pressure)	40, 45, 50	STB
D	Holding Pressure1 (percent of machine max pressure)	75, 80, 85	STB
E	Holding pressure Time (seconds)	6, 8, 10	STB
N	Nuisance factors, injection machines	550A, 550B	STB

Use maximin weight model to choose among the efficient treatments

In DEA method, it is common that more than one DMU would be selected as the efficient one. For the determination of the most efficient DMU, another mathematical modeling should be done. The maximin weight model was presented by Wang et al. (2009) for ranking DEA efficient units. This model is solved for all of efficient DMUs, the model is represented in Equation 6.

$$
\begin{aligned}
& \text{Maximize} \quad w \\
& \text{s.t.} \\
& \sum_{i=1}^{m} v_i = 1, \\
& \sum_{r=1}^{s} u_r \hat{y}_{ro} - \sum_{i=1}^{m} v_i \hat{x}_{i0} = 0, \\
& \sum_{r=1}^{s} u_r \hat{y}_{rj} - \sum_{i=1}^{m} v_i \hat{x}_{ij} \le 0, \quad j = 1, ..., n \\
& u_r \ge w, \quad r = 1, ..., s, \\
& v_i \ge w, \quad i = 1, ..., m,
\end{aligned}
\tag{6}
$$

where w is the maximin weight that could keep DMU_0 efficient, $\hat{x}_{ij} (i = 1, ..., m; j = 1, ..., n)$ is the normalized input obtained using Equation 7.

Table 3 Summary of experimental results for injection molding process example

Run	L$_{18}$					Y$_1$		Y$_2$	
	A	B	C	D	E	N_1	N_2	N_1	N_2
1	40	55	40	75	6	483.36	483.32	483.96	483.98
2	40	55	45	80	8	483.34	483.38	483.74	484.06
17	50	65	45	75	10	483.4	483.42	483.84	484.14
18	50	65	50	80	6	483.38	483.32	483.88	484.06

Table 4 Algorithm and its parameters considered for training the neural network

Training algorithm	Maximum number of epochs to train	Performance goal	Maximum validation failures	Minimum performance gradient	Initial *mu*	*mu* decrease factor	*mu* increase factor	Maximum *mu*
Levenberg-Marquardt back propagation	100	0	5	1e-10	0.001	0.1	10	1e10

$$\hat{x}_{ij} = \frac{x_{ij}}{\sum_{j=1}^{n} x_{ij}}, i = 1, ..., m; \; j = 1, ..., n \qquad (7)$$

If there would be k efficient units, by solving maximin weight model represented in Equation 7 for each efficient unit, a group of maximin weights, $w_{i1}^*, w_{i2}^*, ...w_{ik}^*$, are obtained and the DMU with the largest value of w is considered as the most efficient treatment.

Illustrative example

For verifying the applicability of the proposed method, a real case study of an injection molding process, in a TV and monitor production line, is conducted. In the case under study, two responses are of interest and the responses are the sizes of the upper side and the lower side of a cabinet front of a monitor. The specification limits of responses are represented in Table 1. In this table, the response types are determined so that we would be able to choose among the formula of computing SN ratios from Equations 2, 3, and 4.

According to the initial experiments and also operational experiences, five controllable factors are screened to set up the experimental design. The description of each controllable factor and also their corresponding levels are illustrated in Table 2.

According to the information given in Table 2, the L_{18} Taguchi design is selected for design of experiments. Since two similar machines can be used for production of the mentioned part, two nuisance factors are considered.

As it can be easily seen in Table 2, the controllable factors are mainly from the type of injection pressure and injection speed or processing time. According to maintenance considerations of injection machines, a production process with less needed injection pressure and less needed injection speed would incur less erosion to the machines, so all pressure-related factors and speed-related factors are STB type. Besides, as much as the required time for a process is shorter, the production rate would increase, so a shorter processing time is more desired. By considering these aspects, all the mentioned controllable factors can be considered as STB-type inputs. An efficient treatment is a setting in which not only the quality requirements are met, but also the abovementioned preferences of the operation are satisfied. Some of the L_{18} orthogonal Taguchi experimental design results are reported in Table 3.

Tune-effective parameters of ANN for estimation of responses

In the training phase, factor level values and noise factor levels are considered as inputs and corresponding response values are considered as outputs of neural networks. As results, there are 36 data divided into three groups of training, testing, and validation data. An important point in choosing testing and training data is that considering one replicate as training and other replicate as a testing data causes erroneous judgment. In this study, the 5th, 10th, and 16th experiments in condition of the first noise factor and second noise factor are considered as testing and validation data, respectively, and others as the training data. The algorithm and its significant parameters considered for training the neural network are reported in Table 4. In the training phase of the ANN, the linear function is selected as the transfer function for the last layer and the tangent sigmoid is used for other layers. By considering the dimension of training data, axial points for designing of experiments by CCD is determined in Table 5. For more details of the tuning method, see Hsieh and Tong (2001).

By performing such an experiment, the optimal setting for the parameters of the neural network are determined as reported in Table 6.

Train the ANN using tuned parameters and estimate unpracticed treatments

In this step, data are divided into testing, training, and validation data subsets as mentioned in the previous section, and finally, the training of ANN is conducted using the tuned parameters obtained from the previous step.

It should be noted that the initial conducted experiment whose results are reported in Table 3, just included 4 treatment results from the 18 practiced treatments,

Table 5 Parameters and their levels studied in experiments based on CCD design in ANN parameter tuning

Factors	Cube points		Central point	Axial points	
	Low	High		Low	High
The number of neuron in first layer	4	8	6	2	10
The number of neuron in second layer	1	3	2	0	4

Table 6 Optimum values of effective parameters in performance of ANN

Parameter	Optimum value
The number of neurons in first layer	5
The number of neurons in second layer	3
RMSE of test data	14.32
RMSE of validation data	14.70

Table 8 Estimated values by neural networks for actual experiments for the injection molding process example

Combination	L_{18}					Y_1		Y_2	
	A	B	C	D	E	N_1	N_2	N_1	N_2
C1	40	55	40	75	6	483.36	483.34	483.95	484.02
C2	40	55	40	75	7	483.35	483.35	483.9	484.05
C1874	50	65	50	85	9	483.32	483.33	483.78	484.13
C1875	50	65	50	85	10	483.32	483.35	483.79	484.1

while by using ANN, many other unpracticed treatments and corresponding results can be estimated in the current step. For this regard, new levels are defined between initial levels. These levels are shown in Table 7.

According to the levels defined in Table 7, there are $5^4 \times 3 = 1,875$ possible treatments (because of the five levels for factors B, C, D, E, and three levels for factor A) which their responses can be estimated by the trained ANN. The responses of these factor level combinations are estimated using the trained neural network. The estimated responses for new defined treatments are reported in Table 8.

Evaluate the efficiency of all treatments using DEA

In this step, we assume each of the estimated treatments reported in Table 8 as a certain DMU. The SN ratios of estimated responses are computed and then normalized in each treatment using Equations 3 and 5. In this step, evaluation of the efficiency for each treatment can be conducted by solving Equation 1.

Matlab software (MathWorks, Inc., Natick, MA, USA) is used to solve 1,875 linear models to determine the efficiency of each treatment. Finally, the efficiency value for 12 of them is obtained as equal to 1. In the next step, the most efficient treatment would be selected among these 12 treatments.

Use maximin weight model to choose among the efficient parameter combinations

In this step, minimum weighted restriction is applied to determine the most efficient treatment among all. Efficient treatments and weights which are obtained by solving Equation 6 are illustrated in Table 9. As mentioned before, the treatment with maximum weight is selected as the most efficient solution of the case study. The most efficient treatment is obtained as $A = 45\%$, $B = 55\%$, $C = 40\%$, and $D = 80\%$ and $E = 6$ s.

The combination and corresponding values with the most efficient treatment are denoted in italics. For verifying the obtained results, the real results of the most efficient treatment are performed for a batch of 30 parts and the most efficient and the extracted results are compared with previous experimented data. Results of the comparisons are illustrated in the Table 10, in which SN_1 and SN_2 are signal-to-noise ratios of the first and second responses, respectively. Note that the most efficient treatment result has been considered as the 19th

Table 7 Factors and their new defined levels for the injection molding process example

Row	Factors	Levels
A	Injection pressure (percent of machine max pressure)	40, 45, 50
B	Injection speed (percent of machine max speed)	55, 57.5, 60, 62.5, 65
C	Holding pressure 1 (percent of machine max pressure)	40, 42.5, 45, 47.5, 50
D	Holding pressure 1 (percent of machine max pressure)	75, 77.5, 80, 82.5, 85
E	Holding pressure time (seconds)	6, 7, 8, 9, 10
N	Nuisance factors, injection machines	550A, 550B

Table 9 The maximin calculated weight for efficient combinations

Combination	A	B	C	D	E	w_j
C100	40	55	47.5	85	10	0.000507
C530	40	65	42.5	75	10	0.000508
C636	*45*	*55*	*40*	*80*	*6*	*0.000508*
C675	45	55	42.5	85	10	0.000426
C748	45	55	50	85	8	0.000494
C1008	45	62.5	40	77.5	8	0.000502
C1201	45	65	47.5	75	6	0.000508
C1420	50	57.5	42.5	82.5	10	0.000477
C1505	50	60	40	75	10	0.000485
C1691	50	62.5	45	82.5	6	0.000077
C1768	50	65	40	82.5	8	0.000487
C1815	50	65	45	80	10	0.000505

The combination and corresponding values with the most efficient treatment are denoted in italics.

Table 10 Validating results

Run	SN ratio of the results		Relative efficient	w_j
	SN₁	SN₂		
1	84.6542	90.6861	0.94	
2	84.6545	66.6024	0.63	
3	84.6549	70.6859	0.67	
4	76.6955	63.8360	0.00	
5	81.1332	67.7633	0.35	
6	81.1329	66.0764	0.33	
7	90.6753	71.6012	1.00	0.0316
8	84.6549	68.4083	0.57	
9	84.6552	73.7852	0.68	
10	78.6346	63.8382	0.15	
11	90.6757	69.8589	1.00	0.0297
12	81.1332	69.8579	0.37	
13	90.6746	70.6848	1.00	0.0256
14	84.6549	67.7630	0.57	
15	81.1332	66.0764	0.32	
16	84.6545	67.1635	0.57	
17	90.6760	67.1646	1.00	0.0000
18	81.1325	71.6012	0.39	
19	83.5709	106.2443	1.00	0.0348

The combination and corresponding values with the most efficient run and corresponding maximin weight is denoted in italics.

run. At first, efficient treatments among these 19 treatments are determined, then the most efficient treatment is selected. As it is observed, in this stage, the previous most efficient treatment is selected as the most efficient one. It shows that the obtained treatment as the most efficient one is theoretically and practically the most possible efficient treatment.

For increasing the applicability of the final results, one-way and two-way sensitivity analyses of efficiency under different values of related factors have been studied. For this purpose, lots of experimented data is necessary to track the sensitive parameters; also, efficiency calculation requires lots of computations for the estimated response values. So, an artificial neural network is applied as the efficiency estimator of different treatments. For this purpose, an ANN is trained by using level values of 1,875 treatments as inputs and efficiency of them as outputs. Then, the sensitivity analysis is done under different values of controllable factors in one or two ways using the trained efficiency estimator neural network. Results of the sensitivity analyses are illustrated in Figure 3. It can be observed that controllable factors A and D can be changed in intervals of about (42 to 45) and (79 to 81), respectively, without changing the selected treatment efficiency score; however, factors B, C, and E have inverse relation to the efficiency score in the analyzed amplitude.

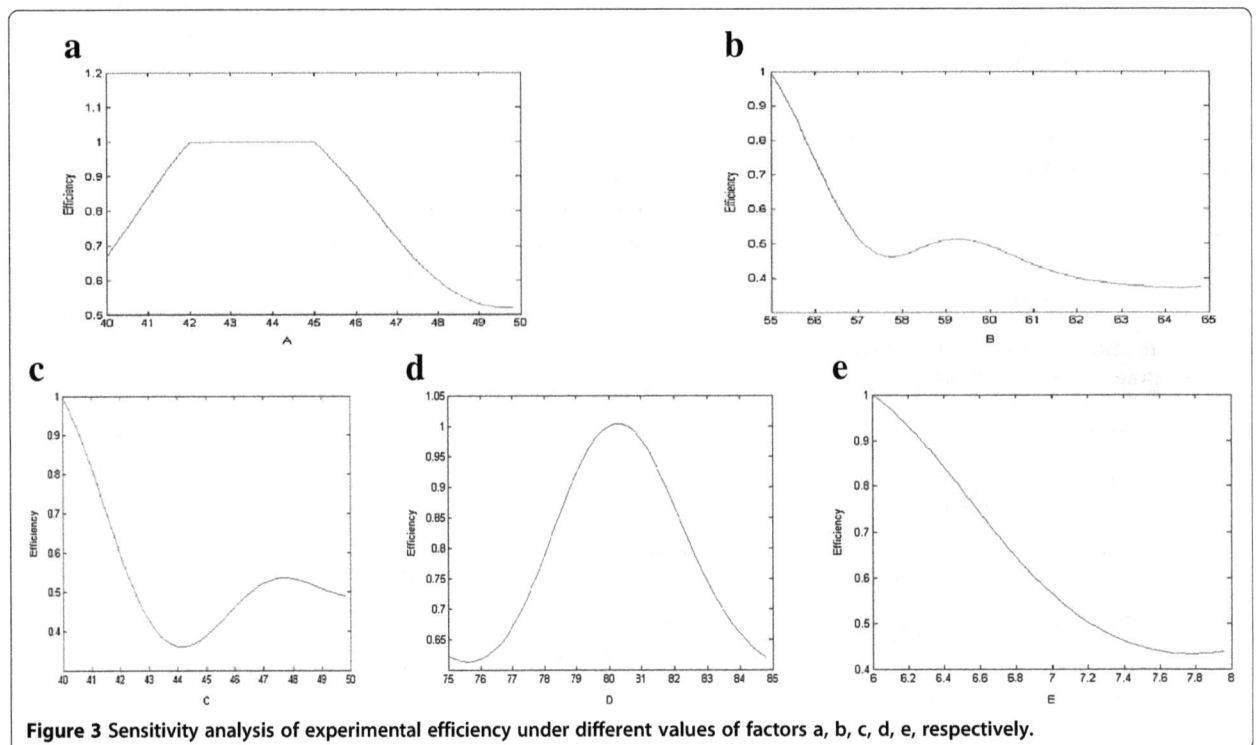

Figure 3 Sensitivity analysis of experimental efficiency under different values of factors a, b, c, d, e, respectively.

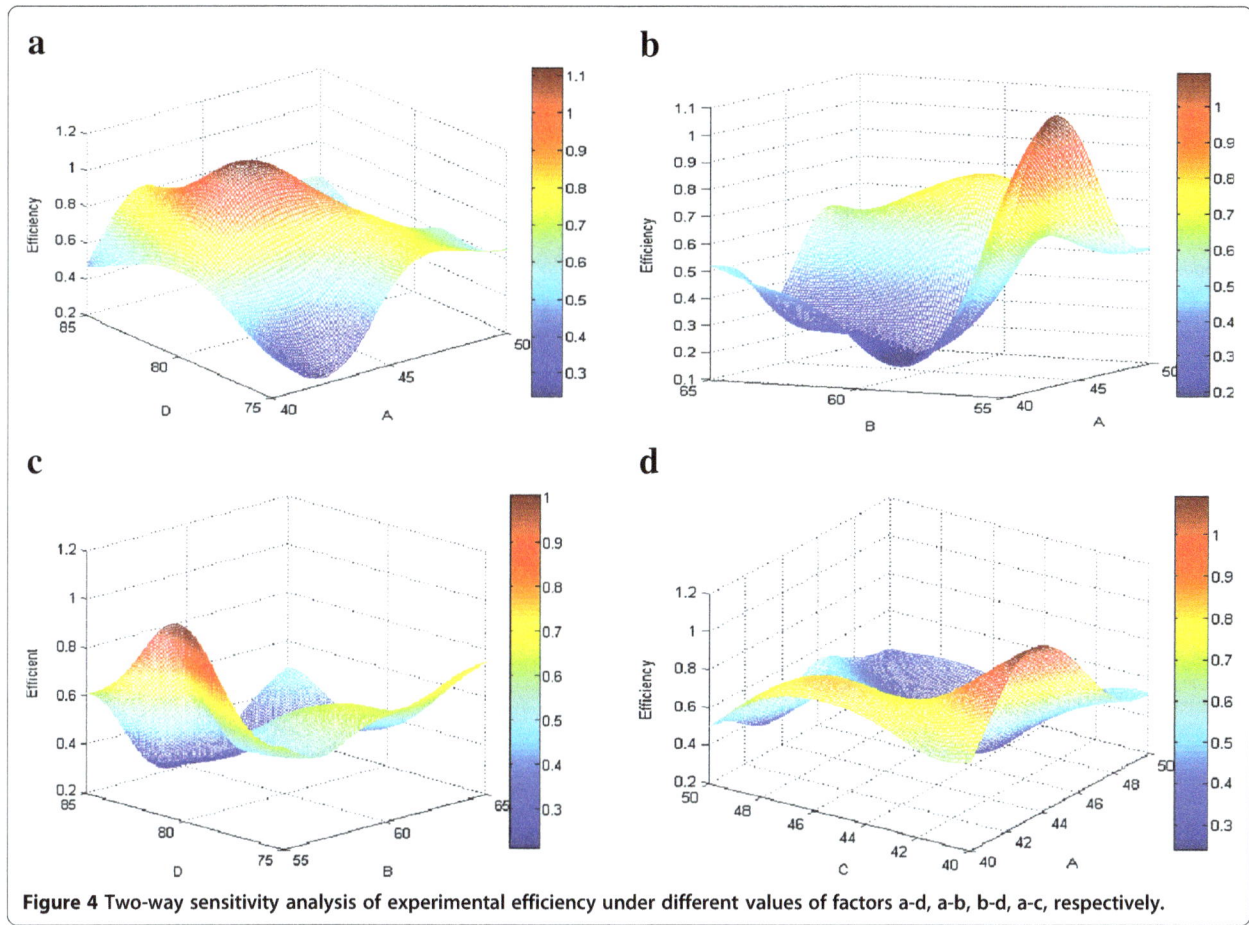

Figure 4 Two-way sensitivity analysis of experimental efficiency under different values of factors a-d, a-b, b-d, a-c, respectively.

Two-way sensitivity analysis has been illustrated in Figure 4; it shows that there are interactions between any pairs of controllable factors which affect the efficiency score and that simultaneous analysis of controllable factors instead of individual analysis is more desirable. This verifies that the proposed efficiency analysis of treatments in multiple-response optimizations is a preferred technique than others especially when the controllable factors are as smaller-the-better type.

Conclusions

In this study, a four-step approach was presented to find the optimal treatment in multiple-response optimization problems. After conducting a Taguchi-designed experiment and collecting the response data, by using a tuned neural network, the responses of unpracticed treatments were estimated considering new levels defined for the controllable factors. Each treatment was assumed to be as a DMU and the smaller-the-better-type controllable factors were assumed as the input variables, whereas the computed SN ratios of responses were assumed to be as the outputs for the DEA modeling. Then data envelopment

analysis was applied to obtain the efficiency of each treatment. Finally, the maximin weight model was applied to find the most efficient treatment.

In the proposed method, for assisting the economic aspects of the process improvement, the type of controllable factors was accounted for, and using this approach, the overall output quality of the process was maximized while the usage of input variables was minimized. For validation of the study, the proposed method was applied in a plastic molding process as a real case and the results were compared and analyzed. Sensitivity analysis of the efficiency deviation, which was the useful tool for analyzing results, was presented by applying another neural network. The analysis showed that the proposed approach was a proper tool in discrete multiple response optimization especially for the STB controllable factors. As illustrated in this paper, the proposed approach was based on the response values and it did not consider the variations of responses; so, using other approaches which consider the variation can be future studies of STB-type controllable factors in MRO problems.

Competing interests
The authors declare that they have no competing interests.

Authors' contributions
MB managed the study, completed the main idea and was responsible for integrating and revising the manuscript. AFG developed the methodology of the proposed approach. Besides HM contributed in developing the methodology as well as verifying it in the case study. All authors read and approved the final manuscript.

Acknowledgements
The authors would like to acknowledge the management of Iran office machine industry plastic factory for contributing in the case study.

Author details
[1]Department of industrial Engineering, Faculty of Engineering, Shahed University, Khalij Fars Highway, Tehran P.O. BOX: 3319118651, Iran. [2]Faculty of industrial engineering and systems, college of engineering, University of Tehran, Enghelab street, Tehran P.O. BOX: 11155-4563, Iran.

References
Amiri A, Bashiri M, Mogouie H, Doroudynan MH (2012) Non-normal multiple response optimization using process capability index. Scientia Iranica 19(6):1894–1905

Bashiri M, Geranmayeh AF (2011) Tuning the parameters of an artificial neural network using central composite design and genetic algorithm. Scientia Iranica 18(6):1600–1608

Bashiri M, Moslemi A (2013) Simultaneous robust estimation of multi-response surfaces in the presence of outliers. Int J Ind Eng Int 9(7):2–12

Caporaletti LE, Dulá JH, Womer NK (1999) Performance evaluation based on multiple attributes with nonparametric frontiers. Omega 27:637–645

Chang H (2006) Dynamic multi-response experiments by backpropagation networks and desirability functions. J Chin Inst Ind Eng 23(4):280–288

Chang H (2008) A data mining approach to dynamic multiple responses in Taguchi experimental design. Expert Syst Appl 35(3):1095–1103

Chang H, Chen Y (2011) Neuro-genetic approach to optimize parameter design of dynamic multiresponse experiments. Appl Soft Comp 11:436–442

Charnes A, Cooper WW, Rhodes E (1978) Measuring the efficiency of decision making units. Eur J Oper Res 2(6):429–444

Desai K, Survase S, Saudagar P, Lele S, Singhal R (2008) Comparison of artificial neural network (ANN) and response surface methodology (RSM) in fermentation media optimization: case study of fermentative production of scleroglucan. Biochem Eng J 41(3):266–273

Erzurumlu T, Oktem H (2007) Comparison of response surface model with neural network in determining the surface quality of moulded parts. Mater Des 28(2):459–465

Gauri SK, Pal S (2010) Comparison of performances of five prospective approaches for the multi-response optimization. Int J Adv Manuf Technol 48(9):1205–1220

Goel T, Vaidyanathan R, Haftka TR, Shyy W, Queipo VN, Tucker K (2007) Response surface approximation of Pareto optimal front in multi-objective optimization. Comp Methods Appl Mech Eng 196(1):879–893

Gutierrez E, Lozano S (2010) Data envelopment analysis of multiple response experiments. Appl Math Model 34(5):1139–1148

Hesieh K, Tong L (2001) Optimization of multiple quality responses involving qualitative and quantitative characteristics in IC manufacturing using neural networks. Comp Ind 46(1):1–12

Hsu C (2001) Solving multi-response problems through neural networks and principal component analysis. J Chin Inst Ind Eng 18(5):47–54

Hsu C (2004) An integrated approach to enhance the optical. Int J Prod Econ 92(3):241–254

Khaw JF, Lim B, Lim LE (1995) Optimal design of neural networks using the Taguchi method. Neurocomputing 7(3):225–245

Kim Y, Yum B (2004) Robust design of multilayer feedforward neural networks: an experimental approach. Eng Appl Artificial Intell 17(3):249–263

Li T, Su C, Chiang T (2003) Applying robust multi-response quality engineering for parameter selection using a novel neural-genetic algorithm. Comp Ind 50(1):113–122

Liao H (2004) A data envelopment analysis method for optimizing multi-response problem with censored data in the Taguchi method. Comp Ind Eng 46(4):817–835

Liao HC (2005) Using N-D method to solve multi-response problem in Taguchi. J Intell Manuf 16:331–347

Liao HC, Chen YK (2002) Optimizing multi-response problem in the Taguchi method by DEA based ranking method. Int J Qual Reliability Manage 19(7):825–837

Maghsoodloo S, Ozdemir G, Jordan V, Huang CH (2004) Strengths and limitations of Taguchi's contributions to quality, manufacturing, and process engineering. J Manuf Syst 23(2):73–126

Myers R, Montgomery D (2002) Response surface methodology: process and product optimization using designed experiments, 2nd edition. Wiley, New York

Namvar-Asl M, Soltanieh M, Rashidi A, Irandoukht A (2008) Modeling and preparation of activated carbon for methane storage I. Modeling of activated carbon characteristics with neural networks and response surface method. Energy Conv Manage 49(9):2471–2477

Niaki STA, Hoseinzade S (2013) Forecasting S&P 500 index using artificial neural networks and design of experiments. J Indust Eng 9(1):1–9

Noorossana R, Davanloo Tajbakhsh S, Saghaei A (2009) An artificial neural network approach to multiple-response optimization. Int J Adv Manufact Technol 40(11):1227–1238

Ortiz F, Simpson J, Pignatiello J, Heredia-Langner A (2004) A genetic algorithm approach to multiple-response optimization. J Qual Technol 36:432–450

Ozcelik B, Erzurumlu T (2006) Comparison of the warpage optimization in the plastic injection molding using ANOVA, neural network model and genetic algorithm. J Mat Proces Technol 171(3):437–445

Packianather M, Drake P, Rowlands H (2000) Optimizing the parameters of multilayered feedforward neural networks through Taguchi design of experiments. Qual Reliability Eng Int 16(6):461–473

Peterson G, St Clair D, Aylward S, Bond WE (1995) Using Taguchi's method of experimental design to control errors in layered perceptrons. IEEE Transac Neural Networks 6(1):949–961

Robinson TJ, Borror CM, Myers RH (2004) Robust parameter design: a review. Qual Reliability Eng Int 20:81–101

Sibalija TV, Majstorovic VD (2012) An integrated simulated annealing-based method for robust multiresponse process optimisation. Int J Adv Manufact Technol 59:1227–1244

Sukthomya W, Tannock J (2005) The optimisation of neural network parameters using Taguchi's design of experiments approach: an application in manufacturing process modelling. Neural Comp Appl 14:337–344

Taguchi G, Chowdhury S (2000) Robust design engineering. McGraw-Hill, New York

Tbanassoulis E (2001) Introduction to the theory and application of data envelopment analysis. Kluwer, Norwell

Tortum A, Yayla N, Çelik C, Gökdağ M (2007) The investigation of model selection criteria in artificial neural networks by the Taguchi method. Physica A: Stat Mech Appl 386:446–468

Tsao C (2008) Comparison between response surface methodology and radial basis function network for core-center drill in drilling composite materials. Int J Adv Manufact Technol 37(11):1061–1068

Wang YM, Luo Y, Liang L (2009) Ranking decision making units by imposing a minimum weight restriction in the data envelopment analysis. J Comput Appl Math 223(1):469–484

Zang C, Friswell MI, Mottershead JE (2005) A review of robust optimal design and its application in dynamics. Comp Struct 83:315–326

16

Solving the vehicle routing problem by a hybrid meta-heuristic algorithm

Majid Yousefikhoshbakht[1*] and Esmaile Khorram[2]

Abstract

The vehicle routing problem (VRP) is one of the most important combinational optimization problems that has nowadays received much attention because of its real application in industrial and service problems. The VRP involves routing a fleet of vehicles, each of them visiting a set of nodes such that every node is visited by exactly one vehicle only once. So, the objective is to minimize the total distance traveled by all the vehicles. This paper presents a hybrid two-phase algorithm called sweep algorithm (SW) + ant colony system (ACS) for the classical VRP. At the first stage, the VRP is solved by the SW, and at the second stage, the ACS and 3-opt local search are used for improving the solutions. Extensive computational tests on standard instances from the literature confirm the effectiveness of the presented approach.

Keywords: Ant colony system, NP-hard problems, Sweep algorithm, Vehicle routing problem

Background

One of the parameters that is always in mind in production and services is the reduction of the product's expense since lower product costs yield increase in the company's competitive advantage in terms of production and services, so the company's profit is growth. Furthermore, minimizing the transportation cost is one of the cost reduction methods in which the goods are transported from one place to other places with minimum cost. Therefore, effectively resolving the vehicle routing problem (VRP) is incredibly important in a distribution network of a logistics system.

A lot of research has been carried out in the field of logistics, from the traveling salesman problem to complex dynamic routing problems. Among the prominent problems in distribution and logistics, the vehicle routing problem and its extensions have been widely studied for many years, mainly because of their applications in real world logistics and transportation problems. The VRP has been proven to be a non-deterministic polynomial time problem (Bodin & Golden 1981). This means that the time of the VRP solution grows exponentially with increasing the distribution points.

A large number of techniques in the literature deal with the VRP and its variations such as VRP with pickup and delivery, time windows VRP, stochastic VRP, multi-depot VRP and others. Generally, the techniques used for solving the VRP can be categorized into exact, heuristic and hybrid methods. Although exact algorithms are appropriate for instances with small size, they are not often suitable for real instances owing to the computational time required to obtain an optimal solution. There have been many papers proposing exact algorithms for solving the VRP. These algorithms are based on column generation approach (Choi & Tcha 2007), implicit enumeration algorithm (Ruiz et al. 2004), branch-and-bound method (Valle et al. 2011), etc.

The VRP is an NP-hard problem and difficult to solve by exact methods within acceptable computing times (Lee et al. 2003). In other words, when the problem size of the VRP is increased, the exact methods cannot solve it. So, heuristic methods are necessary to be used for solving instances with large size in a reasonable amount of time. These algorithms can be divided into two main groups including classical heuristics and meta-heuristics. Since heuristic approaches are not so efficient for escaping local optimum values, the obtained solutions by them may have a large disparity compared to the best known or sub-optimal solutions. Some of the well-known heuristic algorithms are gravitational emulation search (Clarke

---
* Correspondence: khoshbakht@iauh.ac.ir
[1]Young Researchers Club, Hamedan Branch, Islamic Azad University, Hamedan, Iran
Full list of author information is available at the end of the article

& Wright 1964), local search (Gillett & Miller 1974) and Lin-Kernighan (Christofides et al. 1979).

A new kind of emerged algorithm basically tries to combine basic heuristic methods in higher level frameworks aimed at efficiently and effectively exploring a search space in the last 30 years. These methods are nowadays commonly called *meta-heuristics*. Since the meta-heuristic approaches are very efficient for escaping from local optimum, they are one of the best group algorithms for solving combinatorial optimization problem. A great number of meta-heuristic algorithms have been applied to the VRP, including genetic algorithm (Baker & Ayechew 2003), large neighborhood search (Hong 2012; Pisinger & Ropke 2010), tabu search (TS) (Leung et al. 2011), simulated annealing (Osman 1993), memetic algorithm (Ngueveu et al. 2010), neural networks (Su & Chen 1999) and particle swarm optimization (Ai & Kachitvichyanukul 2009). On the other hand, a limited amount of research addressing the VRP has used the ant colony optimization (ACO), with candidate lists and ranking techniques to improve the ability of a single ACO to solve the VRP (Bullnheimer et al. 1998; Bullnheimer et al. 1999).

Recently, many researchers have found that the employment of hybridization in optimization problems can improve the quality of problem solving in comparison with heuristic and meta-heuristic approaches. Since hybrid algorithms such as genetic algorithms with a local search (Prins 2009; Lima et al. 2004), genetic algorithms with sweep algorithm (SW) and nearest addition method (Wang & Lu 2009), ACO and greedy heuristics (Zhang & Tang 2009), simulated annealing and TS (Lin et al. 2009), neural networks and genetic algorithms (Potvin et al. 1996a; Potvin et al. 1996b), genetic algorithms and ACO (Reimann et al. 2001), etc. have greater ability to find an optimal solution, they have been considered by researchers and scientists in recent years.

The ACO approach is one of the famous meta-heuristic algorithms that simulates the ants' food-hunting behavior and is used for solving combinatorial optimization problems that do not have a known efficient algorithm. This technique, one of the most powerful methods compared to other meta-heuristic algorithms nowadays, was introduced by Dorigo et al. in (Dorigo 1992) (Dorigo et al. 1991; Dorigo 1992). Because a limited amount of research addressing the VRP has used the ACO combined with heuristic algorithms, a hybrid meta-heuristic algorithm based on ant colony system is proposed for the VRP in this paper. In this algorithm, the VRP is solved by the SW firstly, and then the ant colony system (ACS) and 3-opt local search are used for improving the solutions.

In the following parts of this paper, in 'VRP mathematical model', we will explain the formulation of VRP.

Then, at the beginning of 'Methods', we specially explain the SW and ACS, and then, the combination of ACS and SW is extendedly explained in this section. In 'Numerical calculations', the proposed algorithm will be compared with some of the other algorithms on standard problems, which are included in the VRP library. Finally in the last section, the conclusions are presented.

VRP mathematical model

The VRP is one of the applied problems in the real world. This problem consists of a large number of customers in which everyone has a known demand level. These customers must be supplied from node 0 called the depot. Delivery routes which start and finish at the depot are required for vehicles so that all customer demands are satisfied and each customer is visited by just one vehicle. Vehicle capacities are given and limited by Q. So, the VRP can be explained mathematically as follows.

Let G (V, A) be a perfect undirected connected graph with a vertex set $V = \{0, 1, \ldots, n\}$ and an edge set $A = \{(i,j):i,j \in V, i \neq j\}$. If the graph is not perfect, we can replace the lack of any edge with an edge that has an infinite size. Vertex 0 represents the depot; each of the other vertex $i \in V - \{0\}$ is a customer with a non-negative demand q_i, and a non-negative distance c_{ij} is associated with each arc $(i,j) \in A$ and represents the travel cost from node i to node j. The cost matrix is symmetric, i.e., for all $i,j \in V$, $c_{ij} = c_{ji}$. The use of the loop arc (i,i) is not allowed and defining $c_{ii} = +\infty$ for all $i \in V$.

So, the solution for the VRP determines a set of delivery routes that satisfies the requirements of distribution points and obtains the minimum total cost for all vehicles. In practice, minimizing the total cost is equivalent to minimizing the total distance traveled by $m > 1$ vehicles.

An example of a single solution consisting of a set of routes constructed for a classical VRP is presented in Figure 1, where $m = 4$ and $n = 12$. So, the solution of this example is 0-1-2-3-4-0; 0-5-6-7-0; 0-8-9-10-0; 0-11-12-0.

For presenting the integer linear programming model for VRP, the variables below are introduced:

n = the number of nodes for each instance.
m = the number of vehicles used for each instance.

$$x_{ij} = \begin{cases} 1 & \text{if the vehicle travels directly from node } i \text{ to node } j \\ 0 & \text{otherwise.} \end{cases}$$

$$y_{ik} = \begin{cases} 1 & \text{if the node } i \text{ is visited by the vehicle } k \\ 0 & \text{otherwise.} \end{cases}$$

Hence, one of the integer programming formulations for the VRP can be written as follows:

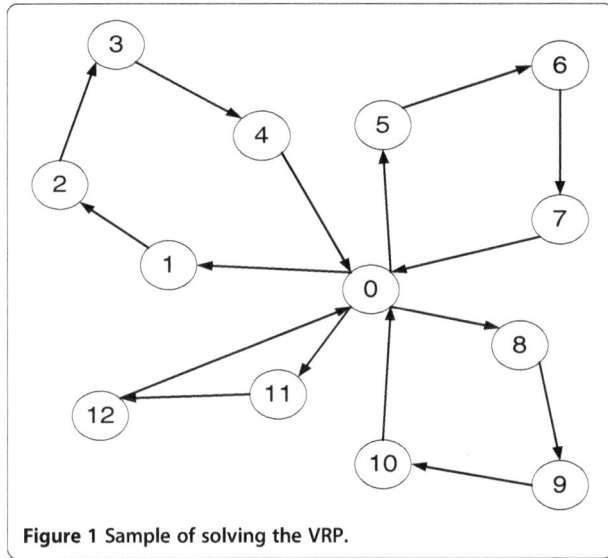

Figure 1 Sample of solving the VRP.

$$\min \sum_{i=0}^{n} \sum_{j=0}^{n} c_{ij} x_{ij} \tag{1}$$

subject to

$$\sum_{i=0}^{n} x_{ij} = 1 (j = 1, \ldots, n) \tag{2}$$

$$\sum_{i=0}^{n} x_{i0} = m \tag{3}$$

$$\sum_{j=0}^{n} x_{ij} = 1 (i = 1, \ldots, n) \tag{4}$$

$$\sum_{j=0}^{n} x_{0j} = m \tag{5}$$

$$\sum_{i=0}^{n} q_i y_{ik} \leq Q(k = 1, \ldots, m) \tag{6}$$

$$\sum_{i \in S} \sum_{j \in N-S} x_{ij} \geq 1 (\phi \neq S \subset \{1, \ldots, n\}, |S| \geq 2 \tag{7}$$

$$\sum_{i \in N-S} \sum_{j \in S} x_{ij} \geq 1 (\phi \neq S \subset \{1, \ldots, n\}, |S| \geq 2 \tag{8}$$

$$x_{ij} \in \{0, 1\} (i = 0, \ldots, n; j = 0, \ldots, n) \tag{9}$$

$$y_{ik} \in \{0, 1\} (i = 0, \ldots, n; k = 1, \ldots, m) \tag{10}$$

The objective function (1) minimizes the total distance traveled in a tour. Constraint sets (2) and (3) ensure that the vehicle arrives once at each node and m times at the depot. Constraint sets (4) and (5) ensure that the vehicle leaves each node once and leaves the depot m times. Constraint set (6) prevents vehicles from carrying loads more than their capacity. Constraint sets (7) and (8) avoid the presence of sub-tours for each vehicle. Finally, Constraint sets (9) and (10) define binary conditions on the variables.

Results and discussion

Numerical calculations

At the first stage in this section, sensitivity analyses of parameters in the proposed algorithm (PA) are performed, and at the second stage, the SW + ACS, which was discussed in the previous section, is analyzed using several benchmarking problems of Christofides et al. (Christofides et al. 1979) and Taillard (Taillard 1993). The proposed algorithm was coded using MATLAB language executed on a computer with a 1.00-GB RAM and an Intel, 2.80-GHz CPU.

Sensitivity analyses of parameters

In this section, the parameters in our algorithm are tuned, in which there are four parameters including α, β, ρ and q_0. The ranges of the four parameters were set to $\alpha \in \{1, 2, 3\}$, $\beta \in \{2, 4, 6\}$, $\rho \in \{0.1, 0.2, 0.3\}$ and $q_0 \in \{0.9, 0.95, 0.99\}$. When tuning the parameters, the instance E-n51-k5 was determined as the test problem. Then, the algorithm with each parameter combination for this instance was tested five times.

Based on the gained results, the algorithm with the smaller weight parameter (α) of pheromone trails possesses higher performance. This may be attributed to the fact that in the SW + ACS, the initial pheromone trails are large values. If using the large control factor of the pheromone trail, the effect of visibility value is weakened and results in a premature convergence. In addition, the qualities of the solutions of the algorithms with $\beta = 2$ are better than those of 4 and 6.

From the test results, it can be found that by setting the evaporation factor to 0.1, the proposed algorithm can yield better solutions. This can be attributed to the fact that if pheromone evaporation is too rapid, it more easily results in the search to be trapped in local minimum. In other words, the smaller evaporation factor can ensure the sufficient diversity of search space and guide following ants to explore better solutions.

Finally, the SW + ACS, in which q_0 is set to 0.99, can provide better solutions in comparison with other values. Thus, the combinations of optimal parameters are determined: $\{\alpha = 1, \beta = 2, \rho = 0.1, q_0 = 0.99\}$.

Benchmark instances

In the instances of Christofides et al. (Christofides et al. 1979), which can be downloaded from the DEIS -

Table 1 Comparing algorithms for standard VRP problems

Example	PA	ACO (Mazzeo & Loiseau 2004)	SA + 3-opt	SA (Clark & Wright 1964)	TS (Leung et al. 2011)	BKS (Mazzeo & Loiseau 2004)
E-n51-k5	521	521	578.56	584.64	524.61	521
E-n76-k10	838	877	888.04	900.26	844	832
E-n101-k8	839.2	845	878.70	886.83	835	815
E-n101-k10	823.74	838	824.42	833.51	820	820
E-n121-k7	1050	1,189	1,049.43	1,071.07	1,042.11	1,042.11
M-n151-k12	1,030.46	1,105	1,128.24	1,133.43	1,052	1,028.42
M-n200-k17	1,325.62	1,606	1,336.84	1,395.74	1,378	1,291.45

ACO, ant colony optimization; BKS, best known solution; PA, proposed algorithm; SA, saving algorithm; SA + 3-opt, saving algorithm + 3-opt; TS, tabu search.

Operations Research Group Library of Instances (DEIS - Operations Research Group 2012), the results obtained from calculations by the proposed methods are compared with other algorithms including saving algorithm (SA), saving algorithm + 3-opt (SA + 3-Opt), TS search and ACO. The first two of these algorithms are traditional, and the next two are relatively new. These algorithms were executed on standard instances containing between 51 and 200 nodes of the VRP problem including E-n51-k5, E-n76-k10, E-n101-k8, E-n101-k10, E-n121-k7, M-n151-k12 and M-n200-k17. The number at the right of each instance shows the number of vehicles, and the middle number indicates the number of related customers.

Table 1 summarizes the computational results of the proposed, compared and other mentioned algorithms. Moreover, to show the method's performance more clearly, we present the best known solutions (BKSs) that have been published in the related literature in this table.

Besides, Figure 2 shows a comparison between the gap values of the heuristic and meta-heuristic algorithms, where the gap is defined as the percentage of deviation from the best known solution in the literature. The gap is equal to $100[c(s^{**}) - c(s^{*})]/c(s^{*})$, where s^{**} is the best solution found by the algorithm for a given instance, and s^{*} is the overall best known solution for the same

instance on the web. A zero gap indicates that the best known solution is found by the algorithm.

As can be seen from Table 1, the proposed algorithm finds the optimal solution for one out of seven problems that are published in the literature. The results indicate that SW + ACS is a competitive approach compared to the traditional and new meta-heuristics. For instances E-n101-k8 and M-n200-k17, the gap is relatively as high as 3%. However, in other instances, the proposed algorithm finds nearly the best known solution, i.e., the gap is below 0.76%, and overall, the average difference is 1.1%. The performance comparison of results shows that the SW + ACS method clearly yields better solutions than the ACO, saving algorithm and its modified version (SA + 3-opt).

Moreover, computational results of the proposed algorithm and TS show that these algorithms have a close competition, and the proposed algorithm gives four better solutions than the TS. In other words, the performance of the SW + ACS is better in reaching the suboptimal solution than the TS (Table 1).

From the statistical viewpoint in Table 2, there is no statistical significant difference between the PA and other meta-heuristic algorithms (p value = 0.9881). Furthermore, it can be observed from both the medians plot (Figure 3) and the means plot (Figure 4) that the PA is

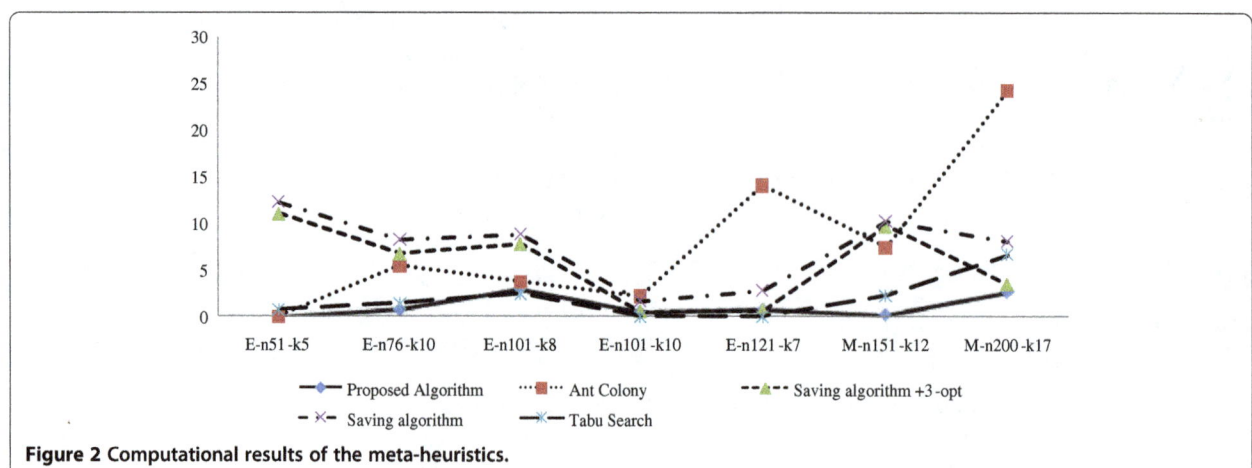

Figure 2 Computational results of the meta-heuristics.

Table 2 ANOVA table

Source	SS	DF	MS	F	The *p* value
Columns	41,994.6	5	8,398.9	0.12	0.9881
Error	2,607,392.3	36	72,427.6		
Total	2,649,386.9	41			

Df, degrees of freedom; F, F ratio; MS, mean square; SS, sum of the squares.

better than the ACO, SA + 3-Opt and SA algorithms. Also, by observing these figures, we can yield the approximate equality of the PA and TS methods.

In Table 3, the values obtained by the common ACO and proposed algorithm are shown. The results indicate that these two algorithms have very good ability for small problems, have approximately similar behavior and can converge to the best solutions. However, this ability of ACO against the proposed algorithm decreases when the number of nodes and feasible solutions increases.

Table 4 shows the results obtained for the second problem instances including 12 instances proposed by Taillard and presents the comparison between the best results of our algorithm and the BKS. As shown in Table 4, in three cases, the BKS can be found by our algorithm. In other cases, the deviation between the BKS reported and the solution found by the proposed algorithm is very low. In addition, the proposed algorithm finds closely the BKS for most of the instances, i.e., the gap is below 1.10%.

Conclusions

In this paper, a hybrid algorithm combining ACS, SW and 3-opt local search was proposed for solving the VRP. The SW + ACS is more efficient than the ACO.

Especially for large-size problems, this algorithm yields better solutions compared with previous algorithms. Another benefit of this algorithm is that when ACS is used, the number of nodes and complexity of the problem are decreased. As a result, the algorithm can solve these small problems easily. So, it is not necessary for the algorithm to be executed many times for finding the best solutions since the SW and 3-opt local search do not have great time requirements. It also seems that the combination of the proposed algorithm and genetic or simulated annealing algorithms will yield better results. Using the proposed algorithm for other versions of the VRP is suggested for future research.

Methods

Because the VRP is one of the most important combinatorial optimization problems, a hybrid algorithm called SW + ACS is proposed in this paper. In this section, the SW is presented first then the ACS will be explained, and finally, the hybrid algorithm will be analyzed in more detail.

Sweep algorithm

Gillett and Miller (Gillett & Miller 1974) proposed a SW in 1974 for Euclidean networks, which ranks and links demand points by their polar coordinate angle (Figure 5). The polar coordinate angle is calculated as follows:

$$An(i) = \arctan\{(y(i) - y(0))/(x(i) - x(0))\}$$

It is noted that if $y(i) - y(0) < 0$ and $y(i) - y(0) \geq 0$ then $-\pi < An(i) < 0$ and $0 \leq An(i) \leq \pi$, respectively. In this

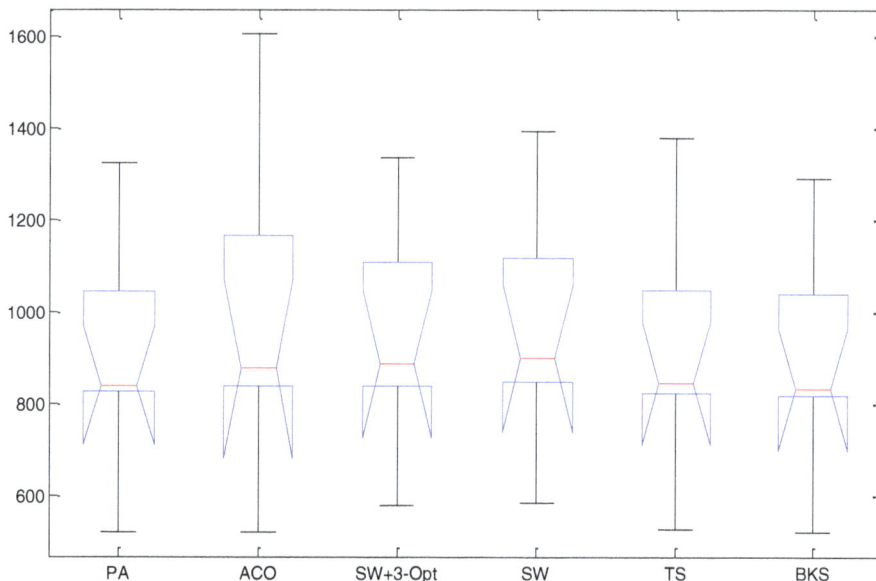

Figure 3 Medians plot.

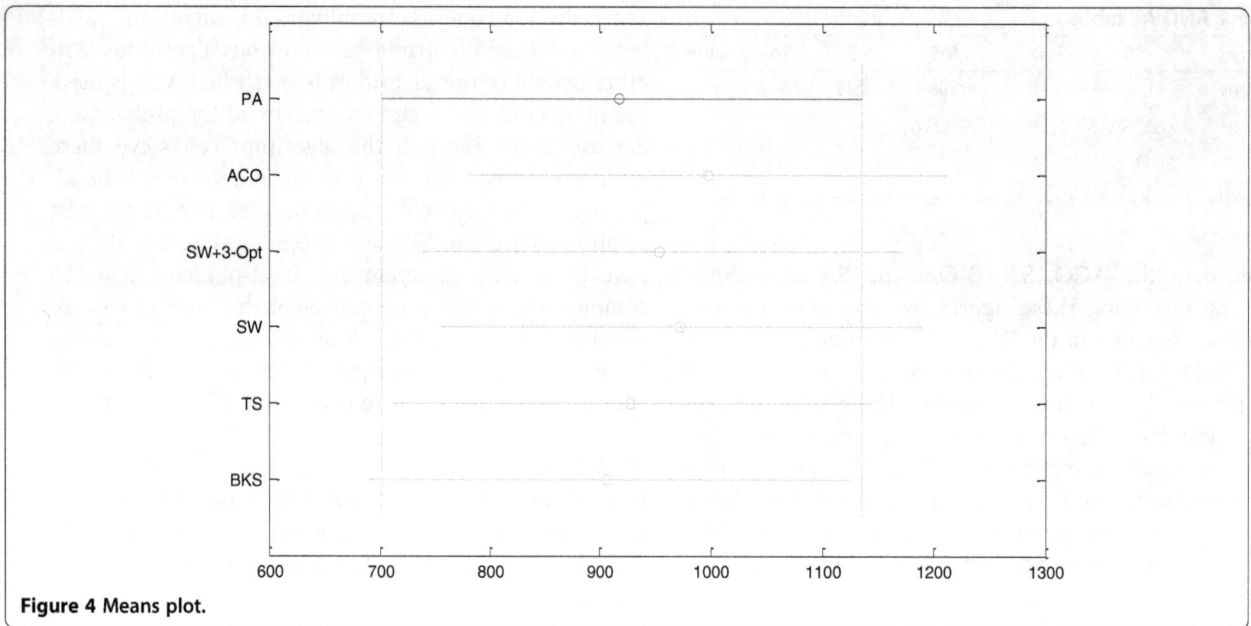

Figure 4 Means plot.

algorithm, which is one of the famous and powerful heuristic algorithms, at first, the polar coordinates of all n customers are calculated, where the depot is node 0 and so $An(1) = 0$.

Then, sweeping (clockwise or counterclockwise) is started from customer i, which has not been visited, from the smallest angle to the largest angle until all customers are included in a tour.

$$0 = An(1) \leq An(2) \leq \ldots \leq An(n).$$

Ant colony system

ACS is one of the most powerful versions of ACO. This meta-heuristic algorithm was introduced in 1996 by Dorigo and Gambardella (Dorigo & Ganbardella 1997), which was strongly inspired by the ant system (AS). So, this algorithm achieves performance improvements because of the introduction of new mechanisms based on ideas not included in

the original AS. The ACS algorithm has two major changes to the rules employed in the AS algorithm, namely:

(1) A novel transition rule that favors either exploitation or exploration is introduced. Node j that is next to node i, among the unvisited nodes J_i^k, is selected by ant k in the route. According to the following transition rule, the probability of each node being visited is

$$P_{ij}^k(t) = \begin{cases} 1 & \text{if}\{q \leq q_0 \text{ and } j = j^*\} \\ 0 & \text{if}\{q \leq q_0 \text{ and } j \neq j^*\} \\ \dfrac{\tau_{ij}^\alpha(t)\eta_{ij}^\beta(t)}{\sum_{r \in J_i^k} \tau_{ir}^\alpha(t)\eta_{ir}^\beta(t)} & \text{otherwise} \end{cases} \quad (11)$$

Where

j^* = the unvisited node j in J_i^k for which $[\tau_{ir}(t)][\eta_{ir}(t)]^\beta$ is maximized.
$\tau_{ij}(t)$ = the amount of pheromone that is on the edge joining nodes i and j.
$\eta_{ij}(t)$ = the heuristic information for the ant visibility measure defined here as the reciprocal of the distance between node i and node j.
α, β = two control parameters that represent the relative importance of the amount of pheromone on the edge between node i and node j compared to the ant visibility value respectively.
q = a random variable in $[0, 1]$.
q_0 = a given arbitrary parameter fixed before the program is started such that when q is less than q_0, the ant employs exploitation to select node j^* as the

Table 3 Mean gap values of ACO and the SW + ACS

Example	ACO	PA
E-n51-k5	0	0
E-n76-k10	5.41	0.72
E-n101-k8	3.68	2.97
E-n101-k10	2.20	0.46
E-n121-k7	14.10	0.76
M-n151-k12	7.45	0.20
M-n200-k17	24.36	2.65

ACO, ant colonization optimization; PA, proposed algorithm.

Table 4 Computational results for the benchmark of Taillard

Problem instance	Size	PA	BKS	Gap
Tai75a	75	**1,618**	1,618	0
Tai75b	75	**1,344**	1,344	0
Tai75c	75	1,295	1,291	0.31
Tai75d	75	1,380	1,365	1.10
Tai100a	100	2,050	2,041	0.44
Tai100b	100	1,940	1,939	0.052
Tai100c	100	1,411	1,406	0.36
Tai100d	100	1,585	1,581	0.25
Tai150a	150	**3,055**	3,055	0
Tai150b	150	2,732	2,656	2.86
Tai150c	150	2,364	2,341	0.98
Tai150d	150	2,660	2,645	0.57

Values in bold are the three cases where the best known solution can be found by our algorithm. PA, proposed algorithm; BKS, best known solution.

next node in its tour, whereas if q exceeds q_0, the ant uses probabilistic exploration to select the next node in its tour.

(2) There are two ways for updating pheromone trail as follows:

- Local updating: When an ant moves from node i to node j, it updates the amount of pheromone on the traversed edge using the following formula:

$$\tau_{ij}(t+1) = (1-\rho).\tau_{ij}(t) \\ + \rho\tau_0 \quad \text{if}(i,j) \in T_k \quad (12)$$

where τ_0, the initial amount of pheromone, is calculated as $\tau_0 = (nC_i)^{-1}$, n is the number of nodes, C_i is the cost of the initial tour produced by a construction heuristic such as the nearest neighbor

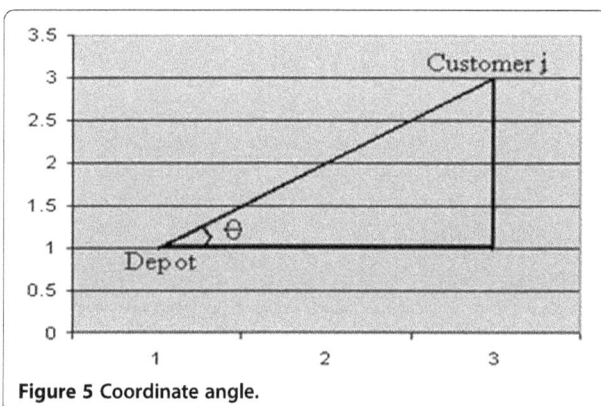

Figure 5 Coordinate angle.

heuristic, SW, etc., and ρ is a parameter called evaporation rate in the range $[0, 1]$ regulating the reduction of pheromone on the edges. It should be noted that local updating has an important effect, such that whenever an ant traverses an edge (i, j), its pheromone trail τ_{ij} is reduced. So, the edge becomes less desirable for the ants in future iterations. Local updating not only encourages an increase in the exploration of edges that have not been visited yet but also helps avoid poor stagnation situations.

- Global updating: When all tours are generated by ants, the edges that belong to the best tour are updated using the following formula:

$$\tau_{ij}(t+1) = (1-\rho).\tau_{ij}(t) \\ + \rho(1/C_b) \quad \text{if}(i,j) \in T_b \quad (13)$$

where C_b is the cost of the best tour T_b found yet. It is important to note that the pheromones of the edges belonging to the best tour are only updated in the global updating. This encourages ants to search in the vicinity of this best tour in future iterations.

Hybrid algorithm

From another viewpoint, there are also two heuristic groups for solving combinatorial optimization problems: construction algorithms that produce a feasible solution themselves, and improvement algorithms that can improve the solutions with having a feasible solution. Here, a construction algorithm, namely SW, and two improvement algorithms, namely ACS and 3-opt local search, are used. In this algorithm, the ACS is applied for improving every route of the vehicle; however, the nodes of each vehicle should be unchanged. On the other hand, the 3-opt local search is used for changing nodes and improving each vehicle. In other words, if this algorithm is separated into two parts, the first one to improve the route of the vehicle without changing the vehicle's nodes and the second one to improve it only by changing the vehicle's nodes, then the ACS does operation 1 and the 3-opt local search does operation 2 (Figure 6).

In this method, first, the nodes that should be visited by vehicles are ordered with respect to the depot by SW, and then, they are set in an array. Second, all of the

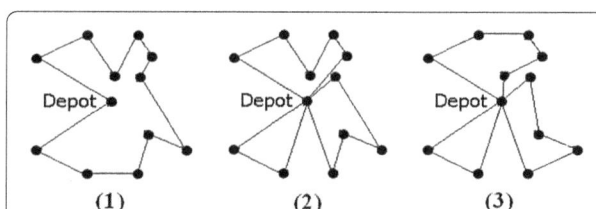

Figure 6 SW (1), (2) and implementing ACS regarding the vehicle's constraints (3).

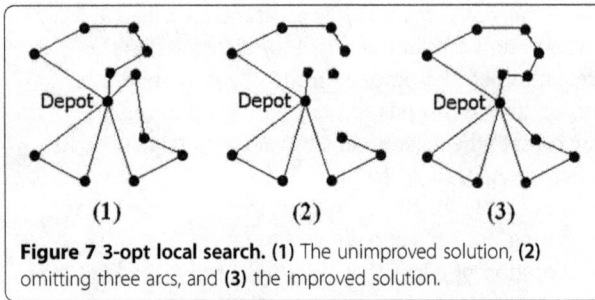

Figure 7 3-opt local search. (1) The unimproved solution, **(2)** omitting three arcs, and **(3)** the improved solution.

states in which the above-mentioned order is preserved in spite of a change in their locations are obtained. For example, if 1 2 3 is the order obtained for a feasible solution, then orders 2 3 1 and 3 1 2 are set in other arrays. Then for each array, the vehicle starts to move from the depot and visits the nodes in the arrays in the order described until it is not possible to add a further node. This means that if the load of a vehicle is more than its capacity, it returns to the depot and repeats these steps until there is no more node to be visited. When vehicle

routes are obtained, the ACS is implemented for every route until the best route is obtained.

Furthermore, the vast literature on meta-heuristics tells us that a promising approach for obtaining high-quality solutions is to couple a local search algorithm with a mechanism to generate initial solutions. The ACO's definition includes the possibility of using local search as daemon actions. Daemon actions are used to implement centralized actions that cannot be performed by a single ant. An example of daemon actions is the activation of a local optimization procedure. In implementation, 3-opt local search procedure is used to obtain more improvement in the algorithm's performance.

This algorithm, which is shown in Figure 7, is based on omitting three arcs of the tour that are not adjacent and connecting them again by another method. It can be noted that there are several routes for connecting nodes and producing the tour again, but a state that satisfies the problem's constraints is acceptable. So, this unique tour will be accepted only if, first, the above constraints are not violated specially regarding each vehicle's capacity,

```
procedure the proposed algorithm
    n is number of nodes;
    m is number of vehicles;
    s* is the random solution;           // s* is the best solution found //
    f* = ∞;
    Set parameters of the proposed algorithm;
    Initialize pheromone trails;
    Use SW to order nodes that should be visited by vehicles with respect to the depot
    and called it R
    gain all of the situations for R and set them to the matrix T
        for i := 1 to length(T) do      // main cycle //
            begin
                Use SW for T(i,:) to produce a solution called T*(i,:)
                For the route of each vehicle in T*(i,:), execute the ACS;
                If new solution is better than T*(i,:), replace it;
                Apply 3-Opt local search algorithm for T*(i,:)
                if f(T*(i,:)) < f* then
                begin
                f* := f(T*(i,:)) ;
                s* = T*(i,:);
                    end                        // save the best so far solution //
                end                            // main cycle //
            show s* and f*
        end                                    // procedure //
```

Figure 8 Description of the proposed algorithm.

and second, the new tour produces a better value for the problem than the previous solution. Besides, omitting three arcs and reconnecting them are repeated until no improving 3-opt is found.

When all of these n arrays are gained, the best one is selected as the best solution, and the algorithm will be finished. So, it must be noted that although the hybrid algorithm is composed of two heuristic and one meta-heuristic method, the strengths of this algorithm in comparison with other algorithms such as ACO or TS are, first, the less time taken to obtain the solution and, second, the algorithm not needing many iterations to gain a better solution. In other words, when the ACS is activated, the number of nodes is decreased, and the algorithm can reach a good solution during the execution. In addition, because of the lack of division in other meta-heuristic algorithms, the quality of solutions obtained are usually less stable compared to those of this algorithm in various executions. A pseudo-code of the proposed algorithm for the VRP is presented in the Figure 8, while in the current section, the procedures of the algorithm are explained in detail.

Competing interests
The authors declare that they have no competing interests.

Authors' contributions
MY performed the numerical experiment and carried out sensitivity analysis by taking an illustration. EK managed the study and participated in its design, drafted the manuscript and performed the statistical analysis. Both authors read and approved the final manuscript.

Acknowledgements
We are thankful to the reviewers for their valuable comments.

Author details
[1]Young Researchers Club, Hamedan Branch, Islamic Azad University, Hamedan, Iran. [2]Faculty of Mathematics and Computer Science, Amirkabir University of Technology, Hafez Avenue, Tehran, Iran.

References
Ai J, Kachitvichyanukul V (2009) Particle swarm optimization and two solution representations for solving the capacitated vehicle routing problem. Comput Ind Eng 56:380–387

Baker BM, Ayechew MA (2003) A genetic algorithm for the vehicle routing problem. Comput Oper Res 30:787–800

Bodin L, Golden B (1981) Classification in vehicle routing and scheduling. Networks 11:97–108

Bullnheimer B, Hartl RF, Strauss C (1998) Applying the ant system to the vehicle routing problem. In: Voss S, Martello S, Osman IH, Roucairol C (eds) Meta-heuristics: advances and trends in local search paradigms for optimization. Kluwer, Boston

Bullnheimer B, Hartl RF, Strauss C (1999) An improved ant system algorithm for the vehicle routing problem. Ann Oper Res 89:319–328

Choi E, Tcha DW (2007) A column generation approach to the heterogeneous fleet vehicle routing problem. Comput Oper Res 34(7):2080–2095

Christofides N, Mingozzi A, Toth P (1979) The vehicle routing problem. Combinatorial optimization. Chichester, Wiley, pp 315–338

Clark G, Wright JW (1964) Scheduling of vehicles from a central depot to a number of delivery points. Oper Res 12:568–581

Clarke G, Wright JW (1964) Scheduling of vehicles from a central depot to a number of delivery points. Oper Res 12:568–581

DEIS - Operations Research Group (2012) Library of Instances., http://www.or.deis.unibo.it/research_Pages/ORinstances/VRPLIB/VRPLIB.html

Dorigo M (1992) Optimization, learning and natural algorithms (in Italian), PhD thesis, Dipartimento di Elettronica., Politecnico di Milano, Italy, p p 140

Dorigo M, Ganbardella L (1997) Ant colony system: a cooperative learning approach to the traveling salesman problem. IEEE Trans Evolutionary, Computing, pp 53–66

Dorigo M, Maniezzo V, Colorni A (1991) Positive feedback as a search strategy, Technical Report 91–106. Dipartimento di Elettronica, Politecnico di Milano, Italy

Gillett BE, Miller LR (1974) A heuristic algorithm for the vehicle dispatch problem. Oper Res 22:340–349

Hong L (2012) An improved LNS algorithm for real-time vehicle routing problem with time windows. Comput Oper Res 39(2):151–163

Lee ZJ, Su F, Lee CF (2003) Efficiently solving general weapon-target assignment problem by genetic algorithms with greedy eugenics. IEEE Transactions on Systems, Man and Cyberkinetics, Part B 33:121–133

Leung SCH, Zhou X, Zhang D, Zheng J (2011) Extended guided tabu search and a new packing algorithm for the two-dimensional loading vehicle routing problem. Comput Oper Res 38(1):205–215

Lima CMRR, Goldberg MC, Goldberg EFG (2004) A memetic algorithm for the heterogeneous fleet vehicle routing problem. Electronic Notes in Discrete Mathematics 18:171–176

Lin SW, Lee ZJ, Ying KC, Lee CY (2009) Applying hybrid meta-heuristics for capacitated vehicle routing problem. Expert Syst Appl 36:1505–1512

Mazzeo S, Loiseau I (2004) An ant colony algorithm for the capacitated vehicle routing problem. Electronic Notes in Discrete Mathematics 18:181–186

Ngueveu SU, Prins C, Wolfler-Calvo R (2010) An effective memetic algorithm for the cumulative capacitated vehicle routing problem. Comput Oper Res 37:1877–1885

Osman LH (1993) Metastrategy simulated annealing and tabu search algorithms for the vehicle routing problem. Ann Oper Res 41:421–451

Pisinger D, Ropke S (2010) Large neighborhood search. In: Gendreau M, Potvin J-Y (eds) Handbook of metaheuristics of International series in operations research & management science, 146th edn. Springer, New York, pp 399–419

Potvin JY, Dube D, Robillard C (1996a) A hybrid approach to vehicle routing using neural networks and genetic algorithms. Appl Intell 6:241–252

Potvin JY, Duhamel C, Guertin F (1996b) A genetic algorithm for vehicle routing with backhauling. Appl Intell 6:345–355

Prins C (2009) Two memetic algorithms for heterogeneous fleet vehicle routing problems. Eng Appl Artif Intel 22:916–928

Reimann M, Shtovba S, Nepomuceno E (2001) A hybrid ACO-GA approach to solve Vehicle Routing Problems. Student Papers of the Complex Systems Summer School, Santa Fe Institute, Budapest

Ruiz R, Maroto C, Alcaraz J (2004) A decision support system for a real vehicle routing problem. Eur J Oper Res 153(3):593–606

Su CT, Chen HH (1999) Vehicle routing design of physical distribution center. Journal of the Chinese Institute of Industrial Engineers 16(3):405–417

Taillard E (1993) Parallel iterative search methods for vehicle routing problems. Networks 23(8):661–673

Valle CA, Martinez LC, Cunha AS, Mateus GR (2011) Heuristic and exact algorithms for a min−max selective vehicle routing problem. Comput Oper Res 38(7):1054–1065

Wang CH, Lu JZ (2009) A hybrid genetic algorithm that optimizes capacitated vehicle routing problems. Expert Syst Appl 36:2921–2936

Zhang X, Tang L (2009) A new hybrid ant colony optimization algorithm for the vehicle routing problem. Pattern Recognit Lett 30:848–855

17

Optimizing the warranty period by cuckoo meta-heuristic algorithm in heterogeneous customers' population

Ali Roozitalab[1*] and Ezzatollah Asgharizadeh[2]

Abstract

Warranty is now an integral part of each product. Since its length is directly related to the cost of production, it should be set in such a way that it would maximize revenue generation and customers' satisfaction. Furthermore, based on the behavior of customers, it is assumed that increasing the warranty period to earn the trust of more customers leads to more sales until the market is saturated. We should bear in mind that different groups of consumers have different consumption behaviors and that performance of the product has a direct impact on the failure rate over the life of the product. Therefore, the optimum duration for every group is different. In fact, we cannot present different warranty periods for various customer groups. In conclusion, using cuckoo meta-heuristic optimization algorithm, we try to find a common period for the entire population. Results with high convergence offer a term length that will maximize the aforementioned goals simultaneously. The study was tested using real data from Appliance Company. The results indicate a significant increase in sales when the optimization approach was applied; it provides a longer warranty through increased revenue from selling, not only reducing profit margins but also increasing it.

Keywords: Free replacement warranty policy; Cuckoo optimization; Heterogeneous population; Warranty period

Introduction

In international competitive markets, the life and profitability of the manufacturers' market are to protect the rights of our clients, in addition to profitability, to earn their satisfaction and loyalty. Today, warranty is an integral part and a strategic competitive tool among producers and is one of the factors that influence the buying decisions of customers. Warranty is very important to initiate the company's commitment to support customers after the sale of the product. Since the cost of warranty obligations is very high and increases with increasing duration length, making profit needs scientific planning and appropriate policy. For example, US$23.6 billion was reported in 2010 to cover worldwide warranty claims of US-based companies (Warranty Week 2011).

Literature review

Based on the research of Murthy and Blischke (2005) 'warranty refers to the obligations of the manufacturer or seller for the occurrence of any possible problem resulting from poor quality materials or poor construction process under the specified conditions and period of time through the repair or replacement of damaged parts or restoring some of the best products'. Many studies have been undertaken in the warranty field. They often equate customers' behavior and ignore performance of different customers and the impact on the failure rate for the purpose of simplification (Murthy and Djamaludin 2002). The conditions that the producers intend to provide for after-sales services are called policies including free replacement warranty (FRW) and PRW.

In earlier researches, different types of warranty policies have been offered to cover specific types. More than 30 sample policies have been introduced by An (1992), and there are numerous others that may be offered depending on a particular product or service. Mathematical models to estimate the cost and analyze expense of the manufacturer's warranty have been studied for a variety of different policies (Blischke and Murthy 1996). This study optimizes the FRW policy in which the manufacturer has commitment to accept the

* Correspondence: Roozitalab@ut.ac.ir
[1]Kish International campus, University of Tehran, Kish Island 71778-5777, Iran
Full list of author information is available at the end of the article

component failure or misusage of the product if it occurs in warranty time. Warranty coverage will end after a time commitment. If an item cannot be repaired and must be replaced during the warranty, date of the swap does not start from the replacement date and the period is nonrenewable.

Sahin and Polatoglu (1996, 1998) offered a random distribution of products based on failure occurrence and variables related to the estimated costs. Murthy and Blischke (2005) studied the terms of warranty obligations of cost models and warranty models based on different platforms. St. John and Cassady (2010) considered a heterogeneous population of customers based on exponentially distributed failures to develop this study and form the basis of theoretical research.

St. John and Cassady (2010) categorized customers based on customer behavior and the effect of usage on the failure rate and considered a different warranty period for each group. In reality, we cannot offer different warranty periods to customers. In addition, his exponential distribution model is not generalized to resolve warranty issues. His model which was developed using Weibull distribution, has great flexibility in solving problems of reliability and warranty due to shape and scale parameters, and it can be extended in various issues.

In recent years, many studies were conducted on evolutionary algorithms inspired from nature and human societies such as genetic algorithms, swarm optimization, and ant colony. One of the most recent evolutionary algorithms that have been recently introduced is cuckoo optimization algorithm with good properties in convergence speed and high accuracy in solving optimization problems and accessing to the optimum overall.

Rajabioun (2011) and Yang and Deb (2009) offered cuckoo nonlinear meta-heuristic optimization algorithm based on cuckoo lives. This algorithm is important in specific issues with speed and precision of convergence to optimality. In this study, we use that to obtain the optimal revenue and optimal duration based on the research model. The use of meta-heuristic optimization techniques in the field is quite new in warranty.

The next section describes the acronyms used in this research. In the third part, we have the theoretical model with Weibull distribution and FRW policy. In the fourth section, we optimize the theoretical model in order to achieve an equal optimization period that maximizes the overall revenue of all groups through COA algorithm. The theoretical results obtained are discussed in the fifth section, and model of the real data obtained from an industrial appliance company are tested to optimize the procedure. Finally, conclusion and suggestions for future research are presented in the sixth section.

Theoretical framework

The theoretical model is presented in this section to form the basis of theoretical research.

Product failure model

Each product has a certain life span, and most products experience defects in their life depending on usage conditions, rate of usage, quality of construction, quality of designing and materials, and many other factors. Higher rate of usage will quicken the occurrence of defects that can be used as an index to budget the after-sales services.

If t is a random variable for the time of failure occurrence of the product t, the failure distribution usage is $F(t)$ and the density usage will be $f(t)$; the reliability usage will be as follows:

$$R(t) = 1 - f(t) \tag{1}$$

When a product is offered by FRW policy, all costs of repair or replacement of defective parts to cope with the after-sales service department is paid by the manufacturer or seller of the product. If you do not repair the defective product or part, it must be replaced with a new piece. We assume that the time of replacement parts and service speed is low and negligible; as a result, the failures in this case are expressed by $F(t)$ (Ross 1972).

Reliability

Reliability refers to the likelihood of appropriate operation of the product based on defined tasks for the product under certain conditions in a period of time, i.e., the probability that a defect or an interruption does not occur in a defined interval under certain circumstances. The reliability function is displayed by $R(t)$. Distribution function of nonfailure or reliability function is complementary to the distribution failure function:

$$R(t) + F(t) = 1. \tag{2}$$

Therefore, we can conclude

$$R(t \geq a) = 1 - F(t \leq a)$$
$$= 1 - \int_0^t f(t)d(t) = \int_t^\infty f(t)d(t) \tag{3}$$

You should consider the product reliability until failure time $t = a$ means that the system will not be experiencing fault at least until a while after this time. Failure rate function $\lambda(t)$ represents the instantaneous failure rate. The equation of this usage is as follows:

$$\lambda(t) = \frac{DR(t)}{D(t)} \times \frac{1}{R(t)} = \frac{f(t)}{R(t)} \tag{4}$$

If we solve the relationship of $R(t)$, the following equation is obtained:

$$R(t) = e^{-\int_0^t \lambda(t)dt'} \tag{5}$$

The warranty cost model under FRW policy

In this policy, if failure occurs, a new one replaces a defective piece. If $\mu(t, u)$ represents the number of expected failures during the commitment period (t) of warranty under usage rate of u, the period time is $[0, t]$ where $0 < t < W$, and $U = u$; we have the following relation based on renewal theory (Hong-Zhong et al. 2008):

$$\mu(T, u) = F(T, U) + \int_0^\tau \mu(t, u)dF(t, U). \tag{6}$$

If it is not conditioned, the equation is as follows:

$$\mu(t) = \int_{u_{\min}}^{U_{\max}} \mu(t, u)dG(u). \tag{7}$$

Repairable products

These products can be fixed or repaired, and the fault will be resolved by replacing the defective part or the entire event. In this case, we assume that maintenance is minimal. The correct model here is based on Blischke and Murthy (1996) with the following equation. Accordingly, it is assumed that $S(t, u)$ represents the expected number of failures in the interval $[0, t]$. If the condition $0 \le t \le W$ and $U = u$, we have

$$S(t, u) = \int_0^t r(t, u)dt. \tag{8}$$

If the condition is eliminated,

$$S(t) = \int_{u_{\min}}^{u_{\max}} dG(u) \int_0^\tau r(t, u)dt. \tag{9}$$

Class model of Weibull failure time

This model has lots of flexibility due to its shape as well as the shape and scale parameters that define it. Many processes of product failures are either modeled by the Weibull distribution. Due to the high flexibility of Weibull distribution parameters, we can have fixed, increased, or decreased failure rates. Weibull belongs to limited failure group in the period with the classification model of Musa and Okumoto (1983) that is of the binomial kind. Thus, the reliability of the distribution will be as follows:

$$r_j(t) = e^{-\left(\frac{\tau-\gamma}{\beta}\right)^\alpha}. \tag{10}$$

Weibull distribution of failure rate is calculated according to the following equation:

$$\lambda(T) = \frac{f(t)}{R(t)} \frac{\alpha}{\beta} \left(\frac{\tau-\gamma}{\beta}\right)^{\alpha-1}. \tag{11}$$

Since we assume that customers use the products in different intensities, all customers are divided into subgroups based on their behavior. This will exert the influence of the intensity use of each subgroup on its failure rate. Therefore, Weibull is considered as a statistical distribution usage for the maximum flexibility due to its shape and scale parameters and can be extended to other distributions. The failure rate of each subgroup of customers is as follows:

$$\lambda(t) = \lambda_0 \times \delta_i. \tag{12}$$

In this equation, $\lambda_0 > 0$ is the basic failure rate, a coefficient that is based on the Weibull distribution, and application intensity δi is based on customer usage rates in subgroup i.

The warranty literature indicates that customer purchases based on the increased warranty period is going to saturate the market; after that, increasing the warranty period is not effective on elevating sales. Let d_i denote the proportion of customers in subgroup i who purchased the product. We use a shifted logistic function, as shown in Equation 13, to describe d_i as a function of the duration of the warranty period:

$$d_i = d_{\max} \left(\frac{1}{1 + \exp\left(2 - \frac{\tau_i}{\gamma}\right)} - \frac{1}{1 + \exp(2)}\right). \tag{13}$$

In this function, $0 < d_{\max} < \frac{1+\exp}{\exp(1)}$ and γ is the parameter of the logistic shift function that is met based on the demand; $d_{\max}$ is the parameter that limits d_I between $[1, 0]$.

τ_i is the warranty period related to subgroup i. According to the statistical distribution of breakdown products, the reliability function will be

$$R_I(t) = e^{-\left(\frac{\tau-\gamma}{\beta}\right)^A}. \tag{14}$$

By knowing the cost per unit for producing and the number of products sold in each subgroup, we can calculate the total income:

$$n_j = d_j \times q_i. \tag{15}$$

n_I represents the number of products expected to be demanded by each subgroup i, and then the total revenue equals to

$$\rho_I = m \times r_i \times n_I. \tag{16}$$

ρ_I represents the income of each i subgroup. Improved earnings in accordance with the above equation are obtained for each subgroup corresponding to warranty period for the group.

Optimization by cuckoo algorithm

Since in reality the duration of different periods cannot be offered for different customers, by cuckoo optimization algorithm, we try to maximize time length and provide a common optimal time for all subgroup charges in order to issue the variables making up the array. These arrays are the same as chromosomes in genetic algorithms which are called *habitat* here. Prior researches (Rajabioun 2011; Yang and Deb 2009, 2010) constitute theoretical studies of this section. In order to determine the primary residence of the cuckoo, we define the problem of optimization variables related to earnings per subgroup, which represents the primary profits of those subgroups (Rajabioun 2011).

$$\text{habitat} = \left[\rho_1, \rho_2, ..., \rho_n\right].$$

Each location with a cuckoo has a special profit that is calculated using the O_P function:

$$\text{Profit} = O_P\left(\rho_1, \rho_2, ..., \rho_n\right).$$

To start this algorithm, inherently seeking maximal profits function, a matrix of the habitat with $N_P \times N_\rho$ dimension is developed, where N_P is the number of cuckoo and N_ρ is the variable income for each subgroup. In this nature, each cuckoo lays about 5 to 20 eggs; the numbers are the maximum and minimum numbers of eggs laid at a time. The cuckoo lays eggs in a particular domain and range (distance from home). This distance is equal to ELR:

$$\text{ELR} = W \times \frac{\text{Number of current cuckoo eggs}}{\text{Total number of eggs}}$$
$$\times \left(\text{Var high} - \text{Var low}\right). \tag{17}$$

W is the ratio of the maximum ELR set up to maximize the convergence.

The cuckoo begins to lay eggs in its coverage range in other bird's nests, with each nest having certain benefits. Locations with less profit are excluded. The rest of the chicks grow in the nest by a huge profit, and merely one bird can grow in every nest.

When the birds grow, they go to the most profitable areas (most likely to survive) and lay eggs. To know each ρ belongs to which group, we use clustering technique of k-means. Clustering is one of the common techniques for data to categorize groups of similar objects, or more precisely, the partitioning of a data set or a subset of a cluster so that each cluster of data in a series has common features. Most of these similarities are defined based on the

Table 1 Basic data related to population, average temperature, and failure rate of subgroups

Subgroups (provinces)	Total population	Ratio of total population to subgroup population	Average of annual temperature (°C)	Product failure rate (λ_1)
Tehran	8,791,378	0:240	$17.3 \simeq 17$	0.0629
Gilan	2,480,874	0:067	$16 \simeq 15.5$	0.0592
Esfahan	4,879,312	0:133	16	0.0592
Fars	4,596,658	0:125	18	0.0666
Khorasan	5,994,402	0:163	14	0.0518
Mazandaran	30,773,943	0:083	15	0.0555
Khoozestan	4,531,720	0:123	$26 \simeq 25.5$	0.0962
Yazd	1,074,428	0:029	19	0.0703
Ghazvin	1,201,565	0:032	13	0.0481
Total	36,624,280	100	-	-

criteria. Due to the large number of data, sometimes, managing this widespread amount of data would be impossible. Therefore, this method can facilitate data analysis by categorizing into groups with common characteristics of the entities.

In this study, we used k-means clustering method that is an unsupervised learning method known to solve clustering problems. This algorithm is a simple way to cluster a data set in a prespecified number k clusters. The main idea is to define the k center for each cluster. These centers should be selected very carefully because different centers create different results.

Therefore, the best choice is putting them (centers) at the farthest possible distance from one another. The next step is to assign each pattern to the closest center. When all the points were assigned to the center, the first phase has been completed and an early grouping is done. Here, we need to calculate a new center cluster level to calculate k for the clusters of the previous levels. After determining the new center k, then we assign the data to the

Table 2 Income of each group and corresponding optimal warranty period

Subgroups (provinces)	Warranty period	Sales
Tehran	52.48	$4/5,025 \times 10^8$
Gilan	14.49	$1/2,898 \times 10^8$
Esfahan	14.49	$2/5,368 \times 10^8$
Fars	95.47	$2/3,194 \times 10^8$
Khorasan	50.46	$3/2,137 \times 10^8$
Mazandaran	77.44	$1/6,227 \times 10^8$
Khoozestan	44.24	$2/415 \times 10^8$
Yazd	47.41	$5/3,423 \times 10^7$
Ghazvin	51.19	$6/5,435 \times 10^7$
Total revenue		1.871498×10^9

Table 3 Total revenue results according to different policies

Kind of population	Warranty period length	Total revenue
Homogeneous population	51	1.1707×10^9
Heterogeneous population with different periods of warranty for each subgroup	-	1.871498×10^9
Heterogeneous population with optimized warranty period length (optimized by cuckoo algorithm)	$856/50 \approx 51$	1.9782×10^9
Homogeneous population with average usage intensity	$836/49 \approx 50$	1.9360×10^9

appropriate centers again. This process is repeated so many times that k of the center does not move.

In every turn, average earnings are calculated for cuckoo groups (clusters) to obtain the optimization of the place, and the group that has the most benefits attracts other cuckoos. The maximum number of cuckoos that live in a moment to kill cuckoos living in inappropriate areas is 200 (Rajabioun 2011).

After moving several times and laying eggs with high convergence, cuckoos go to an optimal point, which has a higher profit than elsewhere, and this is the optimal solution (Rajabioun 2011; Yang and Deb 2009).

Numerical example

Numerical examples of fan failures for an industrial production company have been studied in nine provinces in Iran.

The failure rate that was based on Equation 11: $\lambda_0 = 0.037$. Since the intensity of a fan's usage is depended on temperature of each zone, the warmer region has a higher intensity of usage. The mean of annual temperature was taken from the Iran weather organization, and average temperatures of the regions were considered as a usage of the effective failure rate. Each fan is sold at US$103. Table 1 shows the statistics on the population of each subgroup to the total population (Static Center of Iran 2011), the mean annual temperature, and the corresponding risk rate for each group.

Relationship between the demand of fans of the company and duration of warranty period is considered: $d_{max} = 0.75$ and $\gamma = 8$. This number is assigned regarding the problem and the target market; however, the maximum saturation of the market's share is taken at 75%. Table 2 represents the amount of each group's income and corresponding optimal warranty period.

To estimate the parameters of the Weibull distribution, we have different ways like the method of least squares or graphics in this study. Weibull distribution parameters are estimated using the R software and values of $\alpha = 1274,262$ and $\beta = 44/199,218$ were obtained, respectively. Since the shape parameter is greater than one, it can be concluded that the damages are more likely due to frazzle and long length of product usage, which are rare in new products.

The results of the testing data without optimization and with optimization by cuckoo algorithm are shown in Table 3. By optimizing the models obtained by cuckoo optimization algorithm, optimal time was considered at 50.856 months,

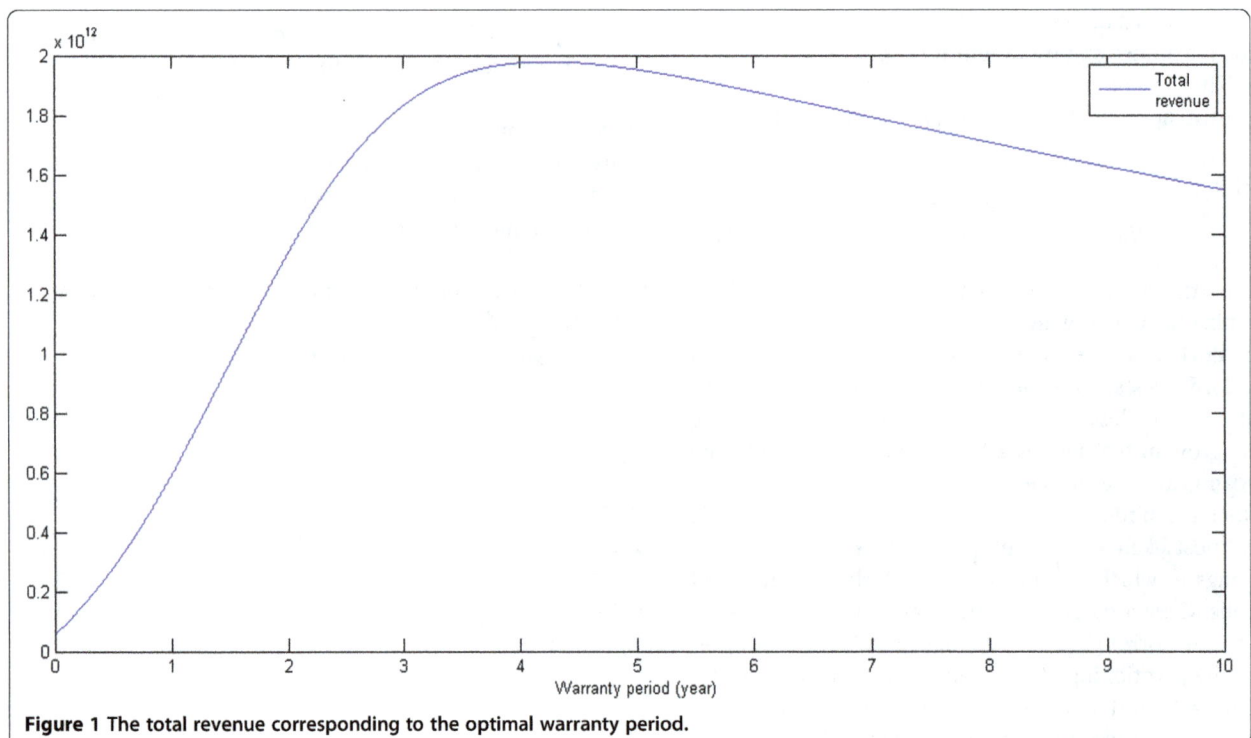

Figure 1 The total revenue corresponding to the optimal warranty period.

the longest period that the revenue is maximized as shown in Figure 1.

Total revenue results according to different policies are presented in Table 3. In the case of a homogeneous population (without the effect of usage intensity), the total income for the period of 51 months has been calculated as equal to FRW policy optimization. Heterogeneous population with different periods is used for each group in the model of St. John and Cassady (2010). Heterogeneous population with the same optimization period of the developed model and optimization point are presented in this study. In the last case, instead of considering the different usage intensity for each group, an average failure rate for all groups was selected to better compare the optimum performance with and without the effect of application intensity.

Conclusions and recommendations for future research

One of the key points in providing an appropriate policy of warranty is to estimate the optimized warranty coverage. This feature must be able to get more customers to buy these products and gain more trust from them. The profit of the company should not be threatened by the cost of repairing or replacing the defective products.

In this paper, FRW policy based on heterogeneous population of customers was developed based on different application intensities with statistical distribution of Weibull. To determine the optimal duration of warranty, optimal time was estimated using the cuckoo optimization algorithm. By comparing the amount of income earned during the cuckoo optimized period, heterogeneous population without optimization, homogeneous population without optimization, and the length of warranty period in numerical data, a significant increase was observed in the amount of the manufacturer's revenue. Furthermore, with the extended warranty period, we expect that costs imposed upon customers decrease and their satisfaction increases which will also bring about social and economic benefits for the producer in the long run.

The theoretical model using actual data of a manufacturing plant in the field of home appliances was tested. Results show a dramatic increase in revenue due to the longer period of optimization; we have reached customers' satisfaction, trust, and indirect benefits of the proposed policy. The cuckoo meta-heuristic algorithm was used for the first time in the field to optimize the warranty that was quite new. We suggest a more frequent and wider use of such algorithms in warranty policies. Using these algorithms, we are able to take into account the dynamic price and inflation that will lend special attraction to the research studies in competitive markets.

This model was discussed in a manufacturing company in the field of home appliances; nonetheless, services can be covered under the same warranty policy, too. In this study, FRW policy is selected as the base policy to optimize.

Optimizing other models of policies from previous studies with meta-heuristic algorithms can help industry managers and decision-makers in warranty policies to select and apply the most appropriate method and policy. Some of these problems are currently under investigation by the authors.

Competing interests
The authors declare they have no competing interest

Authors' contributions
EA provide the warranty literature and participated in data calculating. AR creates the mathematic model and has share in collecting raw data, also Case study analyses and the Conclusions and recommendations for future research part do by him. All authors read and approved the final manuscript.

Author details
[1]Kish International campus, University of Tehran, Kish Island 71778-5777, Iran. [2]Tehran University, Gisha bridge, Jalale ale ahmad blvd, Tehran 6311-14155, Iran.

References
An DL (1992) Asset management: remedy for addressing the fiscal challenges facing highway infrastructure. Int J Transp Manag 1:41–54

Blischke WR, Murthy DNP (1996) Warranties: concepts and classification. In: Blischke WR, Murthy DNP (eds) Product warranty handbook. Marcel Dekker, New York

Hong-Zhong H, Zhi-Jie L, Yanfeng L, Yu L, Liping H (2008) A warranty cost model with intermittent and heterogeneous usage. Eksploatacja I Niezawodnosc-Maintenance and Reliability 40:9–15

Murthy DNP, Blischke WR (2005) Warranty management and product manufacture. Springer Series in reliability engineering, USA. ISBN-10: 1852339330, chapter 1, 5, 6.

Murthy DNP, Djamaludin I (2002) New product warranty: a literature review. Int J Prod Econ 79(3):231–260

Musa JD, Okumoto K (1983) Software reliability models: concepts, classification, comparisons, and practice. In: Skwirzynski JK (ed) Electronic systems effectiveness and lifecycle costing, NATO ASI Series, F3. Springer-Verlag, Heidelberg, pp 395–424

Rajabioun R (2011) Cuckoo optimization algorithm. Appl Soft Comput J 11:5508–5518

Ross SM (1972) On optimal assembly of systems. Nav Res Logist 19:569–574

Sahin I, Polatoglu H (1996) Manufacturing quality, reliability and preventive maintenance. Prod Oper Manag 5:132–147

Sahin I, Polatoglu H (1998) Probability distributions of cost, revenue and profit over a warranty cycle. Eur J Oper Res 108:170–183

St. John D, Cassady RC (2010) Customized warranty policies for heterogeneous populations. In: 2010 Proceedings - Reliability and Maintainability Symposium (RAMS). San Jose, 25–28 Jan 2010

Static Center of Iran (2011) Population estimates until September of 2011. http://www.amar.org.ir/. Accessed 31 Sep 2012.

Warranty Week (2011) Eighth Annual Warranty Report., Totals and Averages http://www.warrantyweek.com/archive/ww20110401.html. Accessed 01 May 2012.

Yang X-S, Deb S (2009) Cuckoo search via Lévy flights. In: World Congress on Nature & Biologically Inspired Computing, Coimbatore, 9–11 Dec 2009. IEEE, Piscataway, pp p. 210–214

Yang X-S, Deb S (2010) Engineering optimization by cuckoo search. Int J Mathematical Modeling and Numerical Optimization 1(4):330–343

18

Bi-product inventory planning in a three-echelon supply chain with backordering, Poisson demand, and limited warehouse space

Maryam Alimardani[*], Fariborz Jolai[*] and Hamed Rafiei

Abstract

In this paper, we apply continuous review (*S-1, S*) policy for inventory control in a three-echelon supply chain (SC) including *r* identical retailers, a central warehouse with limited storage space, and two independent manufacturing plants which offer two kinds of product to the customer. The warehouse of the model follows (*M/M/1*) queue model where customer demands follow a Poisson probability distribution function, and customer serving time is exponential random variable. To evaluate the effect of considering bi-product developed model, solution of the developed model is compared with that of the two (*M/M/1*) queue models which are separately developed for each product. Moreover, and in order to cope with the computational complexity of the developed model, a particle swarm optimization algorithm is adopted. Through the conducted numerical experiments, it is shown that total profit of the SC is significantly enhanced using the developed model.

Keywords: Bi-product multi-echelon inventory planning; Markov model; Backordering; Exponential lead time; Particle swarm optimization

Background

Today, increasing competitive pressure and market globalization directly results in an environment in which supply chains (SCs) are organized in such a way to be able to respond to the customer needs (Gumus and Guneri 2007). In this regard, diverse competitive metrics are elicited as cost, quality, delivery, and flexibility; while some are noted with respect to the production process and planning, such as technology, capacity, facilities, and workforce planning (Sana 2011). To do so, integrated multi-echelon inventory management is employed in SCs in order to obtain inventory cost reduction as well as customer service improvement of the total SC (Gumus and Guneri 2007). Upon the principles of integrated inventory management, two main concerns are considered: coordination of different SC elements and arranging their inventory levels according to the interrelationship among the elements.

Most of today's manufacturing firms are organized as a network of manufacturing plants and distribution centers, which consists of a value chain including providing raw material, processing them into end-products, and distributing/delivering the finished goods to the customers. In this regard, the term 'multi-echelon' or 'multi-level' production/distribution network is adopted in order to refer to a network during which a good is delivered to the end customer after more than one step. In multi-echelon supply networks, there might be various forms of inventories: raw material, work in process, and/or finished products. Procured raw materials from suppliers are processed to finished products in manufacturing plants. Then, the finished products are sold to distributors, retailers, and/or end customers. As mentioned above, a multi-echelon SC is a network when an item moves towards the final customer after passing more than one stage (Ganeshan 1999). The echelon stock of a stockpoint comprises all stock at this stockpoint, plus in-transit to or on-hand at any of its downstream stockpoints, minus the backorders at its downstream stockpoints (Diks and de Kok 1998).

Upon multi-echelon inventory systems, distribution of products to customers is widely conducted over extensive geographical areas. Given the importance of these systems, many studies have been presented based upon

* Correspondence: m.alimardani@ut.ac.ir; fjolai@ut.ac.ir
School of Industrial and Systems Engineering, College of Engineering, University of Tehran, Tehran, P.O. Box: 11555-4563, Iran

the various conditions and assumptions (Moinzadeh and Aggarwal 1997). On the other hand, numerous studies have been devoted to the inventory control policy of multi-echelon systems with stochastic demands (Gumus and Guneri 2007). In this paper, a bi-product, three-echelon SC is considered which comprises r identical retailers, a central warehouse with limited storage space, and two independent manufacturing plants. The assumed network presents two types of product to the customers who arrived into the retailer nodes. Customer demands for each product j enter retailer i upon Poisson probability distribution function with rate λ_{ij}. Also, central warehouse responds the needs upon exponential probability distribution function with respect to the on-hand inventory. When a retailer places an order and faces inventory shortage in the warehouse, she/he waits up until product availability (backordering). The warehouse holds inventory and applies a continuous review ($S-1$, S) inventory policy; that is, a replenishment order is forwarded to the manufacturer after each order enters the warehouse. Therefore, the purpose of this study is to seek an optimal S^* for inventory control in order to minimize both inventory holding cost and backordering cost. The detailed descriptions of the problem are elaborated in 'Single-product model' and 'Bi-product model' sections.

The aim of the paper is the improvement of the total network costs in a stochastic environment. On the other hand, increasing the profit of the SC is the purpose of this study. In this regard, we suppose a bi-product three-echelon SC which is modeled as an ($M/M/1$) queue model for each type of product offered through the developed network. In addition, to show the performance of the proposed bi-product SC, a network including two Markov models ($M/M/1$) for each type of product is also considered. The procedure of the comparison of both proposed SCs is executed applying a particle swarm optimization (PSO) algorithm.

Sherbrooke's (1968) paper can be introduced as one of the oldest researches conducted in the field of continuous review multi-echelon inventory systems, leading to several instances of the papers which addressed continuous review models of multi-echelon inventory systems during 1980s, such as Graves (1985), Moinzadeh and Lee (1986), Lee and Moinzadeh (1987), Deuermeyer and Schwarz (1981), and Svoronos and Zipkin (1988). In addition, other authors have studied the continuous review ($S-1$, S) policy in multi-echelon inventory systems with the assumption of stochastic environments (Axsäter 1990; Axsäter 1993; Andersson and Melchiors 2001). Besides, other inventory control policies such as continuous review (R, Q) and (s, Q) policies and one-for-one have been investigated in this era

under different assumptions on echelon number, product number, demand probability distribution function, transportation times, and backlogging assumptions (Kevin Chiang and Monahan 2005; Thangam and Uthayakumar 2008; Haji et al. 2009; Teimoury et al. 2010).

Kevin Chiang and Monahan (2005) developed a two-echelon dual-channel inventory model which was controlled upon a one-for-one policy. They supposed stochastic demands of two customer segments. Thangam and Uthayakumar (2008) studied a two-level supply chain consisting of a single supplier and several identical retailers. In their considered problem, retailers received Poisson demands while partial backordering was permitted. The finished goods were delivered to the customers in a network with constant transportation times. In another paper, a modified one-for-one policy was extended to inventory control of one central warehouse in a two-echelon inventory system with a number of non-identical retailers. The authors considered that the central warehouse was facing uniformly distributed demands from each retailer (Haji et al. 2009).

Since problems of stochastic inventory systems have a significant level of complexity, queueing approach has been applied to obtain a simple performance result. In this regard, Chao (1987) developed a Markov decision process model to describe the optimal control policy for an inventory system with zero lead time and market interruptions. Gupta (1996) presented an exact cost function for a lost-sale (s, Q) inventory system with Poisson demand, constant lead time, and exponentially distributed supplier's on/off periods, while a similar paper was presented by Mohebbi (2003) in this context. He developed an analytical model for computing the stationary distribution of the on-hand inventory in a continuous review inventory system with lost sales in which demands and lead times were stochastic. In the study, demand process was performed upon Poisson probability distribution function and replenishment lead time based on Erlang (E_k). Under a continuous-time Markov chain, the author assumed two available and unavailable states at any points of time. Moreover, a modified (s, Q) policy was applied for the inventory system.

Some authors have developed queuing models which are related to production authorization strategy for supply chain inventory analysis and optimization (Dong 2003; Teimoury et al. 2011a, 2011b). Teimoury et al. (2011a) modeled an inventory system of a three-echelon multi-product supply chain including a manufacturing plant, several warehouses, and several retailers as queue model $GI/G/1$. In their network, after manufacturing products in batches, each type of products which are demanded by several retailers is held in separate warehouses with different holding and backorder costs. Moreover, they studied a multi-product multi-echelon manufacturing supply chain network with batch ordering in which performance evaluation had been analyzed upon the developed queuing models. The purpose of their study was to determine the optimal inventory level at the warehouse of each product so as to minimize total expected cost (Teimoury et al. 2011b). Some other papers have been published towards another aspect of the integrated inventory systems of SCs and disruption effects. One of the notable instances is the paper by Bozorgi Atoei et al. (2013). The authors designed a reliable capacitated supply chain network considering random disruptions in both distribution centers and suppliers.

To the best of our knowledge, there is no attention to extend equilibrium equations in multi-item multi-echelon supply chain in order to achieve one-stage transition probabilities in the related literature. To do so, we developed a bi-product three-echelon SC with full backordering. The considered SC comprises one central warehouse with limited storage space, several identical retailers, and two independent manufacturing plants. We modeled the central warehouse which is investigated upon a continuous review (S-1, S) inventory policy as a (M/M/1) queue model. Besides, a network including two Markov models (M/M/1) for each type of product is also considered.

The remaining of the paper is organized as follows. The proposed three-echelon supply chain structure is elaborated in the 'Bi-product model' section. Steps of the proposed structure including proposed assumptions and equilibrium equations are described thoroughly as well. 'Single-product model' section describes single-product inventory model in order to compare its performance with that of the developed bi-product model. Proposed PSO algorithm is stated in details in 'Solution methodology' section, while experimental results are reported in the 'Results and discussion' section. Eventually, remarkable conclusions and future research directions are provided in the 'Conclusion' section.

Bi-product model

The notations used in two coming sections are as follows:

- r Number of retailers.
- λ_{ij} Demand rate at retailer i for product j, $i = 1, 2,..., r$; $j = 1,2$.
- λ_j Demand rate for product j, $j = 1,2$ ($\lambda_j = \sum \lambda_{ij}$).
- h_j Holding cost rate at central warehouse for product j.
- b_j Backordering cost rate at central warehouse for product j.
- Ch Holding cost at central warehouse in steady state for each product.

- Cb Backordering cost at central warehouse in steady state for each product.
- TC Total cost at central warehouse in steady.

Problem assumptions

In order to describe the developed model further in subsequent sections of the paper, the following assumptions are made:

- Two kinds of products are offered to the customer in the supposed SC.
- Customer demands enter the retailers upon Poisson distributions with rates λ_{i1} and λ_{i2} for the two product kinds.
- The accumulation of retailer demands enters the central warehouse with rate λ_1 and λ_2 for the products.
- The corresponding service time and the on-hand inventory at the warehouse are random exponential variables with means $1/\mu_j$ and $1/\mu_j'$ respectively.
- The central warehouse has limited storage space.
- Continuous review (S-1, S) inventory policy controls the warehouse inventory system.

Model description and assumptions

In this study, a bi-product three-echelon supply chain network which consists of r identical retailers, a central warehouse with limited storage space, and two independent manufacturing plants are supposed. The system receives stochastic demands from retailers. Two types of products are offered to the customer who arrived into the retailer nodes. The whole accumulation of customer needs which arrived into the retailers depletes the on-hand inventory at warehouse, if any. Otherwise, stockout situation occurred in the warehouse, and the backlogged orders have to wait to be fulfilled. Moreover, it is assumed that the retailer orders are filled using FIFO policy by the single server. The demand for each product j, $j = 1, 2$, at each retailer i; $i = 1, 2, ..., r$, which is independent, and distributed upon Poisson distribution with rates λ_{i1} and λ_{i2} enter to the central warehouse. The demand rate arrived to the warehouse equals λ_1 and λ_2 for each product j, $j = 1, 2$, respectively, and the warehouse places orders of products 1 and 2 with rates λ_1 and λ_2, respectively, to the corresponding manufacturing plant, where, $\lambda_1 = \sum \lambda_{i1}$ and $\lambda_2 = \sum \lambda_{i2}(i = 1,, ..., r)$. Whenever the retailers face available on-hand inventory at the warehouse, the service time (including transportation time from the warehouse to each retailer and warehouse service time) for each product is considered as random exponential variables with means $1/\mu_j(j = 1, 2)$.

Otherwise, when a retailer places an order and faces inventory shortage at the warehouse (backordering conditions), the service time (including transportation time from the warehouse to each retailer, warehouse service time, transportation time from the manufacturer to the central warehouse, and manufacturing time) is considered a random exponential variable with mean $1/\mu_j'$ (j = 1, 2). With respect to the above described network, the space Markov system is as the one depicted in Figure 1. Figure 1a shows the demand rate of the described network, while the corresponding response rates are depicted in Figure 1b.

The central warehouse holds inventory and applies a continuous review ($S-1$, S) inventory policy; that is, a replenishment order is forwarded to the manufacturer

after each order enters the warehouse. The supposed model seeks a S^* in order to minimize a cost function which captures two different operational cost factors including inventory holding and backordering costs. We modeled the central warehouse as a ($M/M/1$) queue model with the state space (N_1, N_2), where, n_1 and n_2 represent stocks on hand of product 1 and product 2 at central warehouse, respectively. Moreover, $-S \leq \sum_{j=1,2} n_j \leq S$ is satisfied. Figure 2 demonstrates the transition diagram of the system. As mentioned earlier, the objective function seeks to minimize the total cost including holding cost of on-hand inventory at the central warehouse and backordering cost of unsatisfied demands based upon the inventory controlling policy ($S-1$, S) by Equation 1.

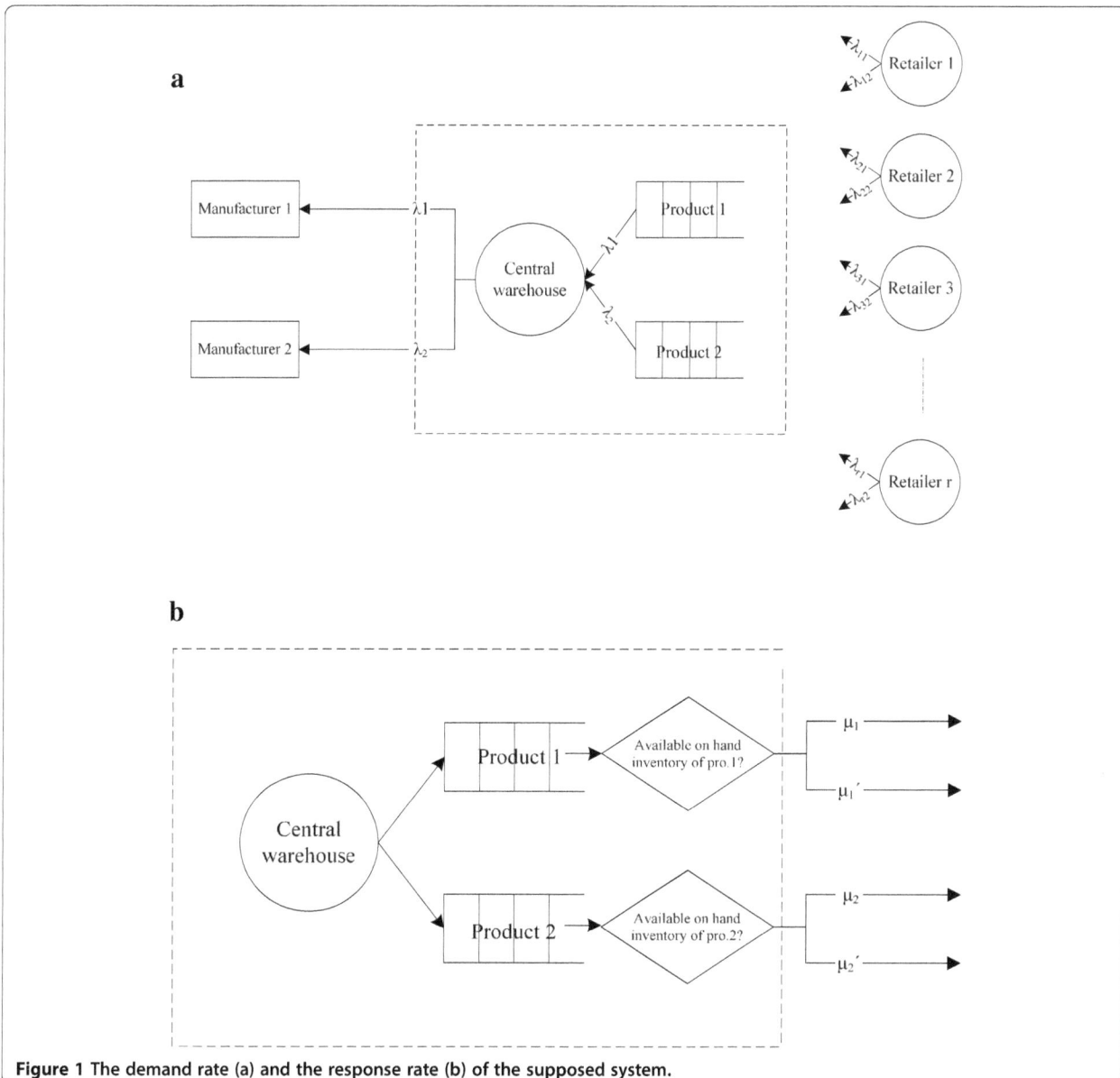

Figure 1 The demand rate (a) and the response rate (b) of the supposed system.

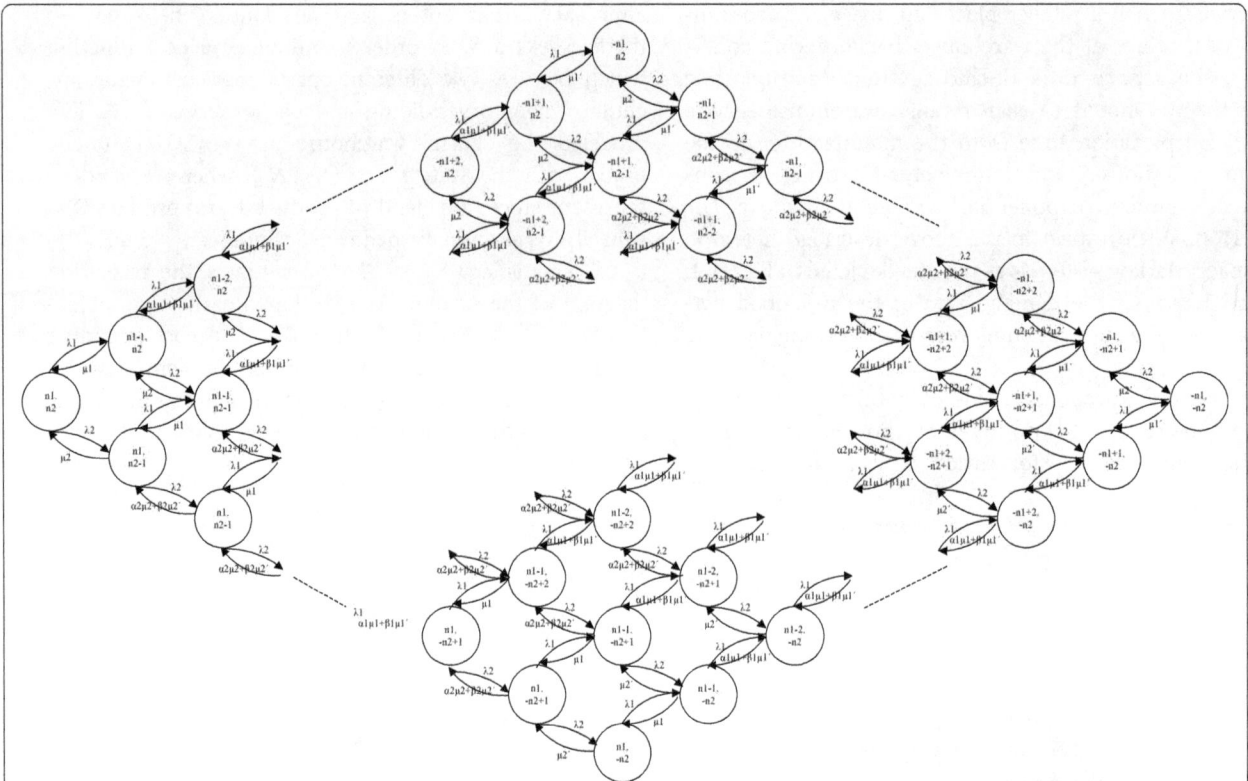

Figure 2 Transition diagram of the described system in this paper.

Objective function:

$$\text{Min} \quad z : TC = Ch + Cb. \tag{1}$$

Subject to:

$$Ch = I.\Pi(p,q); \tag{2}$$

$$Cb = b.\Pi(p,q); \tag{3}$$

$$I = p^+ + q^+; \tag{4}$$

$$b = p^- + q^-; \tag{5}$$

$$n_1 + n_2 \leq S_{max} \tag{6}$$

$$n_1 + n_2 = S^* \tag{7}$$

In the above model, inventory holding costs and backordering costs are estimated by Equations 2 and 3,

respectively. Equation 4 calculates on-hand inventory of both products at the warehouse as well as Equation 5 for the unsatisfied demands. The limited storage space of the central warehouse is considered using Equation 6. Finally, Equation 7 shows that the sum of the two kinds of products equals to the optimal on-hand inventory at the supposed warehouse to respond the customers' needs.

Determining one-stage transition probabilities

The state of the system is defined by (N_1, N_2), where n_1 and n_2 denote the number of on-hand stock of product 1 and product 2 at central warehouse, respectively. (N_1, N_2) can be depicted as a matrix. In this regard, we introduce the states matrix that is a $((2 \times n_1 + 1) \times (2 \times n_2 + 1))$ matrix as in Figure 3. The states matrix has one-stage transitions among states of the system. In states matrix, the states of the system are divided into four different sections: A, B, C, and D. The central warehouse on-hand stock of both types of products are not available in section A. In section B, on-hand inventory of product 1 is not available while it is sufficient for product 2. In section C, on-hand inventory of product 2 is not available while it is sufficient for product 1. Finally, and in section D, on-hand stock of both types of products are available at the central warehouse.

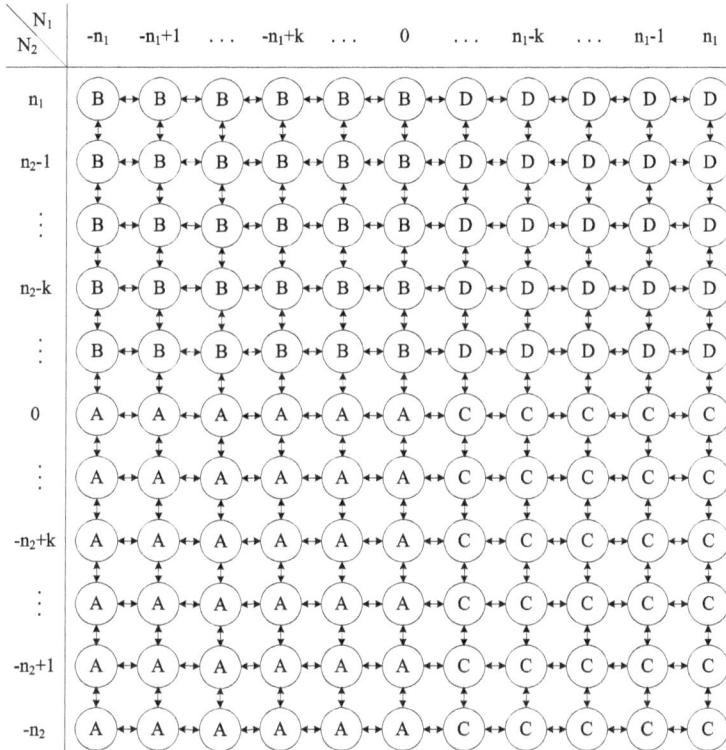

Figure 3 States matrix.

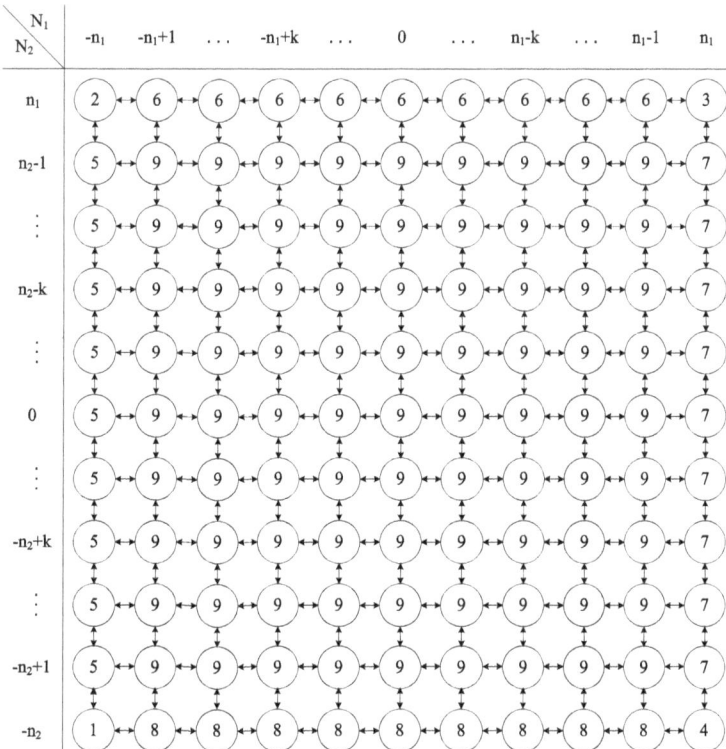

Figure 4 Different regions of various equilibrium equations.

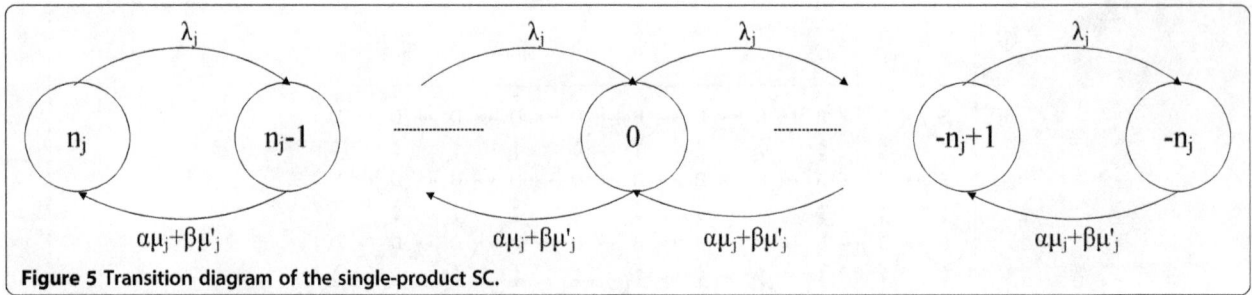

Figure 5 Transition diagram of the single-product SC.

Equilibrium equations

In order to denote the equilibrium equation in a given state (p, q), all the departure states with state (p, q) as their corresponding destination state are specified. To do so, the system states are divided into nine categories according to their similarities among departure states. States in one category are the ones which have similar departure states. The nine mentioned categories are illustrated in Figure 4. All states which are classified in a category have a similar form of equilibrium equations. The equilibrium equations for the states located in each of the nine categories presented in Figure 3 are computed upon the one-stage transition probabilities. These equations are presented through (8 to 16) for regions 1 to 9, respectively.

$$(\lambda_1 + \lambda_2)\Pi_{n1,n2} = \mu_1\Pi_{n1-1,n2} + \mu_2\Pi_{n1,n2-1} \tag{8}$$

$$\left(\lambda_2 + \mu_1{}'\right)\Pi_{-n1,n2} = \lambda_1\Pi_{-n1+1,n2} + \mu_2\Pi_{-n1,n2-1} \tag{9}$$

$$\left(\lambda_1 + \mu_2{}'\right)\Pi_{n1,-n2} = \lambda_2\Pi_{n1,-n2+1} + \mu_1\Pi_{n1-1,-n2} \tag{10}$$

$$(\mu_1{}' + \mu_2{}')\Pi_{-n1,-n2} = \lambda_1\Pi_{-n1+1,-n2} + \lambda_2\Pi_{-n1,-n2+1} \tag{11}$$

$$\left(\lambda_1 + \lambda_2 + \alpha_1\mu_{1+}\beta_1\mu_1{}'\right)\Pi_{n1-k1,n2}$$
$$= \lambda_1\Pi_{n1-k1+1,n2} + \left(\alpha_1\mu_{1+}\beta_1\mu_1{}'\right)\Pi_{n1-k1-1,n2}$$
$$+ \mu_2\Pi_{n1-k1,n2-1} \tag{12}$$

$$\left(\lambda_1 + \lambda_2 + \alpha_2\mu_{2+}\beta_2\mu_2{}'\right)\Pi_{n1,n2-k2}$$
$$= \lambda_2\Pi_{n1,n2-k2+1} + \left(\alpha_2\mu_{2+}\beta_2\mu_2{}'\right)\Pi_{n1,n2-k2-1}$$
$$+ \mu_1\Pi_{n1-1,n2-k2} \tag{13}$$

$$\left(\lambda_2 + \mu_1{}' + \alpha_2\mu_{2+}\beta_2\mu_2{}'\right)\Pi_{-n1,n2-k2}$$
$$= \lambda_2\Pi_{-n1,n2-k2+1} + \left(\alpha_2\mu_{2+}\beta_2\mu_2{}'\right)\Pi_{-n1,n2-k2-1}$$
$$+ \lambda_1\Pi_{-n1+1,n2-k2} \tag{14}$$

$$\left(\lambda_1 + \mu_2{}' + \alpha_1\mu_{1+}\beta_1\mu_1{}'\right)\Pi_{n1-k1,-n2}$$
$$= \lambda_2\Pi_{n1-k1,-n2+1} + \left(\alpha_1\mu_{1+}\beta_1\mu_1{}'\right)\Pi_{n1-k1-1,-n2}$$
$$+ \lambda_1\Pi_{n1-k1+1,-n2} \tag{15}$$

$$\left(\lambda_1 + \lambda_2 + \alpha_1\mu_{1+}\beta_1\mu_1{}' + \alpha_2\mu_{2+}\beta_2\mu_2{}'\right)\Pi_{n1-k1,n2-k2}$$
$$= \lambda_2\Pi_{n1-k1,n2-k2+1}$$
$$+ \left(\alpha_1\mu_{1+}\beta_1\mu_1{}'\right)\Pi_{n1-k1-1,n2-k2}$$
$$+ \left(\alpha_2\mu_{2+}\beta_2\mu_2{}'\right)\Pi_{n1-k1,n2-k2-1}$$
$$+ \lambda_1\Pi_{n1-k1+1,n2-k2} \tag{16}$$

where:

$$k_j = 1, 2, ..., 2n_j-1; j = 1, 2.$$

If $p \le n_1$ $\alpha_1 = 1$, $\beta_1 = 0$; Otherwise $\alpha_1 = 0$, $\beta_1 = 1$.
If $q \le n_2$ $\alpha_2 = 1$, $\beta_2 = 0$; Otherwise $\alpha_2 = 0$, $\beta_2 = 1$.

In order to solve the above formulated problem with regard to the various numbers of the equilibrium equations as well as their complexity, all equilibrium equations are coded in the form of a matrix in MATLAB software for the first time in this context. In the proposed program, all equation components are transferred

Table 1 Network's data

Parameters/problem number	1	2	3	4	5	6	7	8	9	10
μ_1	5	5	8	5	5	6	9	7	5	10
μ_2	8	8	5	8	8	4	3	8	9	4
μ_1'	6	4	6	6	6	8	4	6	4	7
μ_2'	10	6	10	10	10	7	6	3	5	3
λ_1	3	3	3	4	3	5	4	3	1	3
λ_2	4	4	4	3	4	3	2	1	5	1
h1	5	5	5	5	8	5	5	5	5	5
h2	8	8	8	8	5	8	8	8	8	8
b1	7	7	7	7	7	7	7	7	7	7
b2	6	6	6	6	6	6	6	6	6	6

Table 2 Parameters and their levels

P factor	Inertial weight	Individual learning	Plural learning	Population size	Maximum iteration
Lower limit	0.5	1	1	10	50
Upper limit	1	2	2	50	200

Table 3 Tuned parameters of the proposed PSO

P factor	Inertial weight	Individual learning	Plural learning	Population size	Maximum iteration
Tuned value	1	2	2	20	100

to the left side, and then, the equation coefficients, the one-stage transition probability, and the quantities of right-hand side appeared as matrices A, X, and b (i.e., $AX = b$), respectively. In addition, the constraints of the storage space at the warehouse are added as a violation coefficient to the objective function. It is noted that initial solution is generated randomly in the proposed solution methodology, while one-stage transition probabilities are computed exactly. Problem solution includes a twofold string that represents optimal on-hand inventory of each product at the warehouse.

Single-product models

In order to compare SC performance of the proposed bi-product multi-echelon SC with the single product SC, a network including two Markov models $(M/M/1)$ for each type of product is considered in this section. Note that in this network, the common warehouse holds inventory and applies a continuous review (S_{j-1}, S_j) inventory policy for each type of product; that is, a replenishment order is forwarded to the manufacturer after each order enters the warehouse. The supposed model seeks a S_j^* in order to minimize a cost function which captures two different operational cost factors including inventory holding and backordering costs. Here, the corresponding Markov model $(M/M/1)$ for product j has the state space (N_j) in which n_j represents on-hand stock of product j at the common warehouse $(-S_{max} \leq S_j^* \leq$

S_{max}). The transition diagram of this system is obtained as shown in Figure 5.

In order to denote the equilibrium equation in a given state (p_j), all the departure states with state (p_j) as their corresponding destination state are specified. To do so, system states are divided into three categories according to their similarity among departure states. All states which are classified in a category have a similar form of equilibrium equations. The equilibrium equations for the states located in each of the three categories are computed upon the one-stage transition probabilities. These equations are as the ones in (17 to 19) for regions 1 to 3, respectively.

$$\lambda_j \Pi(n_j) = \mu_j \Pi(n_{j-1}) \tag{17}$$

$$\left(\lambda_j + \alpha\mu_j + \beta\mu_j'\right) \Pi(n_{j-h})$$
$$= \lambda_j \Pi(n_{j-h+1}) + \left(\alpha\mu_j + \beta\mu_j'\right) \Pi(n_{j-h-1}); h$$
$$= 1, 2, ..., 2n_j - 1 \tag{18}$$

$$\left(\alpha\mu_j + \beta\mu_j'\right) \Pi(-n_j) = \lambda_j \Pi(-n_j + 1) \tag{19}$$

while:

If $p_j \leq n_j \alpha = 1, \beta = 0;$ Otherwise $\alpha = 0, \beta = 1.$

The objective function and constraints of the second condition are constructed as the bi-product model. Here, Ch_j, Cb_j, I_j, and b_j denote holding cost, backorder cost, on-hand inventory, and unsatisfied demand of product j, respectively. Optimal solution of the developed model represents optimal on-hand inventory of each product at the warehouse.

Objective function:

$$\text{Min} z: \ \text{TC} = \Sigma_j Ch_j + Cb_j \tag{20}$$

Table 4 Computational results of the single-product model

Problem number	Product 1			Product 2				
	n_1	OFV	CPU time (s)	n_2	OFV	CPU time (s)	$n_1 + n_2$	Total cost of SC
1	36	46.27	6.17	93	46.27	4.40	129	92.54
2	35	43.02	5.95	80	44.42	4.65	115	87.44
3	63	42.02	6.22	92	47.59	4.43	155	89.61
4	43	42.02	6.15	61	47.59	4.52	104	89.61
5	31	46.09	6.58	83	46.09	4.42	114	92.18
6	76	31.69	7.27	44	37.09	5.72	120	68.78
7	67	43.76	8.33	41	43.77	5.59	108	87.53
8	57	41.06	7.32	79	44.61	5.09	136	85.67
9	32	41.43	6.08	82	45.02	6.14	114	86.45
10	79	45.32	5.36	35	47.02	4.86	114	92.34

Table 5 Computational results of the bi-product

Problem number	n_1	n_2	OFV	CPU time (s)
1	32	6	52.12	1,493.08
2	41	10	55.45	1,603.65
3	22	20	48.87	1,569.91
4	33	9	44.83	1,486.68
5	21	23	61.55	1,406.48
6	48	25	27.1	1,979.46
7	31	3	43.8	1,874.17
8	39	46	142.4	1,957.23
9	39	5	60.55	1,899.17
10	19	2	118.86	1,938.01

Subject to:

$$Ch_j = I_j . \Pi \left(p_j \right); \tag{21}$$

$$Cb_j = b_j . \Pi \left(p_j \right); \tag{22}$$

$$I_j = p^+{}_j; \tag{23}$$

$$b_j = p^-{}_j; \tag{24}$$

$$n_j \leq Smax \tag{25}$$

The developed bi-product and single-product models are non-linear integer programming models. In the next section, a solution algorithm is proposed to cope with complexity of the models.

Solution methodology

Having the problem defined, solution methodology is developed to tackle computational complexity of the problem. To do so, a metaheuristic algorithm (i.e., particle swarm optimization) is developed because of two major reasons:

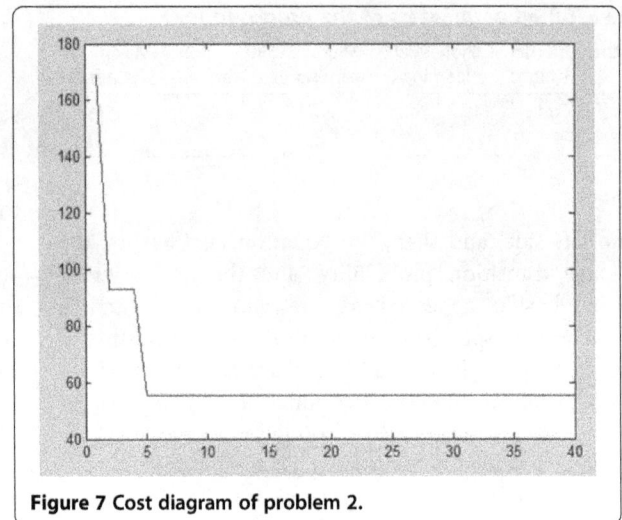

Figure 7 Cost diagram of problem 2.

numerous numbers of the equilibrium equations as well as their levels of complexity. Moreover, there is no compact mathematical formulation available for the index of the system steady state.

The introduced PSO by Kennedy and Eberhart (1995) and Eberhart and Kennedy (1995) is an evolutionary calculations technique. Particle swarm concept has been driven from the simulation of social behavior. The PSO algorithm works as the behavior of flying birds which exchange their information to solve optimization problems. This algorithm is an optimization technique in real-world spaces, while many of these problems are set in discrete space (Xia and Wu 2005). The PSO solutions are as particles in the position of search space. In order to achieve the best point in the solution space, they collect both information of their experiences and the others' (Equations 26 and 27). In this regard, they investigate the solution space through exchanging their position and velocity. The velocity is updated by Equation

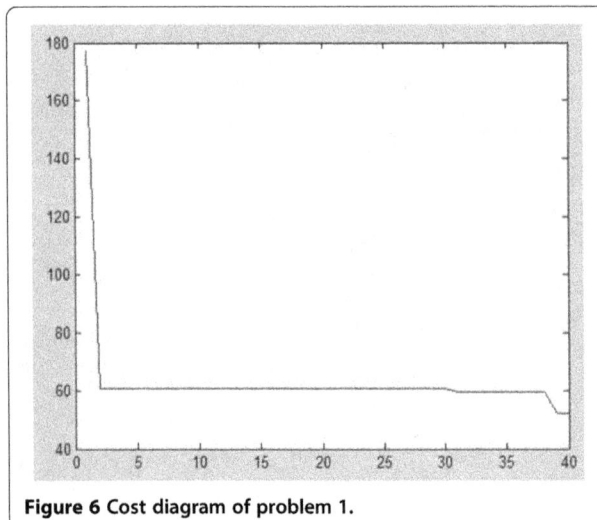

Figure 6 Cost diagram of problem 1.

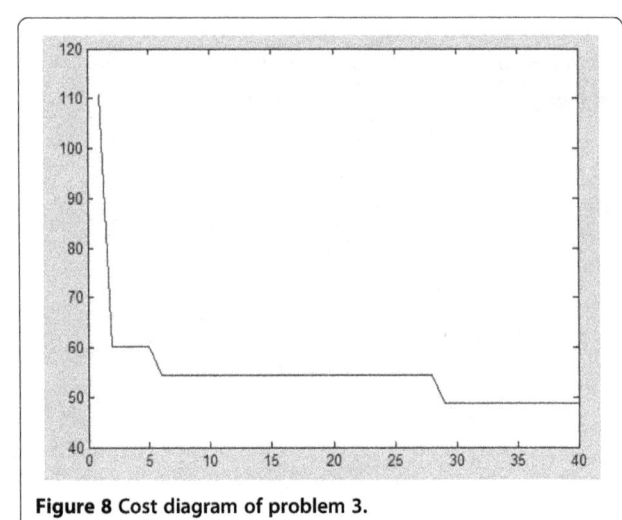

Figure 8 Cost diagram of problem 3.

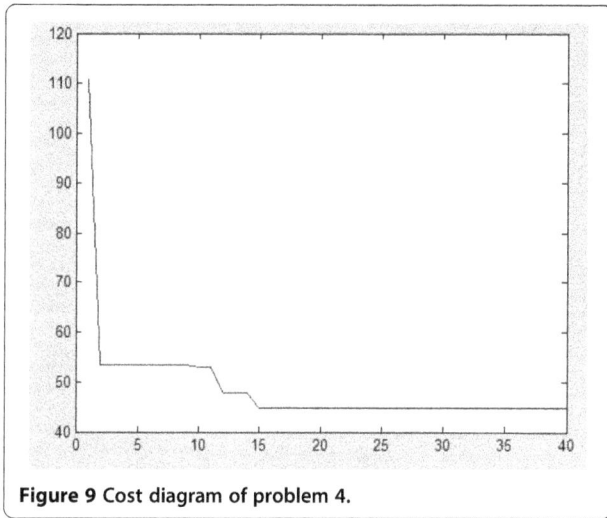

Figure 9 Cost diagram of problem 4.

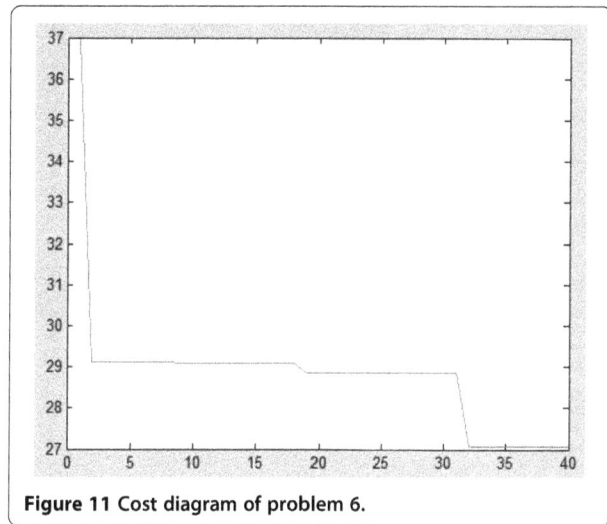

Figure 11 Cost diagram of problem 6.

26. The first part refers to the inertia of previous velocity, while the private particle thinking is denoted by the second part, the 'cognition' part. The third part, the 'social' part, represents the cooperation among the particles.

$$V_i(t + 1) = w \times V_i(t)+$$
$$c_1 \times r_1 \times (x_{i\text{-best}}(t) - x_i(t)) \qquad (26)$$
$$+c_2 \times r_2 \times (x_{\text{gbest}}(t) - x_i(t))$$

$$x_i(t + 1) = x_i(t) + V_i(t + 1) \qquad (27)$$

$V_i(t)$ is called the velocity of particle i in iteration t. In this study, the initial velocity is considered zero. $x_i(t)$ is called the ith particle position, $x_{i\text{-best}}(t)$ is called local best solution that shows the best previous position of each particle, $x_{\text{gbest}}(t)$ represents the best position among all particles in the swarm. W as inertial weight balances the global exploration and the local exploitation abilities of the swarm. c_1 and c_2 represent the weight of the stochastic acceleration terms which is assumed to be equal to 2. r_1 and r_2 are random functions to seek better the space in the range [0,1]. The inertial weight is considered 1 in the first iteration and then, slowly reduces in each other iteration by following Equation 28. *wdamp* is assumed as 0.9:

$$w = w \times wdamp \qquad (28)$$

The $V_{\max}$ equation with search space width that has been presented by Shi and Eberhart (1999) is applied to converge more the search space. As a result of using this approach, particle's velocity and position must be limited through a defined space width.

An initial solution is generated randomly regarding the constraint of the limited storage space at the central warehouse in order to be a feasible solution. Based on the obtained initial solution, we calculate one-stage transition probabilities certainly by Equations 8 to 19. After

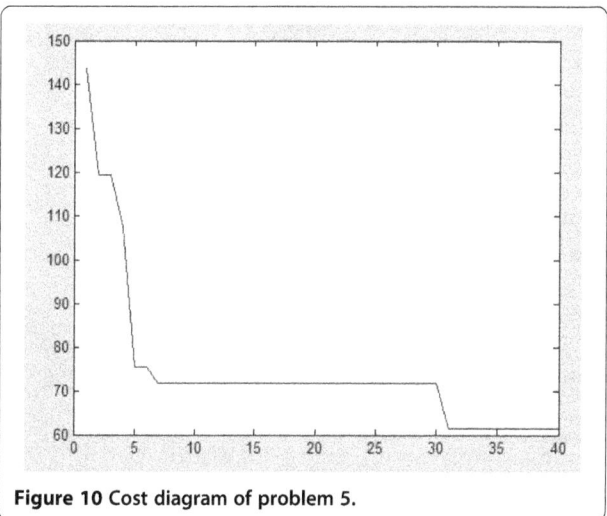

Figure 10 Cost diagram of problem 5.

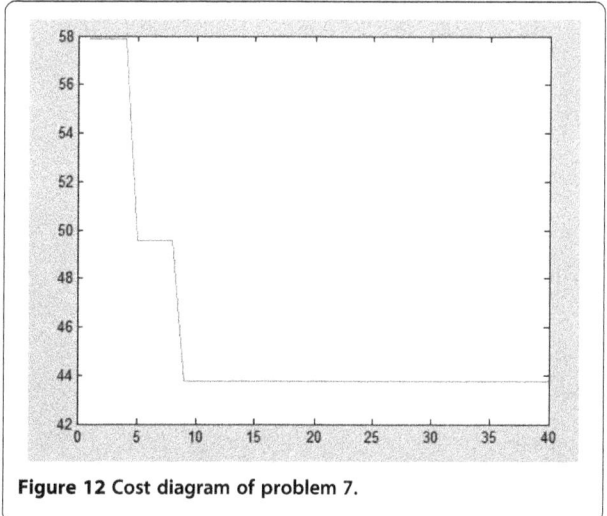

Figure 12 Cost diagram of problem 7.

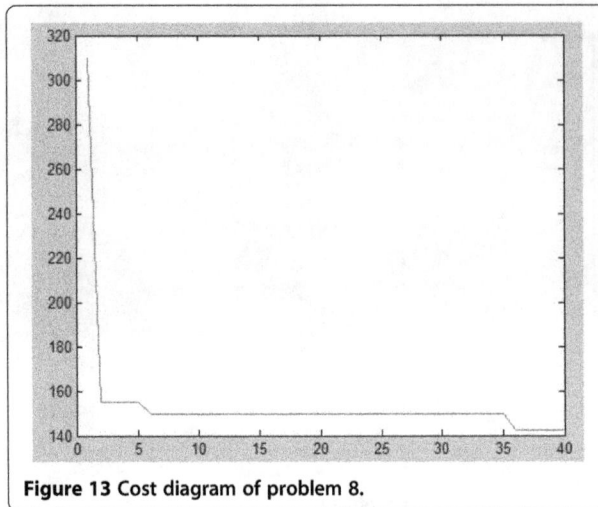

Figure 13 Cost diagram of problem 8.

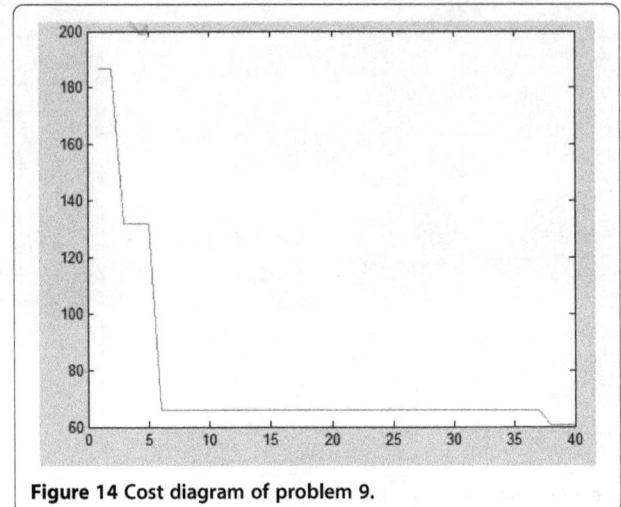

Figure 14 Cost diagram of problem 9.

computing the transition probabilities, the on-hand inventory of each product as problem solution and objective function values are entered to the PSO algorithm to seek neighbor space of the present solution in order to achieve better one. The above procedure will be repeated up to given number of maximum iterations of the algorithm. The PSO algorithm is implemented by considering population size 20, maximum iteration 100, $c_1 = c_2 = 2$. Algorithm 1 shows the structure of the proposed solution methodology as pseudocode.

Results and discussion

This section is developed to evaluate the tractability of the proposed programming model in terms of the solution quality and the required computational time. In this regard, some numerical experiments of the considered supply chain network on a set of randomly generated problem instances are performed. The developed models and the proposed algorithm are coded in MATLAB 7.11.0 (R2010b) software on a PC with 2 GHz processor. The holding cost rate and backordering cost rate at central warehouse for each product are generated from a uniform distribution of [5,10], the demand rate for products is uniformly generated in the interval [1,5], the corresponding service rate with the on-hand inventory at the central warehouse is generated from a uniform distribution of [1,10]. Table 1 shows the values of the parameters of the generated instances. The threshold of the optimal on-hand inventory at the warehouse is considered 100 for all cases.

It is well known that the quality of an algorithm is significantly influenced by the values of its parameters. In this section, for optimizing the behavior of the proposed algorithm, appropriate tuning of the parameters has been realized using response surface methodology (RSM). RSM is defined as a collection of mathematical and statistical method-based experiential, which can be used to optimize processes. Regression equation analysis is employed to evaluate the response surface model. First, the parameters of the supposed algorithm that statistically have significant impact on the algorithm should be recognized. Each factor is measured at two levels, which can be coded as −1 when the factor is at its low level (L) and +1 when the factor is at its high level (H). The coded variable can be defined as follows:

$$x_i = \frac{r_i - \left(\frac{H+L}{2}\right)}{\left(\frac{H-L}{2}\right)} \tag{29}$$

where x_i and r_i are coded variable and natural variable, respectively. H and L represent high level and low level

---

Algorithm 1 Pseudo-code of the proposed PSO optimization algorithm.

Begin

First step:

Initialize parameters value of PSO algorithm, inertial weight, the weight of the stochastic acceleration
terms (individual and plural learning), a spread of searching space and maximum iteration;

Second step:

At first iteration
Generate PSO's initial population;
Locate the particle with lowest fitness on gbest;
Locate each particle on pbest;
Until (the maximum iteration is not met)

Do {

Iteration = Iteration + 1;
Generate particle by Equation (26), (27);
Evaluate particle {

Calculate each particle's fitness function;
Find new gbest and pbest by compration;
Update gbest of the swarm and pbest
of each particle;

}

}

Third step: output optimization results;

End

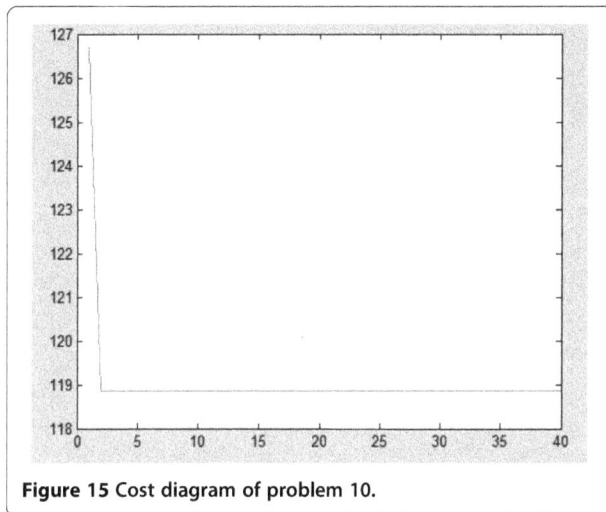

Figure 15 Cost diagram of problem 10.

of the factor, respectively. Factors and their levels are indicated in Table 2.

After developing regression models for each problem size separately, tuned parameters of the proposed PSO are obtained as in Table 3.

Tables 4 and 5 show the solutions and the objective function values of the single and the bi-product models solved by the developed PSO algorithm. Also, the required computational times of the algorithm are reported.

As shown, using bi-product model leads to lower total cost of the network. Also, the solutions of the two single-product models violate the capacity of the common warehouse. This result confirms the necessity of considering multi-product models in the real situation where we have many common sources with limited capacity for products. In these experiments, we also evaluate the convergence rate of the PSO algorithm to obtain a solution for all five considered bi-product problems. Figures 6, 7, 8, 9, 10, 11, 12, 13, 14, and 15 illustrate the convergence trend of the solution for each problems 1 to 10, respectively. These figures show that the PSO algorithm is capable to reach a good solution in its early iterations.

Conclusions

In this paper, a bi-product inventory planning problem is proposed. The problem is developed in a three-echelon SC consisting of r identical retailers, a central warehouse with limited storage space, and two independent manufacturing plants. In this study, demanding process and lead time are stochastic. Customer demands arrived to each retailer following Poisson distribution function with rates λ_{i1} and λ_{i2} for products 1 and 2, and the accumulation of demands with rate $\lambda_1 = \sum \lambda_{i1}$ and $\lambda_2 = \sum \lambda_{i2}$ which enter the central warehouse responded the demand of each product j by the on-hand

inventory at warehouse based upon exponential distribution function with mean $1/\mu_j$, if any. Otherwise, unsatisfied demands are backlogged. In these circumstances, customer demands are satisfied with mean $1/\mu_j'$. Hence, an $(M/M/1)$ queue model has been developed at the warehouse. Then, the problem with two Markov models $(M/M/1)$ has been developed at the warehouse to evaluate the performance of the first problem. Finally, as depicted in the 'Results and discussion' section, the single-product model cannot satisfy the constraint on the warehouse storage space. Moreover, the developed bi-product three-echelon inventory model decreases total network costs significantly, indicating outperformance of the developed bi-product model over the single-product one.

As further research direction, first, it is suggested to consider other assumptions in the considered supply chain to make it more practical, such as more echelons or more products. Second, considering other probability distribution functions for demanding process and service times such as Erlang is recommended. Finally, developing proper exact and inexact algorithm might be of great contribution to the theory of the problem.

Competing interests
The authors declare that they have no competing interests.

Authors' contributions
M. Alimardani has developed the Markov model and the resulted equilibrium equations as well as the proposed mathematical model; while the metaheuristuc algorithm (proposed particle swarm) is designed and applied by H. Rafiei. Both the students have co-worked on the language of the paper. Problem definition, modeling and analysis are conducted under supervision of Dr. Jolai. All authors read and approved the final manuscript.

Acknowledgements
The authors are grateful to the editor of Journal of Industrial Engineering International as well as the anonymous referees for their valuable and constructive comments to enhance quality of the paper.

References
Andersson J, Melchiors P (2001) A two echelon inventory model with lost sales. Int J Prod Econ 69:307–315
Axsäter S (1990) Simple solution procedures for a class of two echelon inventory problems. Oper Res 38:64–69
Axsäter S (1993) Exact and approximate evaluation of batch ordering policies for two-level inventory systems. Oper Res 41:777–785
Bozorgi Atoei F, Teimoury E, Amiri B (2013) A Designing reliable supply chain network with disruption risk. Int J Ind Eng Comput 4:111–126
Chao HP (1987) Inventory policy in the presence of market disruptions. Oper Res 35:274–281
Deuermeyer B, Schwarz LB (1981) A model for the analysis of system service level in warehouse/retailer distribution system: the identical retailer case. In: Schwarz L (ed) Multilevel production/inventory control systems (TIMS studies in management science 16). Elsevier, New York
Diks EB, de Kok AG (1998) Optimal control of a divergent multi-echelon inventory system. Eur J Oper Res 111:75–97
Dong M (2003) Inventory planning of supply chains by linking production authorization strategy to queueing models. Prod Plan Control 14:533–541

Eberhart R, Kennedy J (1995) A new optimizer using particle swarm theory. In: Proceedings of the sixth international symposium on micro machine and human science. , Nagoya, pp 39–43, 4–6 Oct 1995

Ganeshan R (1999) Managing supply chain inventories: a multiple retailer, one warehouse, multiple supplier model. Int J Prod Econ 59:341–354

Graves SC (1985) A multi-echelon inventory model for a repairable item with one for one replenishment. Manage Sci 31:1247–1256

Gumus AT, Guneri AF (2007) Multi-echelon inventory management in supply chains with uncertain demand and lead times: literature review from an operational research perspective. P I Mech Eng B-J Eng 221:1553–1570

Gupta D (1996) The (Q, r) inventory system with an unreliable supplier. INFOR 34:59–76

Haji R, Pirayesh Neghab MA, Baboli A (2009) Introducing a new ordering policy in a two-echelon inventory system with Poisson demand. Int J Prod Econ 117:212–218

Kennedy J, Eberhart R (1995) Particle swarm optimization. In: Proceedings of the 1995 IEEE international conference on neural networks, vol 4, Perth., pp 1942–1948, 27 Nov-1 Dec 1995

Kevin Chiang W, Monahan GE (2005) Managing inventories in a two-echelon dual-channel supply chain. Eur J Oper Res 162:325–341

Lee HL, Moinzadeh K (1987) Operating characteristics of a two-echelon inventory system for repairable and consumable items under batch ordering and shipment policy. Nav Res Log 34:365–380

Mohebbi E (2003) Supply interruptions in a lost-sales inventory system with random lead time. Comput Oper Res 30:411–426

Moinzadeh K, Aggarwal PK (1997) An information based multi-echelon inventory system with emergency orders. Oper Res 45:694–701

Moinzadeh K, Lee HL (1986) Batch size and stocking levels in multi-echelon repairable systems. Manage Sci 32:1567–1581

Sana SS (2011) A production-inventory model of imperfect quality products in a three-layer supply chain. Decis Support Syst 50:539–547

Sherbrooke CC (1968) METRIC: a multi-echelon technique for recoverable item control. Oper Res 36:122–141

Shi Y, Eberhart R (1999) Empirical study of particle swarm optimization. In: Proceedings of congress on evolutionary computation. Washington D.C, pp 1945–1950, 6–9 July 1999

Svoronos AP, Zipkin P (1988) Estimating the performance of multi-level inventory systems. Oper Res 36:57–72

Teimoury E, Modarres M, Ghasemzadehd F, Fathi M (2010) A queuing approach to production-inventory planning for supply chain with uncertain demands: case study of PAKSHOO chemical company. J Manuf Syst 29:55–62

Teimoury E, Mazlomi A, Nadafioun R, Khondabi IG, Fathi M (2011a) A queuing-inventory model in multiproduct supply chains. In: International multiconference of engineers and computer sciences. Hong Kong, 16–18 Mar 2011

Teimoury E, Mazlomi A, Nadafioun R, Khondabi IG, Fathi M (2011b) Inventory planning with batch ordering in multi-echelon multi-product supply chain by queuing approach. In: International multi conference of engineers and computer sciences. Hong Kong, 16–18 Marc 2011

Thangam A, Uthayakumar R (2008) A two-level supply chain with partial backordering and approximated Poisson demand. Eur J Oper Res 187:228–242

Xia W, Wu Z (2005) An effective hybrid optimization approach for multi objective flexible job-shop scheduling problems. Comput Ind Eng 48(2):409–425

19

An application of principal component analysis and logistic regression to facilitate production scheduling decision support system: an automotive industry case

Saeed Mehrjoo[1] and Mahdi Bashiri[2*]

Abstract

Production planning and control (PPC) systems have to deal with rising complexity and dynamics. The complexity of planning tasks is due to some existing multiple variables and dynamic factors derived from uncertainties surrounding the PPC. Although literatures on exact scheduling algorithms, simulation approaches, and heuristic methods are extensive in production planning, they seem to be inefficient because of daily fluctuations in real factories. Decision support systems can provide productive tools for production planners to offer a feasible and prompt decision in effective and robust production planning. In this paper, we propose a robust decision support tool for detailed production planning based on statistical multivariate method including principal component analysis and logistic regression. The proposed approach has been used in a real case in Iranian automotive industry. In the presence of existing multisource uncertainties, the results of applying the proposed method in the selected case show that the accuracy of daily production planning increases in comparison with the existing method.

Keywords: Principal component analysis, Logistic regression, Production planning control, Decision support system

Introduction

Effective planning and control of production processes are usually seen as key to the success of a manufacturing company. During the last 50 years, both academic institutes/universities and industries have put great effort into developing and designing successful approaches and methods for manufacturing planning and control. Indeed, the methods and approaches of how to plan and control production have been changed over time. This occurs in line with changes in customer requirements and technology improvements (Vollmann et al. 2005).

Detailed production scheduling is an extremely complex problem (Brucker 2007) wherein most cases are considered NP-hard (Günther and van Beek 2003). In order to deal with complexities and uncertainties, a detailed production scheduling system should be equipped with all the necessary decision support tools for rendering production problems visible within a planning period

* Correspondence: bashiri@shahed.ac.ir
[2]Industrial Engineering Department, Shahed University, Tehran, Iran
Full list of author information is available at the end of the article

and shift dispatching control from the foremen to the planner (Sotiris et al. 2008). According to Simchi-Levi et al. (2008), the decision support system (DSS) is an analytical tool to aid operations and production planning. The DSS can range from simple tools to expert systems. The DSS helps to solve the problems such as network planning to tactical planning all the way to daily operational problems. Thus, the effective DSS can help managers or production planners to manage uncertainties and achieve better results in daily fluctuations.

The stimulus for this work has been to understand whether or not the historical daily shop floor data can be used for creating more robust daily production plan. Moreover, the paper studies the feasibility of using the multivariate statistical analysis of daily shop floor data as an appropriate solver tool for detailed production scheduling decision support system. In order to answer these, we represent an Iranian automotive case of detailed production planning in applied material requirement planning (MRP) system. The results may not be generalized to JIT and lean manufacturing principles which have a pull approach of planning and control of production. The rest of the paper has been organized as follows: in next section, the related literature has been reviewed, whilst the problem has been defined and the selected case has been presented in the 'Case problem statement' section. The 'Methodology of problem analysis' section has outlined the proposed multivariate DSS as subsequent specifications arising from the 'Case problem statement' section. The 'Proposed multivariate DSS method' section has demonstrated the implementation results, and finally, in the 'Conclusions' section, the conclusions have been drawn and further research efforts have been mapped out.

Literature review

The production planning control models can be characterized in a variety of approaches, but most common categorizations are specific to the application areas (Brucker 2007). In this article, we have classified the production planning problem solving approaches in two groups in terms of their environment and condition: unconditional/ deterministic analytic production planning and production planning under uncertainties (see Table 1).

In the case of unconditional analytic models, we are referring to models which are simplifications of a real system in terms of mathematical expressions and can be solved by exact or heuristic methods. The literature on the exact algorithms in scheduling and production planning problems is extensive. A thorough review of scheduling problems, modeling approaches, and solution methods can be found in Brucker (2007), Framinan and Ruiz (2010), and Ribas et al. (2010). Whereas fluctuations and uncertainties have key roles in real world, deterministic, or exact approaches, such that branch and bound (Yao et al. 2012) and mixed integer linear programming (Maravelias and Sung 2009) are seldom applicable in actual shop floors since they may only solve small-scale problems with distinct parameters and scale of time.

Several heuristics and hybrid methods are recommended in the literature (Jourdan et al. 2009). Ribas et al. (2010) have classified the approximate methods into constructive and improvement heuristics. However, the application of heuristic algorithms is believed to be more applicable for real shop floors due to their lower computations, but they are still limited by the dimension of problems and uncertainties. Therefore, their implementation should be coupled with some decision support tools to aid the production planners (Ross and Bernardo 2011; Sotiris et al. 2008).

To cope with uncertainties in production control, it is worth to investigate a new customized framework for planning and scheduling under uncertainty. Another challenging issue is to investigate the ways of controlling a large number of uncertain parameters. Hence, scheduling

under uncertainty has received a lot of attention in recent years (e.g., Hatzikonstantinou et al. 2012; Vargas and Metters 2011; Torabi et al. 2010; Verderame and Floudas 2009). Uncertainty can be derived from many aspects, such as demand or product orders, alternation or priority of orders, equipment failures, resource changes, and processing time variability. To adapt uncertainties during the manufacturing process, the proposed methods are divided into two main groups: reactive scheduling and preventive scheduling (Aytug et al. 2005). Simulation approach is able to analyze the behavior of the environment when it is characterized by several constraints and uncertainties (Rolo and Martinez 2012; Volling and Spengler 2011; Jahangirian et al. 2010). In these approaches, the outcomes of the simulation software can be used in preventive scheduling and decision support systems, but they need a great deal of efforts to make a practical schedule by some expert production planners.

By the emersion of enterprise resource planning (ERP) systems, the utilization of data becomes more important in production planning and control (PPC). The incredible wealth of available data in SCM and PPC software raises the question of how to help decision makers in harnessing the organization. The answer to this question has defined the production activity control (PAC) subsystem at the lowest level of MRPII (Vollmann et al. 2005). By means of the PAC system, the sequence of the orders is defined with their release and due times. In fact, PAC cannot take into account the real state of the production environment, and it may produce unrealistic or impractical production plan.

Whereas the MRP-based system cannot follow the large number of shop floor fluctuations, production managers bow to the inevitable complex task of scheduling/rescheduling at the shop floor control. Poor production control may cause serious problems to a firm's ability to meet production requirements and constraints. Many researches have focused on developing DSS tools to face this problem (e.g., Ko and Wang 2010; Caricato and Grieco 2009; Mok 2009; Farrella and Maness 2005; McKay and Wiers 2003). These tools are concerned as complementary applications to the ERP/MRP software.

Unfortunately, few success stories have been reported on creating production planning and logistics in a real factory, and there are still many challenges that remain (McKay and Black 2007). In the absence of one sole issue for PPS success or failure (McKay and Wiers 2003, 2004), one potential issue related to the failure of a planning system is the lack of information system and DSS tools for detailed production planning. This was the first insight obtained from this case study (see Table 1).

Meanwhile, combination use of statistical analysis with other methods to control the uncertainties in real condition decision making has been proposed by some

Table 1 Classification of related literatures

	Approach					
	Deterministic production planning			**Production planning under uncertainties**		
	Exact methods	Heuristic methods	Simulation	DSS developing	Statistical analysis	Hybrid/combinational methods
Literatures	Brucker (2007)	Sotiris et al. (2008)	Aytug et al. (2005)	McKay and Wiers (2003)	Mele et al. (2005)	Cunha and Wiendahl (2005)
	Maravelias and Sung (2009)	Jourdan et al. (2009)	Jahangirian et al. (2010)	Sotiris et al. (2008)	Hatzikonstantinou et al. (2012)	Aytug et al. (2005)
				Farrella and Maness (2005)		
	Framinan and Ruiz (2010)	Ribas et al. (2010)	Volling and Spengler (2011)	Mok (2009)	Peidroa et al. (2009)	Volling and Spengler (2011)
	Ribas et al. (2010)	Jahangirian et al. (2010)	Rolo and Martinez (2012)	Caricato and Grieco (2009)	Verderame and Floudas (2009)	
	Yao et al. (2012)	Ross and Bernardo (2011)		Ko and Wang (2010)		
				<————————This paper————————>		
Comments	Seldom applicable in actual shop floors since they may only solve small-scale problems with distinct parameters and scale of time		Simulation needs a great deal of efforts to make a practical schedule by some expert and expensive production planners	DSS tools are concerned with complementary applications to ERP/MRP software. They are practical with lack of ERP system	Statistical and hybrid methods are useful to control uncertainties in real condition and improve the effectiveness of both evaluation and decision making; however, they are not independent and complete tools. They have to be designed for each case problem	

DSS, decision support system; MRP, material requirement planning; ERP, enterprise resource planning.

literatures (Mele et al. 2005). Cunha and Wiendahl (2005) have proposed an evaluation method based on the use of multivariate techniques: principal component analysis (PCA) and cluster analysis (CA) to improve the effectiveness of evaluation and decision making, monitoring and manufacturing control. The idea of using multivariate statistical analysis to develop existing DSS is the second and major contribution of this study. The proposed method in this paper is based on the use of multivariate techniques on shop floor data. We intend to improve the effectiveness of the decision-making tasks undertaken when dealing with detailed production plans in an uncertain condition.

Case problem statement

The case study of detailed production planning has been done in an Iranian automotive manufacturing company. SAIPA Corporation (Tehran, Iran) is a holding company that assembles several types of passenger cars, vans, minibuses, buses, and trucks. As with any other car manufacturing company, the production process followed has a high degree of complexity, coupling the complex bill of materials (BOMs) with equally complex routings that transgress the shop floor boundaries.

The main problem of the selected case has been inferred from logistics staff answers to a set of questions and interviews. It is reported that the accuracy of daily production plans is directly affected by some alternate constraints and probable parameters. Hence, either the rescheduling or planning diversity and related extra material handling or extra/shortage parts and production line stop are enviable tasks every day. By their complaint about the alternate decisions to manage stochastic or abnormal events, we have made inferences about the lack of DSS tools to provide a practical detailed production planning.

Methodology of problem analysis

The following three aspects of the problem have been specified in the analysis of the current situation:

- *Layout and physical constraint.* It focuses on the production flow and is concerned with constraints of layout and any physical limitation in the production lines.
- *Production planning and control system.* It is concerned with the daily activities of the production planners during their detailed production planning in the shop floor control process. Shop floor data in a multi period range was gathered from this aspect of analysis.
- *Uncertainties and stochastic factors.* It is concerned with the source of uncertainties and stochastic factors.

To investigate the mentioned aspects, a mixture of interviews and observation has been applied. The major part of observation and a small part of the meetings were concerned with information about the production process and the production planning control.

In the following two subsections, the basic results concerning the first two aspects of the case problem analysis have been presented. These results are normally used to design the system architecture and functionality as well as the shop floor model of the plant. The results of the third aspect, namely uncertainties, are used to construct a multivariate analysis tool of simplified real production.

Layout and physical constraints

A trim shop is located at the end of the production process. Therefore, it has the highest level of complexity in comparison with subsequent production activity control processes. The main assembly production line is equipped with a conveyor. According to production rate and types of products (seven types), the length of production line is not quite enough for assigning individual locations to keep the minimum stock level of all parts according to the type of product BOM. The logistic area is not available near the line far distance from the main warehouses; thus, the order of completion lead time is long and is influenced by probable accidents.

There is a painted body (PB) stock at the end of the paint shop process. The stock of PB is the same as a single line queue before the entrance of the trim shop, and each PB can be transferred to the trim shop by the sequence of its location. Incapability of selecting the desired PB from the PB stock constrains the production planner to make a daily plan according to the PB color and type sequence. Although the elimination of the layout and physical constraints have been investigated in recent years, due to outstanding required cost and time, the progress of development is not noticeable.

Production planning and control system

Although KANBAN cards and pull production control system have been tried to be applied by production planning and the logistic department, the production control system is still MRP-based. A hierarchical two-level planning framework is used prior to the detailed production scheduling. At the top level, aggregate production planning which controls demand management with a yearly time horizon, has been located. The second planning level which is called midterm planning incorporates a hybrid MRP-PBC approach.

Master production schedule (MPS) outcomes are used to calculate components and material requirements. The final plan is made by revision on a weekly basis using the feedback from the detailed scheduling module. The weekly plan is released by PAC, and detailed production planning is issued as the daily schedule. The daily schedule is derived from a complicated decision-making process which uses shop floor data, inventory status data, BOM, and sequence of PB in stock. Figure 1 demonstrates this process.

PPC suffers from several sources of inconsistencies as a consequence of incomplete ERP implementation. As a result of its complex production process and lack of information technology (IT) infrastructure, the presented case study during the last few years faced numerous problems concerning violated due dates, accumulated late orders, supernumerary production orders, excessive component inventory, poor releasing policies, and low shop floor visibility. The lack of online and integrated information may cause a misunderstanding of the real condition; thus, the production planner faces some unknown parameters in daily scheduling. In this situation, it is not weird if the daily schedule encounters some mistakes. Although the design and implementation of the ERP software is in progress, production planners cannot wait and do not get along with increasing complexity. They really need some practical tools to help them in perfect decision making.

Uncertainties and stochastic factors

Since there are some line-side space constraints, mixed production suffers from lots of problems and obstacles and forces managers to act on the basis of batch production. Meanwhile, there are many sources of stochastic events and uncertainties that batch production such as demand, process, and supply uncertainties (Peidroa et al. 2009). One of the main sources of stochastic factors that have been identified in this study is derived from the paint shop process and PB stock constraint. Due to some small defects on bodies, some of the PBs are selected to go into a touch-up area, and after doing all necessary reworks, they are transferred to the PB stock line. Almost all of the procedures in the defect inspection process are performed manually through human vision and influenced by stochastic factors. On the other hand, the required rework process times depend on the type and the level of defects which are not really exact and deterministic. Hence, the sequence of painted bodies in the queue of stock line cannot be absolutely defined. Meanwhile, supply uncertainties have a key role in unreliability of the production schedule. Each type of products has special parts which are from different suppliers. The availability of all special parts related to the desired type of products is the other vital information for the production planner to make the daily production schedule. According to our observation, the stock levels of these items are not expected to follow exact patterns.

Proposed multivariate DSS method

The complexity of production planning and control process, stochastic factors, physical constraint, uncertainties, and the shortcomings of the underlying IT

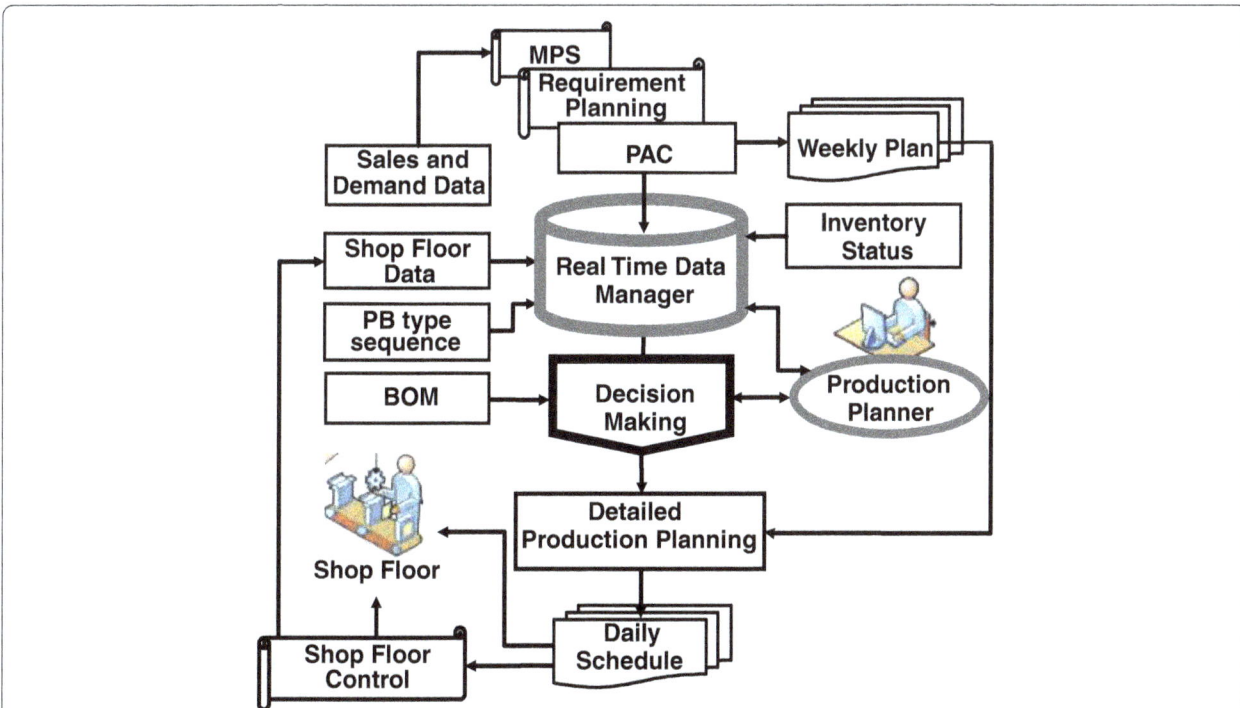

Figure 1 The current MRP-based PPC process.

infrastructure would pose significant drawbacks to the current detailed production scheduling. In this light, to the aforementioned production planning process and fully interoperable, both with the PPC system and existing software package, the proposed approach has been developed on the basis of a custom-built DSS using statistical multivariate techniques.

The integrated approach that will be presented introduces facilities to analyze data which are directly unavailable from the current planning system. This approach is introduced through the use of PCA to decrease the dimension of input-independent variables (Aguilera et al. 2006) and the use of logistic regression (LR) to predict the first priority of available and suitable type of product which can be selected to make a practical and effective detailed production schedule. These are used at different steps as shown in Figure 2.

Shop floor and production plan historical data acquisition

The manner and logical behavior of the production planner to create a weekly plan or change daily detailed scheduling is an important factor through the practical decision-making process which can be used for finding an effective DSS tool. As answer to the main question of this research, the objective has been to find the statistical analysis appropriate for reducing this logical behavior. Hence, it has been required to collect daily shop floor

data and historical data of the daily schedule issued by the production planner. The historical data of PB stock, existing PB quantity, and PB types in paint shop, sale online requests, inventory data, MRP weekly plan, released daily production schedule, and related orders with actual production were the main fields of data that have been collected for this analysis.

Reduction of inventory data by PCA

In this study, the collected inventory data sets (warehouse and line side separately) have at least 40 fields related to each types of products. This high volume and dimension of data matrix increase the complexity of analysis. If a substantial amount of the total variance in these data is accounted for by a few (preferably far fewer) principal components or new variables, then these few principal components can be used for interpretational purposes or in further analysis of the data instead of the original variables. PCA can be viewed as a dimensional reduction technique (Sharma 1996), and it is the appropriate technique for achieving the mentioned objective.

The core idea of PCA is to reduce the dimensionality of a data set comprising a large number of interrelated variables while retaining as much as possible the data set variance (Jolliffe 2002). This is obtained by transforming original variables to a new set of variables or principal

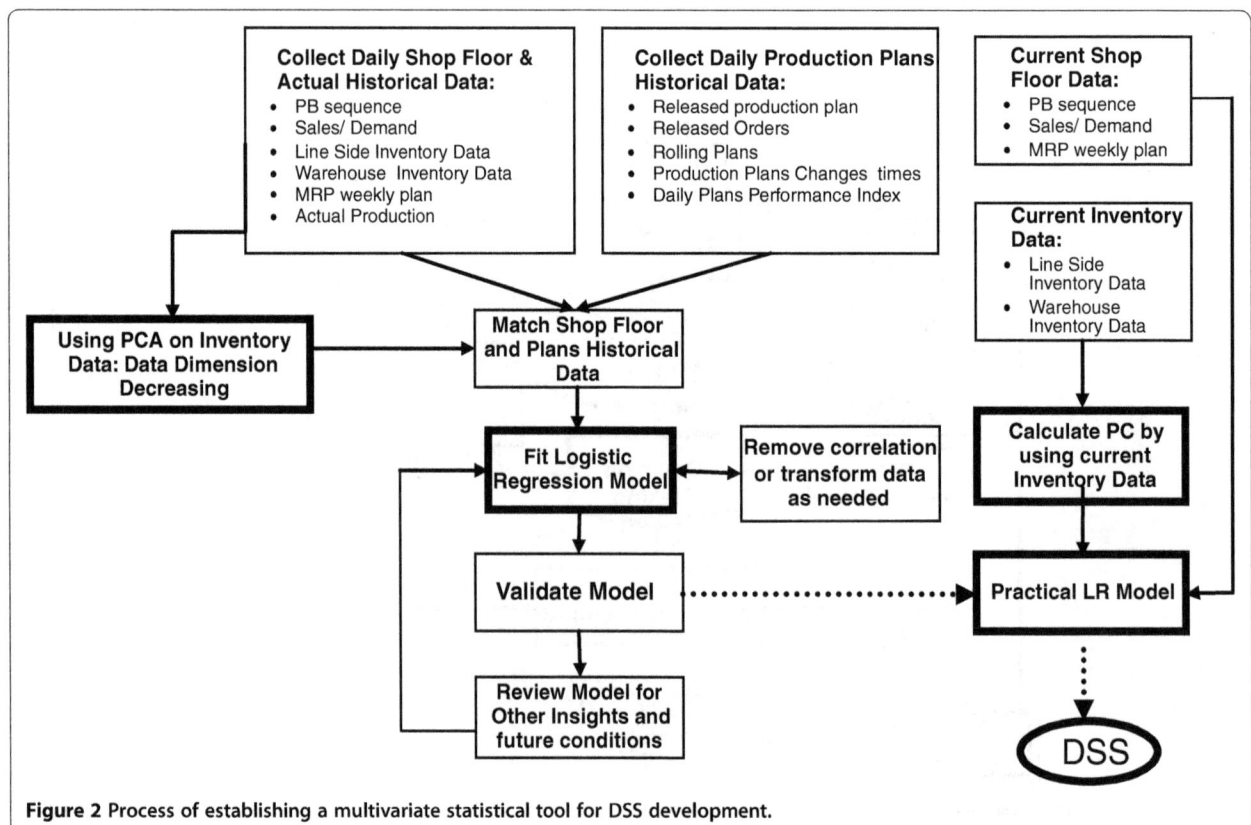

Figure 2 Process of establishing a multivariate statistical tool for DSS development.

components (ξ_i) which are a linear combination of original (p) variables. Due to their properties, they are uncorrelated and are ordered such that the first ($m \leq p$) that are retained contain most of the variation presented in the original data:

$$\text{PC} = \{\xi_1...\xi_m\}; \quad \xi_i = w_{i1}x_1 + w_{i2}x_2 + ... + w_{ip}x_p,$$

where principal component (PC) = $\{\xi_1... \xi_m\}$ are the m principal components and w_{ij} is the weight of the jth variable for the ith principal component.

The reduction in complexity is achieved by performing PCA on collected inventory data. Thus, the original data of inventory can be substituted by PCs, and the new matching table of the shop floor data and corresponding production schedule is established as a contingency table.

Logistic regression model fitting, validation, and review to improvement

The fundamental question in this research motivated us to understand the logical behavior of the production planner in the decision-making process through daily production scheduling. As illustrated in Figure 2, the historical input/output of the decision process is analyzed and the relationship among them is discovered by logistic regression. In the remainder of this section, we briefly discuss about the basic concept and details of developing the logistic regression model and, finally, the validation procedure and review method for the improvement of this model.

Definition of variables

To simplify the discussion and interpretation of estimation model, the notation is introduced and variables are defined which can be recognized from collected data. Table 2 shows a code sheet for definition of preliminary selected variables from collected data.

Basic theory on logistic regression

There are two models of logistic regression to contain binomial/binary logistic regression and multinomial logistic regression. Binary logistic regression is typically utilized when the dependent variable is dichotomous and the independent variables are either categorical or continuous variables (Sharma 1996). Logistic regression is the best to use in this condition. The result of this type of regression can be expressed by a logit function as follows:

$$\text{logit}(p) = \text{Ln}\left(\frac{p}{1-p}\right) = \beta_0 + \beta_1 * x_1 + \beta_2 * x_2 + ... + \beta_k * x_k$$

$$= \beta_0 + XB,$$

$$(1)$$

where $\left(\frac{p}{1-p}\right)$ is the odds.

Table 2 Description of variables

Variable	Description	Code/values
TYP	Compressed natural gas equipped	CNG
	CNG and hydraulic steering wheel	CNG-H
	Hydraulic steering wheel	HYD
	SABA simple injection system	GLXi
	Reinforced new body	X132
	Morvarid hatchback	DM
	Other new models	NEW
CLR	Color production plan	W = white
		G = gray
		B = black
		R = red
		S = silver
PBS	Painted body stock	0–400
SEQ	Compatibility of PB sequence	1 = Low compatibility
		2 = Fair compatibility
		3 = Good compatibility
		4 = High compatibility
INV1	Warehouse and line-side inventory PC 1	0 to 2,000
INV2	Warehouse and line-side inventory PC 2	0 to 1,000
SRT	Scheduled receipt	0 to 1,000
ACP	Available colored parts	0 to 1,000
RWP	Remainder quantity of weekly plan	0 to 5,000
ESD	Emergency sales/demand	1 = Critical
		2 = Urgent
		3 = Normal
		4 = Non-emergency
ADP	Actual daily product	0 to 1,100
DPP	Daily production plan	0 to 1,100
DPC	Daily production plan capability	0 = No
		1 = Yes

The model can either be interpreted using the logit scale, or the log of odds (the relative probability) can be converted back to the probability such that

$$p = \exp\frac{\beta_0 + XB}{1 + \exp(\beta_0 + XB)} \qquad (2)$$

In order to calculate the parameters $\beta_0, \beta_1, \beta_2,..., \beta_k$, the logistic regression transforms the dependent into a logit variable and then uses maximum likelihood estimation. In this paper, logistic regression is used to estimate the daily production planning capability (DPC) from the shop floor data. According to the variables summarized in Table 2, the logit can be defined for this case as follows:

$$
\begin{aligned}
\text{logit} = {} & \beta_0 + \beta_1 * \mathbf{TYP(1)} + \beta_2 * \mathbf{TYP(2)} + \beta_3 * \mathbf{TYP(3)} \\
& + \beta_4 * \mathbf{TYP(4)} + \beta_5 * \mathbf{TYP(5)} + \beta_6 * \mathbf{TYP(6)} \\
& + \beta_7 * \mathbf{INV1} + \beta_8 * \mathbf{INV2} + \beta_9 * \mathbf{CLR(1)} \\
& + \beta_{10} * \mathbf{CLR(2)} + \beta_{11} * \mathbf{CLR(2)} + \beta_{12} * \mathbf{CLR(3)} \\
& + \beta_{13} * \mathbf{CLR(4)} + \beta_{14} * \mathbf{DPP} + \beta_{15} * \mathbf{PBS} \\
& + \beta_{16} * \mathbf{SEQ(1)} + \beta_{17} * \mathbf{SEQ(2)} + \beta_{18} * \mathbf{SEQ(3)} \\
& - \beta_{19} * \mathbf{SRT} + \beta_{20} * \mathbf{ACP} - \beta_{21} * \mathbf{RWP} - \beta_{22} * \mathbf{ESD(1)} \\
& + \beta_{23} * \mathbf{ESD(2)} - \beta_{24} * \mathbf{ESD(3)}.
\end{aligned}
\tag{3}
$$

To find out how effective the model expressed in Equation 3 is, the statistical significance of individual regression coefficients is tested using the Wald chi-square statistic. Goodness-of-fit test assesses the fitness of a logistic model against actual outcomes. Hosmer-Lemeshow test is an inferential goodness-of-fit test which is utilized in this paper. Meanwhile, the consequent predicted probabilities can be revalidated with the actual outcome to determine if high probabilities are indeed associated with events and low probabilities with non-events. The readers are referred to Bewick et al. (2005) and Hosmer and Lemeshow (2000) for more information about the assessment of fitted model.

Predicting capability of daily production planning

The fitted model, which has successfully passed the goodness-of-fit tests, can be used to calculate the predicted Logit (probability) of DPC for a given value of shop floor data. For example, assume that at the end of the working day, the production planner wants to make a decision about tomorrow's production plan and would like to predict the capability of a given production schedule. At first, using the PCA method ('Reduction of inventory data by PCA' section), the inventory level of line-side and related warehouses can be estimated by two principal components (INV1 and INV2), and then, according to the shop floor data, the amount of other independent variables (TYP, PBS, RWP, ESD, and DPP) are defined. Therefore, the probably of response variable (DPC) can be calculated by Equation 4:

or $p = \frac{\exp(\text{logit})}{1 + \exp(\text{logit})}$.

Customized DSS to facilitate detailed production scheduling

Predicting the capability of daily production planning facilitates the decision making of the PPC system, and as a result, the customized DSS can be defined and applied. The new hierarchical planning framework is depicted in Figure 3. The main procedural sequence does not exhibit any remarkable change in comparison with the current PPC process (Figure 1). The MPS calculates long-term end item needs and feeds the PPC system which creates the production order backlog according to MRP procedures. The weekly production plans are issued by the MRP module and feeds detailed scheduling.

The proposed multivariate method contributes in the DSS module which is denoted in Figure 3 in the dashed box. As we have described in the previous sections, the online shop floor data is used by this customized DSS tool and the predicted amount of DPC index is calculated. This production planning capability index can facilitate the decision-making process of detailed scheduling. The production planner can typically run this customized DSS at the beginning of each planning period (commonly one working day), and after making the decision about final changes on the detailed schedule, data of production order is extracted. When detailed scheduling is finalized, the production orders are handed down to the foremen for beginning of production. If dynamic events take place (e.g., a machine breaks down, a rush order arrives, or a subcontractor violates due dates), the planner reschedules to accommodate them.

Numerical experiment and results

According to the defined variables, the 42-week shop floor and inventory data have been collected. Every day, the line-side inventory level of special and important parts as well as warehouse inventory level have been recorded. Table 3 illustrates the sample data which were recorded on the first and second days in the warehouse. Data were collected over a period of 8 months and included PB stock, sale online requests, inventory data, MRP weekly plan, inventory status data schedule, and actual production.

According to the data reduction method described in the 'Reduction of inventory data by PCA' section, the following PC scores can be derived by applying PCA. In this study, the result of PCA shows that the first two principal

$$
p = \frac{e^{\left(\begin{array}{l} \beta_0 + \beta_1 * \text{TYP}(1) + \beta_2 * \text{TYP}(2) + \beta_3 * \text{TYP}(3) + \beta_4 * \text{TYP}(4) + \beta_5 * \text{TYP}(5) + \beta_6 * \text{TYP}(6) + \beta_7 * \text{INV1} + \beta_8 * \text{INV2} + \beta_9 * \text{CLR}(1) + \beta_{10} * \text{CLR}(2) + \beta_{11} * \text{CLR}(3) \\ + \beta_{12} * \text{CLR}(4) + \beta_{13} * \text{DPP} + \beta_{14} * \text{PBS} + \beta_{15} * \text{SEQ}(1) + \beta_{16} * \text{SEQ}(2) + \beta_{17} * \text{SEQ}(3) + \beta_{18} * \text{SRT} + \beta_{19} * \text{ACP} + \beta_{20} * \text{RWP} + \beta_{21} * \text{ESD}(1) + \beta_{22} * \text{ESD}(2) + \beta_{23} * \text{ESD}(3) \end{array} \right)}}{1 + e^{\left(\begin{array}{l} \beta_0 + \beta_1 * \text{TYP}(1) + \beta_2 * \text{TYP}(2) + \beta_3 * \text{TYP}(3) + \beta_4 * \text{TYP}(4) + \beta_5 * \text{TYP}(5) + \beta_6 * \text{TYP}(6) + \beta_7 * \text{INV1} + \beta_8 * \text{INV2} + \beta_9 * \text{CLR}(1) + \beta_{10} * \text{CLR}(2) + \beta_{11} * \text{CLR}(3) \\ + \beta_{12} * \text{CLR}(4) + \beta_{13} * \text{DPP} + \beta_{14} * \text{PBS} + \beta_{15} * \text{SEQ}(1) + \beta_{16} * \text{SEQ}(2) + \beta_{17} * \text{SEQ}(3) + \beta_{18} * \text{SRT} + \beta_{19} * \text{ACP} + \beta_{20} * \text{RWP} + \beta_{21} * \text{ESD}(1) + \beta_{22} * \text{ESD}(2) + \beta_{23} * \text{ESD}(3) \end{array} \right)}},
\tag{4}
$$

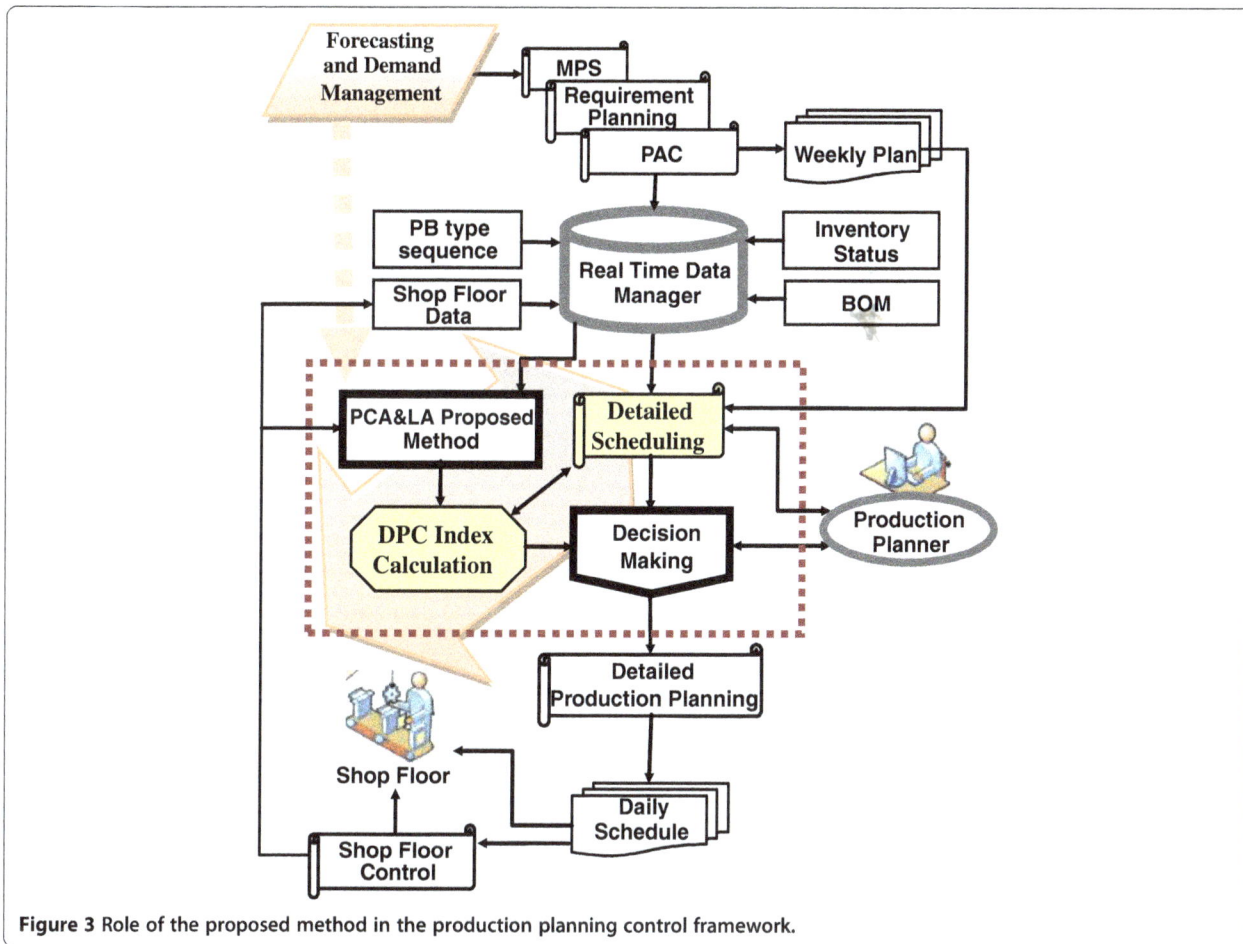

Figure 3 Role of the proposed method in the production planning control framework.

component variables account for about 90% of the total variance of data, and the screen plot shows that the appropriate number of PCs is two. Using these PCA scores, the PC formulas are defined, and then the amount of PCs by the exact amount of each variable every morning can be calculated.

By the results of data reduction, we can make a new shop floor control (SFC) data which can be used for calculating logit more easily and practically. Table 4 shows the sample format of SFC data table which must be created for logistic regression analysis. Using Equation 4,

Table 3 Sample data format of warehouse and line-side inventory levels

Date	TYP	ENGN	AXLE	PIPE	CONT	BODY	DASH	ECUT	STWL	EXST	TRIM	WIRE	SNSR	DMPR	CNGK	HYDK	SUB
09/01	CNG	138	154	129	102	158	96	404	251	157	259	250	451	253	320	-	306
09/01	DM	63	66	67	-	94	41	80	83	46	91	42	88	93	-	49	73
09/01	GLXi	189	159	156	-	150	103	153	-	123	229	266	157	126	-	-	100
09/01	HYD	46	60	72	-	104	48	117	98	69	108	185	205	170	-	27	139
09/01	NEW	27	112	122	-	180	43	174	-	119	164	185	285	219	-	-	142
09/01	X132	127	310	128	-	137	246	256	220	181	250	255	263	268	-	267	302
09/02	CNG	106	153	129	95	152	96	403	250	154	259	259	457	255	300	-	314
09/02	CNG-H	65	62	65	57	75	50	124	126	156	81	167	182	130	236	219	110
09/02	DM	63	64	64	-	139	39	73	81	52	176	49	87	85	-	91	77
09/02	GLXi	171	151	159	-	155	102	158	-	129	266	253	155	126	-	-	104
09/02	NEW	23	117	124	-	184	53	168	-	124	163	182	285	222	-	-	142
09/02	X132	120	243	121	-	139	246	255	223	181	258	259	266	268	-	271	300

ENGN engine; AXLE, rear axle; PIPE, fuel pipe set; CONT, compressed net gas container; BODY, special parts of body; DASH, dashboard or instrument panel; ECUT, electronic central unit; STWL, steering wheel; EXST, exhaust; TRIM, trim parts; WIRE, wiring set; SNSR, sensors set; DMPR, dampers set; SNGK, compressed net gas kit; HYDK, hydraulic kit; SUB, sub assembly required parts.

Table 4 SFC table data format: -input of LR analysis

Week	Working day	TYP	PCs		Daily plan		PBS	SEQ	SRT	ACP	RWP	ESD	ADP	DPC
			INV1	INV2	CLR	DPP								
1	1	CNG	316	166	W	450	180	3	400	385	968	3	446	1
	1	DM	157	44	R	50	50	4	100	86	300	1	51	1
	1	GLXi	127	255	S	100	80	4	100	85	500	3	98	1
	1	HYD	171	36	G	100	80	3	100	65	200	3	105	1
	1	NEW	79	195	S	150	100	1	100	0	430	4	131	0
	1	X132	417	192	W	250	200	2	300	85	1,400	4	212	0
1	2	CNG	293	136	B	200	50	1	100	20	522	2	154	0
	2	CNG-H	208	67	S	150	100	2	150	150	350	3	132	0
	2	DM	253	56	B	250	131	3	130	100	252	3	248	1
	2	GLXi	185	262	W	150	100	4	100	100	395	4	153	1
	2	NEW	66	195	W	100	55	3	100	83	330	4	82	0
	2	X132	409	199	G	250	250	4	400	186	1,105	2	253	1
↓	↓	↓	↓	↓	↓	↓	↓	↓	↓	↓	↓	↓	↓	↓
↓	↓	↓	↓	↓	↓	↓	↓	↓	↓	↓	↓	↓	↓	↓

the probable response variable (DPC) can be calculated by the coefficients derived from logistic regression analysis.

$$logit = 0.93153 - 0.39432 * \mathbf{TYP(1)} - 0.56284 * \mathbf{TYP(2)}$$
$$- 0.7653 * \mathbf{TYP(3)} - 0.56226 * \mathbf{TYP(4)} - 2.28933 * \mathbf{TYP(5)}$$
$$- 1.43129 * \mathbf{TYP(6)} - 0.00502 * \mathbf{INV1} - 0.00157 * \mathbf{INV2}$$
$$+ 0.21325 * \mathbf{CLR(1)} + 0.29959 * \mathbf{CLR(2)}$$
$$- 1.13775 * \mathbf{CLR(3)} - 0.44392 * \mathbf{CLR(4)}$$
$$+ 0.01631 * \mathbf{DPP} - 0.00162 * \mathbf{PBS} + 0.2752 * \mathbf{SEQ(1)}$$
$$+ 0.23753 * \mathbf{SEQ(2)} + 0.20142 * \mathbf{SEQ(3)} - 0.00016 * \mathbf{SRT}$$
$$+ 0.00042 * \mathbf{ACP} - 0.00047 * RWP - 0.01292 * \mathbf{ESD(1)}$$
$$+ 0.48281 * \mathbf{ESD(2)} - 0.02589 * \mathbf{ESD(3)}$$

$$\Rightarrow p = \frac{\exp(logit)}{1 + \exp(logit)}. \tag{5}$$

Table 5 summarizes the test results of null hypothesis in which all the coefficients associated with predictors equal 0. The test statistic $G = 230.037$ with a p-value of 0.000 implies that there is at least one estimated coefficient that is different from 0. The results of Pearson, deviance, and Hosmer-Lemeshow goodness-of-fit tests have been also summarized in Table 5.

In this study, there is insufficient evidence to claim that the LR model does not fit the data adequately because the P-values for all tests are larger than the significance level of 0.05. Therefore, the LR model shown in Equation 5 is appropriate in explaining the DPC prediction.

The association between the response variable and predicted probabilities has been evaluated by some measures such as Somers' D, Goodman-Kruskal's gamma, and Kendall's tau-a in our case; the summary of results is listed in Table 6. The measures indicate that there is a close correspondence between DPC and its predicted probabilities.

The accuracy of the proposed method

Utilizing the discussed PCA and LR model, 42 working weeks of shop floor data (including 1,256 records) were used to evaluate its prediction quality. In addition to real data, Monte Carlo-based simulated data were generated to extend our samples to 100 weeks. The simulation was run under a variety of conditions such as production line, seasonal demand, and probable disruption in production line. Every 4 weeks (1 month), the outcomes of classical detailed planning were compared with the corresponding outcomes of the proposed method. These results have been reported in Table 7, including 10

Table 5 Goodness-of-fit tests

Test		df	P value
All slopes are zero (G)	230.037	23	0.000
Pearson (χ^2)	514.679	584	0.982
Deviance (χ^2)	631.625	584	0.947
Hosmer-Lemeshow (χ^2)	3. 474	8	0.901

Table 6 Measures of association

Pairs	Number	Percent	Summary measures	P value
Concordant	67,781	84.6	Somers' D	0.69
Discordant ties	12,163	15.2	Goodman-Kruskal's gamma	0.70
	151	0.2	Kendall's tau-a	0.30
Total	80,095	100		

Table 7 The comparison of DSS method accuracy

Applied methods		Total observations	Actual daily production capability		Predicted daily production capability			Daily planning accuracy (%)
			DPC = 1	DPC = 0	DPC = 1	DPC = 0	Error	
Instance 1	Classic DSS method	124	81	43	124	0	43	65
	Revised proposed method	124	117	7			7	94
Instance 2	Classic DSS method	136	88	48	136	0	48	65
	Revised proposed method	136	127	9			9	93
Instance 3	Classic DSS method	110	81	32	110	0	32	74
	Revised proposed method	110	105	5			5	95
Instance 4	Classic DSS method	116	93	23	116	0	23	80
	Revised proposed method	116	109	7			7	94
Instance 5	Classic DSS method	84	62	22	64	0	25	74
	Revised proposed method	84	76	8			8	90
Instance 6	Classic DSS method	64	39	25	64	0	25	61
	Revised proposed method	64	55	9			9	86
Instance 7	Classic DSS method	88	75	13	88	0	13	74
	Revised proposed method	88	86	2			2	98
Instance 8	Classic DSS method	101	66	35	101	0	66	74
	Revised proposed method	101	95	6			6	94
Instance 9	Classic DSS method	76	33	43	76	0	43	43
	Revised proposed method	76	69	7			7	91
Instance 10	Classic DSS method	121	80	41	121	0	41	66
	Revised proposed method	121	120	1			1	99

instances which have been selected from the worst to the best states.

From the perspective of daily planning accuracy, the logistic regression model correctly identified 109 of 124 observations (refer to instance 1). The accuracy of each method can be simply calculated by dividing the number of observed actual productions, which are respondents of production plans (DPC = 1), to the total number of production plans. As shown in Table 7, by the proposed DSS method, more reliable detailed production plans can be submitted than by the classic method.

Conclusions

This study presents an application of statistical multivariate method together with the solver module in production activity control of an Iranian automotive manufacturer and introduces a revised decision support system which can provide a productive tool for knowledge workers to offer more reliable detailed production plans.

The proposed method is based on the use of principal component analysis to reduce the extensive dimension of shop floor data and logistic regression analysis to make a predictive tool and pre-check of daily production plan capability to improve the effectiveness of decision

making. In this case study, it is shown that the revised DSS works more reliably and more accurately.

For future studies, either prediction accuracy or data reduction techniques may be improved by applying other specialized models of logistic regression. Manufacturers can also further adjust the proposed prediction models to accord with their production environments and data availability.

Competing interests
The authors declare that they have no competing interests.

Authors' contributions
SM has made substantial contributions to the concept, design, and acquisition of data and analysis and interpretation of data. MB has been involved with the main idea, drafting the manuscript, revising it critically for important intellectual content, and has given final approval of the version to be published. Both authors read and approved the final manuscript.

Authors' information
SM is a PhD student at Payme Noor University. MB is an associate professor at Shahed University.

Acknowledgments
The authors would like to thank the Logistic and Industrial Engineering department and Production Logistics department of SAIPA Company for supporting this case study in data acquisition and serious cooperation.

Author details

[1]Industrial Engineering Department, Payme Noor University (PNU), Tehran, Iran. [2]Industrial Engineering Department, Shahed University, Tehran, Iran.

References

Aguilera AM, Escabias M, Valderrama MJ (2006) Using principal components for estimating logistic regression with high-dimensional multi collinear data. Computational Statistics & Data Analysis 50:1905–1924

Aytug H, Lawley MA, McKay K, Mohan S, Uzsoy R (2005) Executing production schedules in the face of uncertainties: a review and some future directions. European Journal of Operational Research 161(1):86–110

Bewick V, Cheek L, Ball J (2005) Statistics review 14: logistic regression. Critical Care 9(1):112–118. doi:10.1186/cc3045

Brucker P (2007) Production control: a universal conceptual framework. Production Planning and Control 1(1):3–16

Caricato P, Grieco A (2009) A DSS for production planning focused on customer service and technological aspects. Robotics and Computer-Integrated Manufacturing 25:871–878

Maravelias CT, Sung C (2009) Integration of production planning and scheduling: overview challenges and opportunities. Computers Chemical Engineering 33:1919–1930

Cunha PF, Wiendahl HP (2005) Knowledge acquisition from assembly operational data using principal components analysis and cluster analysis. CIRP Annals - Manufacturing Technology 54(1):27–30

Farrella RR, Maness TC (2005) A relational database approach to a linear programming-based decision support system for production planning in secondary wood product manufacturing. Decision Support Systems 40:183–196

Framinan JM, Ruiz R (2010) Architecture of manufacturing scheduling systems: literature review and an integrated proposal. European Journal of Operational Research 205:237–246

Günther HO, van Beek P (eds) (2003) Advanced planning and scheduling solutions in process industry. Springer, Berlin

Hatzikonstantinou O, Athanasiou E, Pandelis DG (2012) Real-time production scheduling in a multi-grade PET resin plant under demand uncertainty. Computers and Chemical Engineering 40:191–201

Hosmer DW, Lemeshow S (2000) Applied Logistic Regression, 2nd edn. John Wiley & Sons, New York

Jahangirian M, Eldabi T, Naseer A, Stergioulas LK, Young T (2010) Simulation in manufacturing and business: a review. European Journal of Operational Research 203:1–13

Jolliffe IT (2002) Principal component analysis, 2nd edn. Springer Series in Statistics, Springer Verlag, New York

Jourdan L, Basseur M, Talbi E (2009) Hybridizing exact methods and metaheuristics: a taxonomy. European Journal of Operational Research 199(3):620–629

Ko CH, Wang SF (2010) GA-based decision support systems for precast production planning. Automation in Construction 19(7):907–916

McKay KN, Black GW (2007) The evolution of a production planning system: a 10-year case study. Computers in Industry 58:756–771

McKay KN, Wiers VCS (2003) Integrated decision support for planning scheduling and dispatching tasks in a focused factory. Computers in Industry 50(1):5–14

McKay KN, Wiers VCS (2004) Practical production control: a survival guide for planners and schedulers. J Ross Publishers, Boca Raton

Mele FD, Musulin E, Puigjaner L (2005) Supply chain monitoring: a statistical approach. Computer Aided Chemical Engineering 20:1375–1380

Mok PY (2009) A decision support system for the production control of a semiconductor packaging assembly line. Expert Systems with Applications 36:4423–4424

Peidroa D, Mulaa J, Polera R, Verdegay JL (2009) Fuzzy optimization for supply chain planning under supply, demand and process uncertainties. Fuzzy Sets and Systems 160:2640–2657

Ribas I, Leisten R, Framin JM (2010) Review and classification of hybrid flowshop scheduling problems from a production system and a solutions procedure perspective. Computers & Operations Research 37:1439–1454

Rolo M, Martinez E (2012) Agent-based modeling and simulation of an autonomic manufacturing execution system. Computers in Industry 63:53–78

Ross JW, Bernardo AL (2011) Single and parallel machine capacitated lotsizing and scheduling: new iterative MIP-based neighborhood search heuristics. Computers & Operations Research 38(12):1816–1825

Sharma S (1996) Applied Multivariate Techniques. John Wiley & Sons, New York

Simchi-Levi D, Kaminsky P, Simchi-Levi E (2008) Designing and managing the supply chain: concepts, strategies and case studies, 3rd edn. McGraw-Hill/Irwin, Boston

Sotiris G, Athanasios S, Ilias T (2008) A decision support system for detailed production scheduling in a Greek metal forming industry. MIBES Transactions 2(1):41–59

Torabi SA, Ebadian M, Tanha R (2010) Fuzzy hierarchical production planning (with a case study). Fuzzy Sets and Systems 161(11):1511–1529

Vargas V, Metters R (2011) A master production scheduling procedure for stochastic demand and rolling planning horizons. Int J Production, Economics 132:296–302

Verderame PM, Floudas CA (2009) Integrated operational planning and scheduling under uncertainty. Computer Aided Chemical Engineering 26:381–386

Volling T, Spengler TS (2011) Modeling and simulation of order-driven planning policies in build-to-order automobile production. International Journal of Production Economics 131(1):183–193

Vollmann ET, Berry WL, Whybark CP, Jacobs CP (2005) Manufacturing planning and control for supply chain management, 5th edn. McGraw-Hill Education, Singapore

Yao S, Jiang Z, Li N (2012) A branch and bound algorithm for minimizing total completion time on a single batch machine with incompatible job families and dynamic arrivals. Computers & Operations 39(5):939–951

20

A hybrid algorithm optimization approach for machine loading problem in flexible manufacturing system

Vijay M. Kumar[1*], A. N. N. Murthy[2] and K. Chandrashekara[3]

Abstract

The production planning problem of flexible manufacturing system (FMS) concerns with decisions that have to be made before an FMS begins to produce parts according to a given production plan during an upcoming planning horizon. The main aspect of production planning deals with machine loading problem in which selection of a subset of jobs to be manufactured and assignment of their operations to the relevant machines are made. Such problems are not only combinatorial optimization problems, but also happen to be non-deterministic polynomial-time-hard, making it difficult to obtain satisfactory solutions using traditional optimization techniques. In this paper, an attempt has been made to address the machine loading problem with objectives of minimization of system unbalance and maximization of throughput simultaneously while satisfying the system constraints related to available machining time and tool slot designing and using a meta-hybrid heuristic technique based on genetic algorithm and particle swarm optimization. The results reported in this paper demonstrate the model efficiency and examine the performance of the system with respect to measures such as throughput and system utilization.

Keywords: Flexible manufacturing system, Production planning, Loading, Hybrid algorithm optimization

Background

In recent years, competitive market conditions coerce manufacturing firms to enhance response times and flexibility in all operations. Flexible manufacturing systems (FMSs) have been proved to respond this challenge positively because of their ability to produce a variety of parts using the same system in the shortest possible lead time. According to Stecke (1983), FMS is characterized as an integrated, computer-controlled, complex arrangement of automated material handling devices and computer numerically controlled (CNC) machine tools that can simultaneously process medium-sized volumes of a variety of part types. The highly integrated FMS offers the opportunity to combine the efficiency of transfer line and the flexibility of a job shop to best suit the batch production of mid-volume and mid-variety of products. However, flexibility has a cost, and the capital investment sustained by firms to acquire such systems is generally

very high. Therefore, adequate planning of FMS during its development phase is pivotal so as to evaluate the performance of the system and justify the investment incurred. Prior to production, careful operational planning is essential to establish how well the system interacts with the operations over time. Hence, successful operation of FMS requires more intense planning as compared to any conventional production system. The decisions related to FMS operations can be broadly divided into pre-release and post-release decisions. Pre-release decisions include the FMS operational planning problem that deals with the pre-arrangement of jobs and tools before the processing begins, whereas post-release decisions deal with the scheduling problems. Pre-release decisions, *viz* machine grouping, part type selection, production ratio determination, resource allocation, and loading problems, must be solved while setting up an FMS. Amongst pre-release decisions, machine loading is considered as one of the most vital production planning problems since operational effectiveness of FMS largely depends on it. Loading problem, in particular, deals with allocation of jobs to various machines under technological

* Correspondence: vijayjss@yahoo.com
[1]Department of Mechanical Engineering, JSS Academy of Technical Education, Bangalore, 560 060, India
Full list of author information is available at the end of the article

constraints with the objective of meeting certain performance measures. Therefore, the problem is combinatorial in nature and happens to be non-deterministic polynomial-time (NP)-hard.

Researchers recognized early on that not all problems can be solved this quickly, and they had a hard time figuring out exactly which ones could and which ones could not. There are several so-called NP-hard problems, which cannot be solved in polynomial time, even though nobody can prove a super-polynomial lower bound.

A decision problem is a problem whose output is a single Boolean value: YES or NO. There are three classes of decision problems:

- *P* is the set of decision problems that can be solved in polynomial time. Intuitively, P is the set of problems that can be solved quickly.
- *NP* is the set of decision problems with the following property: If the answer is YES, then there is a *proof* of this fact that can be checked in polynomial time. Intuitively, NP is the set of decision problems where it can verify a YES answer quickly if we have the solution for it.
- *Co-NP* is the opposite of NP. If the answer to a problem in co-NP is NO, then there is a proof of this fact that can be checked in polynomial time.

The induction of the model is NP. There is a set of *m* input values that produces a TRUE output as a proof of this fact; the proof can be checked by evaluating the model in polynomial time.

Literature review

Formulations of loading problems in FMS and solution techniques have drawn the attention of researchers for quite some time. FMS planning problem was formulated as nonlinear 0–1 mixed integer programming by Stecke (1983), and subsequently, a branch-and-bound algorithm was developed by Berrada and Stecke (1986). Although analytical and mathematical programming-based methods are robust in applications, yet they tend to become impractical when the problem size increases. This motivated the researchers to develop fast and effective heuristics for solving loading problems in large-sized FMSs. One of the important heuristics based on the concept of essentiality ratio for maximization of throughput and minimization of system unbalance simultaneously was proposed by Mukhopadhyay et al. (1992). Later on, Tiwari et al. (1997) developed heuristics using fixed, predetermined job ordering rules as an input while solving loading problems. Moreno and Ding (1993) solved the loading problem using standard sequencing rules such as the shortest processing time (SPT), longest processing time (LPT), first in, first out (FIFO), and last in, first-out (LIFO) and

established that the SPT rule works well in comparison to other rules. The major limitation of heuristics lies in the fact that their inability to estimate the results in a new or completely changed environment as they are generally rule-based and mostly rely on empirical data. Therefore, numerous researchers have used meta-heuristic approaches for solving the machine loading problem. Usually, FMS loading problem seeks a solution that optimizes multiple objectives simultaneously. In this regard, Kumar and Shanker (2000), Tiwari and Vidyarthi (2000), and Swamkar and Tiwari (2004) have addressed a machine loading problem having the bi-criterion objectives of minimizing system unbalance and maximizing the throughput using a hybrid algorithm based on Tabu search and simulated annealing (SA). Genetic algorithm (GA)-based approaches for loading problems is found to ensure an optimal solution with less computational effort (Tiwari et al. 2007).

Since the objective of this paper is to propose an efficient evolutionary search heuristic to solve problems pertaining to job selection and machine loading in random FMS to optimize the system imbalance and throughput simultaneously, only the relevant literature are reviewed in this section. Tiwari and Vidyarthi (2000) proposed a GA-based heuristic to solve the machine loading problem of a random-type FMS. The proposed GA-based heuristic determines the part-type sequence and the operation-machine allocation that guarantee the optimal solution to the problem, rather than using fixed, predetermined part sequencing rules. Swarnkar and Tiwari (Swamkar and Tiwari 2004) proposed a generic 0–1 integer programming formulation and a hybrid algorithm based on Tabu search, and SA is employed to solve the problem. Prakash et al. (Prakash et al. 2008) proposed a special immune algorithm (IA) named 'modified immune algorithm.' This method is capable of learning and memory acquisition, improves some issues inherent in existing IAs, and proposes a more effective IA with reduced memory requirements and reduced computational complexity. Chan et al. (2005) proposed a fuzzy goal programming approach to model the machine tool selection and operation allocation problem of FMSs. The model is optimized using an approach based on artificial immune systems, and the results of the computational experiments are reported. Tripathi et al. (2005) proposed a multi-agent-based approach for solving the part allocation problems in FMSs that can easily cope with the dynamic environment. Kumar et al. (2006) extended the simple GA and proposed a new methodology, a constraint-based GA to handle a complex variety of variables and constraints in a typical FMS loading problem. Yogeswaran et al. (2008, in press) proposed a hybrid algorithm using genetic algorithm and simulated annealing algorithm for their problem. They also proposed efficient machine loading heuristics. The machine loading problem of an FMS is well known for its complexity. This problem encompasses various types of flexibility aspects pertaining to part selection and operation assignments along with constraints ranging from simple algebraic to potentially very complex conditional constraints. In this paper, a hybrid optimization algorithm involving GA and particle swarm optimization (PSO) is proposed to solve this problem. The literature survey clearly supports the proposal of an efficient heuristic to this problem. Besides that, the justification to adopt PSO is mainly due to its performance in solving scheduling problems.

The FMS under consideration in this paper consists of a number of multifunctional CNC machines, tools with the potential to execute several operations. The jobs are available in batches and arrive in random sequences with different requirements for processing. The batch size, number of operations, processing time, and number of tool slots needed for each job are known initially. There are two types of operations accessible for a job, namely essential operation - the job can be performed only in a particular machine - and optional operation - the job can be performed in a number of machines available, which gives the flexibility in the routing of the jobs. The FMS considered has a maximum of six multifunctional machines with each having 960 min of available processing time (8 h = one shift) and six tool slots.

Numerous methods based on mathematics, heuristics, and meta-heuristics have been suggested by the researchers in the pursuit of obtaining quality solutions to loading problems and reducing computational burden. However, these approaches are hardly capable of producing optimal/near optimal solutions or require excessive computational efforts to arrive at quality solutions. In order to alleviate these difficulties, an attempt has been made in this paper to propose a multi-objective meta-heuristic technique based on a hybrid algorithm using genetic and particle swarm optimization (HAO) to solve the machine loading problem of a random FMS with the objective of minimization of system unbalance and maximization of throughput while satisfying the constraints related to available machining time and tool slots. However, GA has an inherent drawback of trapping at local optimum due to appreciable reduction in velocity values as iteration proceeds and hence reduces solution variety. This drawback has been addressed effectively by incorporating mutation, a commonly used operator in GA, to improve the solution quality.

The remainder of this paper is organized as follows: the 'Problem description' section formally defines the problem studied in this paper along with the objectives and assumptions made to solve the problem. In the 'Results and discussion' section, results of benchmark

problems from the open literature are compared with those from the proposed method to illustrate its advantage over other methods. Conclusions drawn from this study are summarized and direction for future research is outlined in the 'Conclusions' section. Finally, the proposed hybrid algorithm based on genetic algorithm optimization (GAO) and PSO is presented in the 'Methods' section.

Problem description

The loading problem in manufacturing deals with selecting a subset of jobs from a set of all the jobs to be manufactured and assigning their operations to the relevant machines in a given planning horizon with the technological constraints in order to meet certain performance measures such as minimization of system unbalance and maximization of throughput. System unbalance can be defined as the sum of unutilized or over-utilized times on all the machines available in the system, whereas throughput refers to the summation of the batch size of the jobs that are to be produced during a planning horizon. Minimization of system unbalance is equivalent to maximization of machine utilization. The processing time and tool slots required for each operation of the job and its batch size are known beforehand. There are two types of operations associated with the part types: essential and optional. Essential operations can be carried out on a particular machine using a certain number of tool slots, while the optional operation can be performed on a number of machines with the same or different processing time and tool slots. The FMS under consideration derives its flexibility in the selection of a machine for optional operation of the job. Generally, the complexity of these problems depends on whether the FMS is of a dedicated type or a random type. A dedicated FMS is designed to produce a rather small family of similar parts with a known and limited variety of processing requirements, while in a random-type system, a large family of parts having a wide range of characteristics with random elements is produced and the product mix is not completely defined at the time of installing the system. This paper addresses the loading problem in a random FMS. The proposed approach has been tested on problems pertaining to three sizes of FMSs (the details are given in Table 1). The details of data related to problem 1 of FMS type 1 (jobs, batch size, unit processing time, machine

Table 1 Details of different FMS scenarios

FMS type	Number of machines	Available time on each machine (min)	Number of tool slots on each machine
FMS 1	4	480, 480, 480, 480	5, 5, 5,5
FMS 2	5	960, 960, 960, 960, 960	10, l2, 10, 12, 10
FMS 3	6	960, 960, 960, 960, 960, 960	14,14,14,14, 14,16

options, number of tool slots, etc.) having four machines are given in Table 2.

In order to minimize the complexities in analyzing the problem for a practical FMS as depicted in Figure 1, the mathematical model is based on the following assumptions:

- Initially, all the jobs and machines are simultaneously available.
- Processing time required to complete an entire job order is known *a priori*.
- Job undertaken for processing is to be completed for all its operation before considering a new job. This is called non-splitting of the job.
- Operation of a job, once started on a machine, is continued until it is completed.
- Transportation time required to move a job between machines is negligible.
- Sharing and duplication of tool slots is not allowed.

Objectives

The overall objective function is represented as: Maximize $'F' = F_1 + F_2$

$$\frac{\sum^{M} F = m = 1 \sum^{J} j = 1 \sum^{O_j} o = 1 \; B_j P_{jom} X_{jom}}{\sum^{M} m = 1 T_m}$$

$$+ \frac{\sum_{j=1}^{J} B_j X_j}{\sum_{j=1}^{J} B_j}, \tag{1}$$

where F_1 indicates minimization of system imbalance which is equivalent to maximizing the system utilization, and F_2 indicates the maximization of throughput which is equivalent to maximizing the system efficiency.

Constraints

$$\sum^{M} m = 1 \sum^{O_j} j = 1 B_j P_{jom} X_{jom} \leq T_m, \tag{2}$$

where $m = 1, 2, \ldots, M$ ensures that overloading of machines is not permitted.

$$\sum_{J=1} \sum_{j_o} G \leq 1, \tag{3}$$

where $J = 1, 2, \ldots, J$ and $o = 1, 2, \ldots, O_j$ ensure that a particular operation of a job is done only on one machine, and $m = 1, 2, \ldots, M$ ensures that the jobs will be loaded only when there is availability of tool slots on each machine.

$$\sum^{O_j} o = 1 \sum^{M} m = 1 X_{jom} = X_j O_j, \tag{4}$$

where $j = 1, 2, \ldots, J$ ensures that the job cannot be split.

Results and discussion

The proposed HAO algorithm for the FMS loading problem is coded in Visual C++ and implemented in a Pentium IV PC. The performance of the HAO algorithms is evaluated using ten benchmark problems available in the open literature representing three different FMS scenarios. In solving the problems, parameters are set as

Table 2 Detailed description of jobs of problem number 1 (FMS 1)

Job number	Batch size	Operation number	Machine number	Unit processing time (min)	Tool slots needed	Total processing time
1	10	1	4	16	1	160
		2	4, 2, 3	7, 7, 7	1, 1, 1	70
2	13	1	12, 3	25	1	325
		2	2,1	17	1	221
		3	1	24	1	312
3	14	1	4, 1	25, 26	2, 2	364
		2	3	11	3	154
4	7	1	3	24	1	168
		2	4	19	1	133
5	9	1	1, 4	25	1	255
		2	4	25	1	255
		3	2	22	1	198
6	8	1	3	20	1	160
7	9	1	2, 3	22, 22	2, 2	198
		2	2	25	1	225

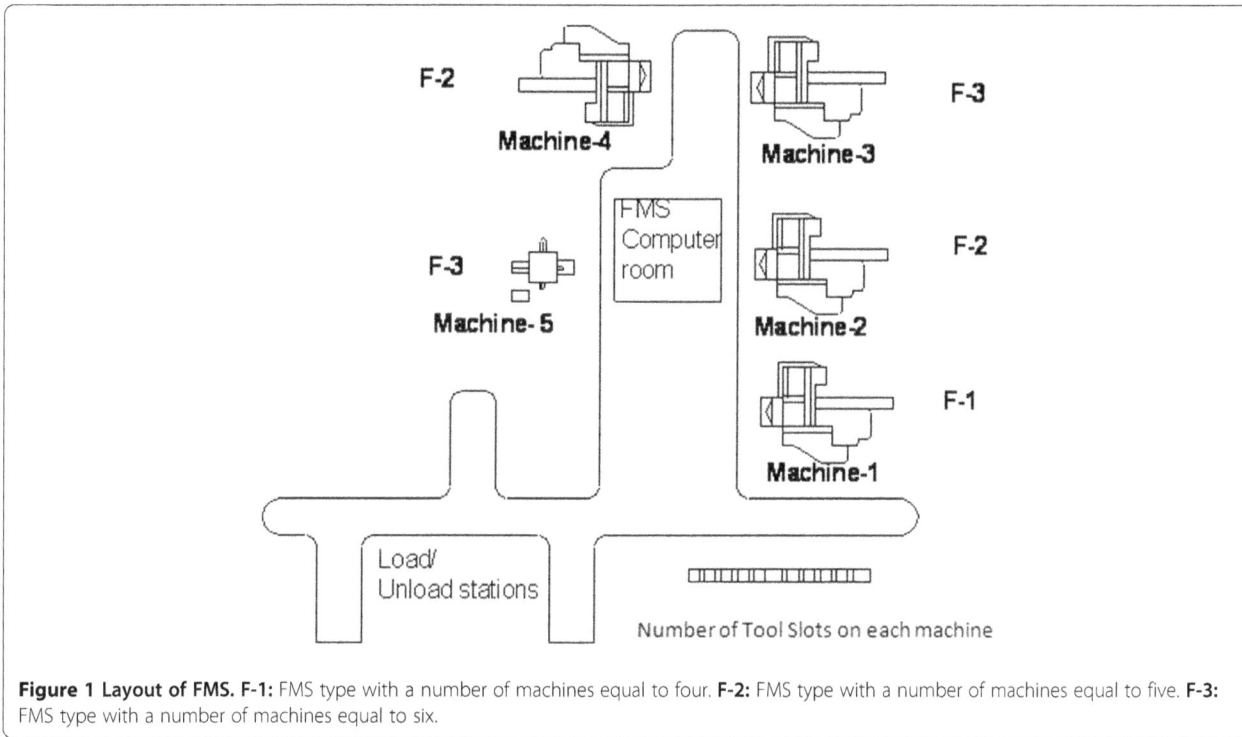

Figure 1 Layout of FMS. F-1: FMS type with a number of machines equal to four. **F-2:** FMS type with a number of machines equal to five. **F-3:** FMS type with a number of machines equal to six.

population size = 50, $w = 0.85$, $\alpha = 0.9$, and $c1 = c2 = 2$ after a thorough examination of the results.

One of the drawbacks of HAO is its premature convergence. In order to alleviate such difficulties and improve solution quality, the mutation operator is adopted from the genetic algorithm. The results of the proposed HAO with mutation (HAOM) are compared with those of the standard HAO, and four standard sequencing rules such as LPT, SPT, FIFO, and LIFO are shown in Table 3. The results are also tabulated for exact solutions using LINDO software for the same data sets. The results indicate that HAOM improves the solution quality and outperforms other techniques in most of the instances. The combined objective of both HAO and HAOM is summarized in Table 4. The last column of Table 4 shows the percentage improvement of HAOM over HAO. The result indicates that the maximum improvement of 4.83 can be made using HAOM.

Figure 2 illustrates the convergence behavior of the HAOM. It can be observed from the figure that the algorithm can achieve the optimal solution after 33 iterations for problem number 7 because no further improvement is observed beyond 33 iterations.

Conclusions

This paper presents an efficient and reliable meta-heuristic-based approach to solve the FMS loading problem. The designed and proposed algorithm based on HAO defined the trade-off regions between the two objectives. Extensive computational experiments have been conducted on different benchmark problems to show the effectiveness of the proposed approach. A comparative study has been carried out for the same problem with similar objective functions and constraints, and the computational experience manifests that the proposed meta-heuristic approach based on HAO outperforms the existing methodologies as far as the solution quality is concerned with reasonable computational efforts. To avoid premature convergence, HAO algorithm is modified in this paper with the introduction of mutation operation. The performance of this algorithm is compared with that of the standard HAO, and the percentage improvement of up to 4.83 is possible using HAOM over HAO.

It is clear from this research that the machine loading problem can be solved using a hybrid algorithm-based heuristic that can tackle the problem in a synergistic way. By combining effective GAO and PSO, resource allocation can be done efficiently. This research had also highlighted the efficiency of the HAOM in the optimization process. The HAOM reduced the time to reach the best fitness value by a considerate amount in most cases. The results presented in Table 5 clearly show that the HAOM algorithm is comparable to the better performing algorithms reported in the literature and it obtains the best results obtained so far at a faster rate. The future work is to fine-tune the parameters of GAO and PSO. In the future, the study can be extended to solve the loading problem by considering more realistic variables and constraints such as availability of

Table 3 Summary of results

Problem number	Number of part types	SPT (SU; TH)	LPT (SU; TH)	LIFO (SU; TH)	FIFO (SU; TH)	Hybrid algorithm (SU; TH)	HAOM (SU; TH)	Branch-and-bound technique using LINDO software
1	7	929; 65	814; 62	459; 36	871; 81	459; 66	449; 65	440; 66
2	6	804; 47	619; 52	467; 62	650; 50	323; 53	313; 51	310; 48
3	6	819; 51	659; 51	819; 51	681; 51	319; 51	314; 51	310; 48
4	8	1,122; 86	950; 81	1,662; 107	932; 86	590; 75	566; 77	555; 71
5	6	896; 54	689; 56	1,398; 110	654; 54	304; 54	301; 56	299; 58
6	5	608; 42	584; 40	2,640; 42	584; 45	296; 46	286; 48	285; 46
7	10	1,388; 92	1,106; 101	1,007; 106	1,118; 80	601; 82	589; 89	585; 82
8	12	1,417; 98	948; 142	707; 127	1,245; 132	790; 115	766; 116	765; 118
9	8	1,154; 87	990; 85	1,699; 88	940; 88	589; 72	569; 78	566; 71
10	14	1,532; 110	1,459; 158	928; 112	1,218; 144	871; 128	845; 136	841; 131
11	6	876; 55	685; 54	1,385; 98	651; 51	301; 52	300; 51	300; 57
12	8	1,110; 84	948; 78	1,658; 101	938; 85	791; 114	764; 114	762; 115
13	10	1,310; 89	1,110; 99	1,001; 102	1,115; 81	602; 83	588; 88	588; 84
14	12	1,410; 95	942; 140	701; 122	1,240; 130	792; 116	762; 114	762; 115
15	14	1,512; 112	1,448; 148	921; 110	1,210; 142	867; 125	841; 131	839; 129

Obtained using different sequencing rules, HAOM, and branch-and-bound technique using the LINDO software for system unbalance (SU) and throughput (TH). SPT, shortest processing time; LPT, longest processing time; LIFO, last in, first-out; FIFO, first in, first out; HAOM, hybrid algorithm with mutation.

pallets, jigs, fixtures, AGVs, etc. in addition to tool slots and machining time.

Methods
Genetic algorithms
GAs are evolutionary programs and adaptive search and optimization algorithms based on the mechanics of natural selection and natural genetics. These are robust in complex search spaces and are versatile in their application. GA is a search and optimization technique operated on Darwin's principle of the 'survival of the fittest,' where weak individuals die before reproducing, while stronger ones survive, bear many offspring, and breed children, which often inherit qualities that are, in many cases,

superior to their parents' qualities. GA evolves with a population of strings created randomly. Each string is evaluated in the population. There are three main GA operators: reproduction, crossover, and mutation. The reproduction is an operator in which the individual strings are copied according to their objective values which results in more highly fit individuals and less weak individuals in the intermediate mating pool. The reproduction operator is followed by the crossover operator which is done in two steps: First, the members of the mating pool are mated at random. Second, each pair of mating individuals undergoes crossover with respect to one or more crossing sites in which portions of the strings are interchanged between pairs. The mechanics of reproduction and crossover, though, seem to be simple; the combined action provides the GA with much of its

Table 4 Comparison of combined objective obtained using HAO and HAOM

Problem number	Number of part types	HAO	HAOM	% Importance of HAOM over HGAO
1	7	1.275	1.275	0.00
2	6	1.581	1.581	0.00
3	6	1.410	1.410	0.00
4	8	1.877	1.882	0.26
5	6	1.645	1.645	0.00
6	5	1.611	1.611	0.00
7	10	1.715	1.776	3.43
8	12	1.696	1.772	1.51
9	8	1.901	1.901	0.00
10	14	1.516	1.593	4.83

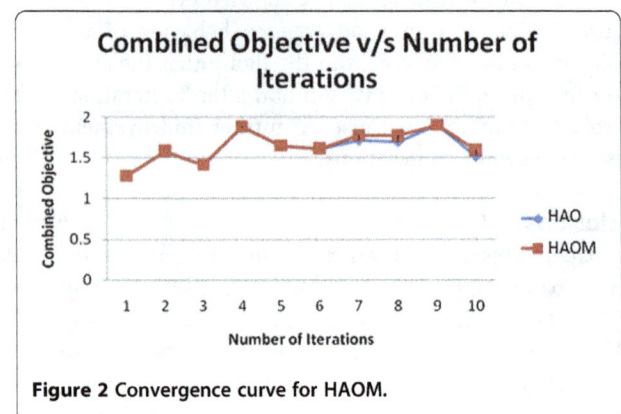

Figure 2 Convergence curve for HAOM.

Table 5 Performance comparison between HAO and HAOM with respect to computation time

Data set	Number of part types	CPU computation time for HAO (s)	CPU computation time for HAOM (s)
1	7	1.627	1.426
2	6	1.423	1.420
3	6	1.428	1.410
4	8	1.781	1.666
5	6	1.424	1.410
6	5	1.395	1.390
7	10	1.923	1.788
8	12	2.122	2.010
9	8	1.788	1.688
10	14	2.315	2.210

CPU, central processing unit; HAO, hybrid algorithm optimization; HAOM, HAO with mutation.

searching power. Mutation plays a secondary role in the GA and tries to ensure potential solutions (individuals). The other operator's reproduction and crossover provide solution convergence, avoiding local optima. The resulting new population is then further evaluated and tested for termination. The termination criteria are designed based on the available response time within which the solution is to be obtained or based on an expected performance level. If the termination criteria are not met, the new population is again produced by the above three genetic operators and evaluated. This procedure is continued and repeated until the termination criteria are met.

Coding

If the binary coding system is considered for making up chromosomes directly, then the digit of the chromosome chain will reach $n \times m$, and the genes will not be independent of each other. This coding system sets the line number of the element valued 1 in each row of matrix X as genes. A natural number is adopted to code the genes since they are independent with each other. The gene code $k1, k2, \ldots, kj, \ldots, kn$, where $k_j \in [1;m]$, is a repeatable positive integer. The gene code is also the number of the machine on which every job is processed; hence, the method is named as genetic algorithm based on the machine code.

Select initial population

Firstly, generate N positive integer-coded n-digit chromosome chains randomly, and set them as the initial population.
   Calculate x(i,j)

$$\forall j \in [1, n] \text{and} \ i \in [1, m], \text{if } kj = i, \text{then} \ x(i, j)$$
$$= 1, \text{else} \ x(i, j) = 0. \tag{5}$$

Reproduce

The fitness function is obtained by means of transforming the objective function, i.e.,

$$\text{let} F(i) = \alpha \exp(-\beta G(i)), \tag{6}$$

where α and β are positive real numbers, and the selection tactics is roulette wheel selection. Assume that $P(i)$ is the selection probability of individual i, then

$$P(i) = \frac{F(i)}{\sum_{j=1}^{n} F(k)}, \text{where } i = 1, 2, \ldots, N. \tag{7}$$

Let $S(0)=0$, then

$$S(i) = P(1) + P(2) + \ldots + P(i), \text{where } i = 1, 2, \ldots, N. \tag{8}$$

Generate N random real numbers ζs which are uniformly distributed between 0 and 1, that is,

$$\zeta s \in U(0, 1), \text{where } s = 1, 2, \ldots, N. \tag{9}$$

If $S(i - 1) < \zeta s < S(i)$, therefore, individual i is sent to the . The objective function is as follows:

$$G(i) = \max\left[\sum x(1, j)t(j), \ldots, \sum x(k, j)t(j), \ldots, \sum x(m, j)t(j)\right]. \tag{10}$$

Crossover

The crossover is carried with the combination of two genes in proper order. The coding method used in this paper makes the crossover completely independent of the genes between each other; if k_j (1,m), the crossover method could adopt a common two-point crossover whose advantages include making message exchange between genes more abundant and obtaining the best solution quickly. Consider a crossover between individuals A and B, randomly select two positions from the chromosome chain, and exchange the chromosome between two positions, thereby generating two child chromosome chains.

   The crossover is as follows:

$$A1|A2|A3 X B1|B2|B3 \Rightarrow A1|B2|A3; B1|A2|B3,$$

where the symbol | represents the position of the crossover point selected randomly, and signal X represents the crossover operation.

Mutation

First, generate a one-digit positive integer, $k_j \in [1; m]$, randomly, then replace the old one when mutating. If k_j is equal to the old one, then select a new positive integer again until they are different; the efficiency of the mutation could be greatly improved using the method. The optimization procedure is explained in Figure 3.

Particle swarm optimization

PSO is a population-based, bio-inspired optimization method. It was originally inspired in the way crowds of individuals move towards predefined objectives, but it is better viewed using a social metaphor. Individuals in the population try to move towards the fittest position known to them and to their informants, that is, the set of individuals of their social circle. The objective is to maximize a fitness function. The structure of the proposed PSO algorithm is as follows:

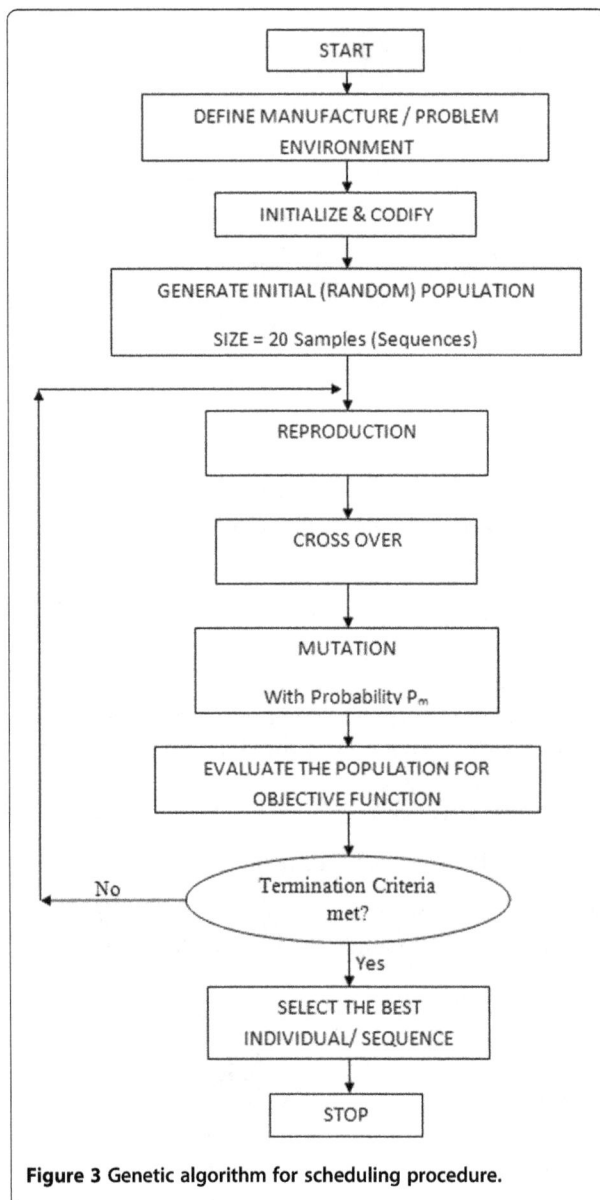

Figure 3 Genetic algorithm for scheduling procedure.

$t \to 0$;
for $(k = 1, N)$
     Generate P_k^t ;
     Evaluate $Z P_k^t$;
e $P_k^t \to P_k^t$
$G^t \to P$ having max $\{Z (^e P_k^t), k\, 1, N\}$
for $(k = 1, N)$
     Initialize v_k^t

//iterative improvement process
do {
     for $(k = 1, N)$
          update Position P_k^{t+1}
          update velocity v_k^{t+1}
     Apply local search on all particle positions;
     Evaluate all particles;
     update e P_k^t and G^{t+1} , $(k = 1, N)$;
     $t \to t + 1$;
     }( ) while $t < t_{max}$
Output G^t

Solution representation

One of the most important issues when designing the HAO lies on its solution representation. In order to construct a direct relationship between the problem domain and the HAO chromosomes for the FMS loading problem, a number of dimensions for n number of jobs are considered. In other words, each dimension represents a typical job. In addition, the chromosome $X_i^t = \left(x_{1i}^t, x_{2i}^t, \ldots, x_{in}^t\right)$ corresponds to the continuous position values for n number of jobs in the loading problem. The particle itself does not present a permutation. Instead, we use the smallest position value (SPV) rule to determine the sequence implied by the position values x_{ij}^t of particle X_i^t . Table 6 illustrates the solution representation of particle X_i^t for the FMS loading problem together with its corresponding velocity and sequence. According to the SPV rule, the SPV is $x_{i1}^t = 0.11$, so the dimension $j = 1$ is assigned to the first job $x_{i1}^t = 4$ in the sequence; the second SPV is $x_{i2}^t = 0.57$, so the dimension $j = 2$ is assigned to be the second job $x_{i2}^t = 6$ in

Table 6 Solution representation of chromosome x_i^t in HAO

Dimension j	x_i^t	v_{ij}^t	Job sequence
1	1.67	2.98	4
2	2.82	−0.87	6
3	1.23	1.51	7
4	0.11	−3.54	3
5	3.47	0.45	1
6	0.57	2.32	2
7	0.98	−1.50	5

the sequence; and so on. In other words, dimensions are sorted according to the SPV rule, i.e., according to the position values x_{ij}^t to construct the initial sequence.

Lack of diversity and mutation operator

HAO schemes described above typically converge relatively rapidly in the first part of the search and then slows down or stops. This behavior has been attributed to the loss of diversity in the population, and a number of researchers have suggested methods to overcome this drawback with varying degrees of success.

As mutation is capable of introducing diversity in the search procedure, two types of mutation have attracted the researchers - mutation of global best and mutation based on sharing information from neighbors. Because the global best individual attracts all members of the swarm, it is possible to lead the swarm away from a current location by mutating a single individual if the mutated individual becomes the new global best. This mechanism potentially provides a means of both escaping local optima and speeding up the search. Looking at the individual components of solution vectors corresponding to the global best function values revealed that it was often only a few components which had not converged to their global optimum values. This suggested the possibility of mutating a single component only of a solution vector. The latter approach introduces diversity by mutating few individuals in the swarm.

In this work, a mutation operator is introduced which mutates position vectors of few particles selected randomly. The mutation operation is not executed in every iteration. HAO algorithm with mutation operation is as follows:

```
// t: time //
// P: populations // n A
// DELTA: the elapsed time of no further
progress //
// MAXT: maximum time of no further
progress // "
t = 0
Initialize P (t)
Evaluate P (t)
While (not—termination—conditi0n) do
t = t + 1
Update swarm according to formulae (ll) and (l2)
If (DELTA > Randi (0, MAXT)
Do mutation
End if
Evaluate the swarm
End
```

Proposed algorithm

The following are the steps of the proposed algorithm:

Step 1. Input the total number of available machines, jobs, batch size, tool slots on each machine operation of all the jobs (both essential and optional), and processing time of every operation of each job.

Step 2. Initialize the parameters. Generate initial population randomly. Construct the initial position values of the particle uniformly: $x_{ij}^t = x_{min} + (x_{max} - x_{min}) \times U(0,1)$, where $x_{min} = 0.0$, $x_{max} = 4.0$, and $U(0, 1)$ is a uniform random number between 0 and l. Generate initial velocities of the particle $v_{ij}^t = v_{min} + (v_{max} - v_{min}) \times U(0,1)$, where $v_{min} = -4.0$, $v_{max} = 4.0$, and $U(0, l)$ is a uniform random number between 0 and l.

Step 3. Get the initial sequence by using the SPV rule. Then, select the first job from that sequence, and do the following:

a. First, load the essential operation on the machine if and only if the available machining time is greater than the time required by the essential operation; otherwise, reject the job.

b. Similarly load the optional operation if and only if the available machining time and tool slot is greater than the time and tool slot required by the optional operation on the basis of the machine having the maximum available time; otherwise, reject the job.

Step 4. Evaluate each particle fitness value, i.e., the objective function.

Step 5. Find out the personal best (pbest) and global best (gbest).

Step 6. If no progress in the pbest value is observed for an elapsed period of DELTA, carry out the mutation of a particle using the mutation strategy as outlined in the 'Particle swarm optimization' section provided that DELTA is greater at random number between zero and maximum time of no progress (MAXT).

Step 7. Update velocity, position, and inertia weight.

Step 8. Compute particle fitness similar to step 3, and find a new pbest and gbest.

Step 9. Terminate if the maximum number of iterations is reached, and store the gbest value; otherwise, go to step 2.

Abbreviations

Notations used for defining the objective function

J, job index, $j = 1, 2,...,J$; M, machine index, $m = 1, 2,...,M$; S_m, tool slot capacity of machine m; O, number of operations for job j, $o = 1, 2,...,O_j$; B_j, batch size of job j; T_m, length of scheduling period for the mth machine; P_{jom}, processing time of operation o of job j on machine m; S_{jom}, number of tool slots required for processing operation 0 of job i on machine m; $B_{(j,o)}$, set of machines on which operation o of job j can be performed.

Variables used for defining the objective function

SU, system unbalance V; TH, throughput ({l, if operation 0 of job j is assigned on machine m).

Parameters used for defining the objective function

$X_{jom} = \{0$, otherwise; $X_j = \{1$, if job is selected; $\{0$, otherwise.

Author details

[1]Department of Mechanical Engineering, JSS Academy of Technical Education, Bangalore, 560 060, India. [2]JSS Academy of Technical Education, Bangalore, 560 060, India. [3]Sri Jayachamarajendra College of Engineering, Mysore, 570 006, India.

References

Berrada M, Stecke KE (1986) A branch and bound approach for machine load balancing in flexible manufacturing systems. Management Science 32(10):1316–1335

Chan FTS, Swamkar R, Tiwari MK (2005) Fuzzy goal programming model with an artificial immune system (AIS) approach for a machine tool selection and operation allocation problem in a flexible manufacturing system. International Journal of Production Research 43(19):4147–4163

Chandrasekaran S, Ponnambalam SG, Suresh RK (2007) (2007) Multi-objective particle swarm optimization algorithm for scheduling in flow shops to minimize make span, total flow time and completion time variance. In: Proceedings of the IEEE international congress on evolutionary computation 2007:25–28

Kennedy J, Eberhart R, Shi Y (2001) Swarm intelligence. Morgan Kaufmann, San Mateo

Kumar N, Shanker K (2000) A genetic algorithm for FMS part type selection and machine loading. International Journal of Production Research 38:3861–3887

Kumar RR, Singh AK, Tiwari MK (2004) A fuzzy based algorithm to solve the machine-loading problems of a FMS and its neuro fuzzy Petri net model. International Journal of Advanced Manufacturing Technology 23(5–6):318–341

Kumar A, Prakash TMK, Shankar R, Baveja A (2006) Solving machine-loading problem of a flexible manufacturing system with constrained-based genetic algorithm. European Journal of Operational Research 175(2):1043–1069

Moreno AA, Ding FY (1993) Heuristics for the FMS loading and part type selection problems. International Journal of Flexible Manufacturing System 5 (4):287–300

Mukhopadhyay SK, Midha S, Muralikrishna V (1992) A heuristic procedure for loading problems in flexible manufacturing systems. International Journal of Production Research 30(9):2213–2228

Nagarjuna N, Mahesh RK (2006) A heuristic based on multistage programming approach for machine loading problem in a flexible manufacturing system. Robotics and Computer Integrated Manufacturing 22(4):342–352

Prakash A, Khilwani N, Tiwari MK, Cohen Y (2008) Modified immune algorithm for job selection and operation allocation problem in flexible manufacturing systems. Advances in Engineering Software 39(3):219–232

Shankar K, Srinivasulu A (1989) Some solution methodologies for loading problems in flexible manufacturing system. International Journal of Production Research 27(6):1019–1034

Srinivas TMK, Allada V (2004) Solving the machine loading problem in a flexible manufacturing system using a combinatorial auction-based approach. International Journal of Production Research 42(9):1879–1893

Stecke KE (1983) Formulation and solution of non-linear integer production planning problem for flexible manufacturing system. International Journal of Management Science 29(3):273–288

Swamkar R, Tiwari MK (2004) Modeling machine loading problem of FMSs and its solution methodology is using a hybrid tabu search and simulated annealing-based heuristic approach. Robotics and Computer Integrated Manufacturing 20:199–209

Tiwari MK, Vidyarthi NK (2000) Solving machine loading problems in a flexible manufacturing system using a genetic algorithm based heuristic approach. International Journal of Production Research 38:3357–3384

Tiwari MK, Hazarika B, Vidyarthi NK, Jaggi P, Mukhopadhyay SK (1997) A heuristic solution approach to the machine loading problem of an FMS and its Petri net model. International Journal of Production Research 35(8):2269–2284

Tiwari MK, Saha J, Mukhopadhyay SK (2007) Heuristic solution approaches for combined job sequencing and machine loading problem in flexible manufacturing systems. International Journal of Advanced Manufacturing Technology 31(7–8):716–730

Tripathi AK, Tiwari MK, Chan FTS (2005) Multi-agent-based approach to solve part section and task allocation problem in flexible manufacturing systems. International Journal of Production Research 43(7):1313–1335

Varadharajan TK, Rajendran C (2005) A multi-objective simulated annealing algorithm for scheduling in flowshops to minimize the makespan and total flowtime of jobs. European Journal of Operational Research 167(3):772–795

Vidyarthi NK, Tiwari MK (2001) Machine loading problem of FMS: a fuzzy-based heuristic approach. International Journal of Production Research 39(5):953–979

Yogeswaran M, Ponnambalam SG, Tiwari MK (2008) An hybrid evolutionary heuristic using genetic algorithm and simulated annealing algorithm to solve machine loading problem in FMS. International Journal of Production Research. Taylor & Francis, Oxford (in press)

Permissions

All chapters in this book were first published in JIEI, by Springer; hereby published with permission under the Creative Commons Attribution License or equivalent. Every chapter published in this book has been scrutinized by our experts. Their significance has been extensively debated. The topics covered herein carry significant findings which will fuel the growth of the discipline. They may even be implemented as practical applications or may be referred to as a beginning point for another development.

The contributors of this book come from diverse backgrounds, making this book a truly international effort. This book will bring forth new frontiers with its revolutionizing research information and detailed analysis of the nascent developments around the world.

We would like to thank all the contributing authors for lending their expertise to make the book truly unique. They have played a crucial role in the development of this book. Without their invaluable contributions this book wouldn't have been possible. They have made vital efforts to compile up to date information on the varied aspects of this subject to make this book a valuable addition to the collection of many professionals and students.

This book was conceptualized with the vision of imparting up-to-date information and advanced data in this field. To ensure the same, a matchless editorial board was set up. Every individual on the board went through rigorous rounds of assessment to prove their worth. After which they invested a large part of their time researching and compiling the most relevant data for our readers.

The editorial board has been involved in producing this book since its inception. They have spent rigorous hours researching and exploring the diverse topics which have resulted in the successful publishing of this book. They have passed on their knowledge of decades through this book. To expedite this challenging task, the publisher supported the team at every step. A small team of assistant editors was also appointed to further simplify the editing procedure and attain best results for the readers.

Apart from the editorial board, the designing team has also invested a significant amount of their time in understanding the subject and creating the most relevant covers. They scrutinized every image to scout for the most suitable representation of the subject and create an appropriate cover for the book.

The publishing team has been an ardent support to the editorial, designing and production team. Their endless efforts to recruit the best for this project, has resulted in the accomplishment of this book. They are a veteran in the field of academics and their pool of knowledge is as vast as their experience in printing. Their expertise and guidance has proved useful at every step. Their uncompromising quality standards have made this book an exceptional effort. Their encouragement from time to time has been an inspiration for everyone.

The publisher and the editorial board hope that this book will prove to be a valuable piece of knowledge for researchers, students, practitioners and scholars across the globe.

List of Contributors

P. C. Tewari
Department of Mechanical Engineering, National Institute of Technology, Kurukshetra, Haryana136119, India

Rajiv Khanduja
Department of Mechanical Engineering, Seth Jai Parkash Mukand Lal Institute of Engineering and Technology (JMIT), Radaur, Yamuna Nagar, Haryana135133, India

Mahesh Gupta
Department of Mechanical Engineering, National Institute of Technology, Kurukshetra, Haryana136119, India

Kandukuri Narayana Rao
Department of Mechanical Engineering, Govt. Polytechnic, Visakhapatnam 530008, India

Kambagowni Venkata Subbaiah
Department of Mechanical Engineering, Andhra University, Visakhapatnam 530003, India

Ganja Veera Pratap Singh
Department of Mechanical Engineering, GITAM University, Visakhapatnam 530045, India

Fatemeh Fardis
Islamic Azad University, South Tehran Branch, Tehran, Iran

Afagh Zandi
Islamic Azad University, South Tehran Branch, Tehran, Iran

Vahidreza Ghezavati
Faculty of Industrial Engineering, Islamic Azad University, South Tehran Branch, Tehran, Iran

Aref Maleki-Darounkolaei
Department of Management and Accounting, South Tehran Branch, Islamic Azad University, Tehran, Iran

Mahmoud Modiri
Department of Management and Accounting, South Tehran Branch, Islamic Azad University, Tehran, Iran

Reza Tavakkoli-Moghaddam
Department of Industrial Engineering, College of Engineering, University of Tehran, Tehran, Iran

Iman Seyyedi
Department of Industrial Engineering, Payame Noor University, Tehran, Iran

Pandian Pitchipoo
Department of Mechanical Engineering, P.S.R. Engineering College, Sivakasi– 626140, Tamil Nadu, India

Ponnusamy Venkumar
Department of Mechanical Engineering, Kalasalingam University, Krishnankoil - 626126, Tamil Nadu, India

Sivaprakasam Rajakarunakaran
Department of Mechanical Engineering, Kalasalingam University, Krishnankoil - 626126, Tamil Nadu, India

Ümit Yüceer
Department of Industrial Engineering, Toros University, Mersin 33140, Turkey

Wu-Lin Chen
Department of Computer Science and Information Management, Providence University, Taichung, Republic of China (Taiwan)

Chin-Yin Huang
Department of Industrial Engineering and Enterprise Information, Tunghai University, Taichung, Republic of China (Taiwan)

Ching-Ya Huang
Department of Industrial Engineering and Enterprise Information, Tunghai University, Taichung, Republic of China (Taiwan)

A Thangam
Department of Mathematics, Pondicherry University – Community College,Lawspet -08, 605 008, Pondicherry, India

Kamlesh Kumar
Department of Mathematics, Indian Institute of Technology Roorkee, Roorkee, Haridwar, Uttarakhand 247667, India

Madhu Jain
Department of Mathematics, Indian Institute of Technology Roorkee, Roorkee, Haridwar, Uttarakhand 247667, India

Mehran Khalaj
Department of Industrial Engineering, Robat Karim Branch, Islamic Azad University, Tehran, Iran

Fereshteh Khalaj
Department of Statistics, Science and Research Branch, Islamic Azad University, Tehran, Iran

Amineh Khalaj
Department of Management, Naragh Branch, Islamic Azad University, Naragh, Iran

Reza Tavakkoli-Moghaddam
School of Industrial & Systems Engineering and Department of Engineering Science, College of Engineering, University of Tehran, Tehran, Iran

Fateme Forouzanfar
School of Industrial Engineering, South Tehran Branch, Islamic Azad University, Tehran, Iran

Sadoullah Ebrahimnejad
Department of Industrial Engineering, Karaj Branch, Islamic Azad University, Karaj, Iran

Zahra Shad
Department of Industrial Engineering, Khaje Nasir Toosi University of Technology, Tehran, Iran

Emad Roghanian
Department of Industrial Engineering, Khaje Nasir Toosi University of Technology, Tehran, Iran

Fatemeh Mojibian
Department of Management and Economics, Tarbiat Modares University, Jalal Ale Ahmad Highway, P.O.BOX: 14115-111, Tehran, Iran

Mirbahador Gholi AriaNezhad
Dep. Of Industrial Management, Science and Research Branch, Islamic Azad University, Tehran, Iran

Ahmad Makuie
Dep. Of Industrial Engineering, University of Science and Technology, Tehran, Iran

Saeed Khayatmoghadam
Dep. Of Management, Mashhad Branch, Islamic Azad University, Mashhad, Iran

Suchismita Satapathy
School of Mechanical Engineering, KIIT University, Bhubaneswar, 751024 Odisha, India

Pravudatta Mishra
Cental electricity supply Utility Dept, Bhubaneswar, 751024 Odisa, India

Mahdi Bashiri
Department of industrial Engineering, Faculty of Engineering, Shahed University, Khalij Fars Highway, Tehran P.O. BOX: 3319118651, Iran

Amir Farshbaf-Geranmayeh
Faculty of industrial engineering and systems, college of engineering, University of Tehran, Enghelab street, Tehran P.O. BOX: 11155-4563, Iran

Hamed Mogouie
Department of industrial Engineering, Faculty of Engineering, Shahed University, Khalij Fars Highway, Tehran P.O. BOX: 3319118651, Iran

Majid Yousefikhoshbakht
Young Researchers Club, Hamedan Branch, Islamic Azad University,Hamedan, Iran

Esmaile Khorram
Faculty of Mathematics and Computer Science, Amirkabir University of Technology, Hafez Avenue, Tehran, Iran

Ali Roozitalab
Kish International campus, University of Tehran, Kish Island 71778-5777, Iran

Ezzatollah Asgharizadeh
Tehran University, Gisha bridge, Jalale ale ahmad blvd, Tehran 6311-14155, Iran

Maryam Alimardani
School of Industrial and Systems Engineering, College of Engineering, University of Tehran, Tehran, P.O. Box: 11555-4563, Iran

Fariborz Jolai
School of Industrial and Systems Engineering, College of Engineering, University of Tehran, Tehran, P.O. Box: 11555-4563, Iran

Hamed Rafiei
School of Industrial and Systems Engineering, College of Engineering, University of Tehran, Tehran, P.O. Box: 11555-4563, Iran

Saeed Mehrjoo
Industrial Engineering Department, Payme Noor University (PNU), Tehran, Iran

Mahdi Bashiri
Industrial Engineering Department, Shahed University, Tehran, Iran

Vijay M. Kumar
Department of Mechanical Engineering, JSS Academy of Technical Education, Bangalore, 560 060, India

A. N. N. Murthy
JSS Academy of Technical Education, Bangalore, 560 060, India

K. Chandrashekara
Sri Jayachamarajendra College of Engineering, Mysore, 570 006, India

9 781682 850534